MODERN MAGNETISM

MODERN MAGNETISM

by

L. F. BATES
Ph.D., D.Sc., F.Inst.P.

Lancashire-Spencer Professor of Physics,
University College, Nottingham

SECOND EDITION
(Revised and Enlarged)

CAMBRIDGE
AT THE UNIVERSITY PRESS

1948

CAMBRIDGE
UNIVERSITY PRESS

University Printing House, Cambridge CB2 8BS, United Kingdom

Cambridge University Press is part of the University of Cambridge.

It furthers the University's mission by disseminating knowledge in the pursuit of education, learning and research at the highest international levels of excellence.

www.cambridge.org
Information on this title: www.cambridge.org/9781316601860

© Cambridge University Press 1948

First edition 1939
Second edition 1948
First paperback edition 2015

A catalogue record for this publication is available from the British Library

ISBN 978-0-521-09162-6 Hardback
ISBN 978-1-316-60186-0 Paperback

PREFACE

Experience as a university teacher has firmly convinced me that the average student finds far more difficulty in acquiring a satisfactory knowledge of magnetism than of any other branch of physics. In my view this is almost entirely because the subject is usually treated far too much from a strict theoretical standpoint and far too little from an experimental one. It is no exaggeration to state that some of the most important experiments whose results have been used very many times in theoretical discussions have not been adequately described in English. An attempt has been made to remedy this state of affairs by writing this book, in which prominence is given to the description of fundamental experiments. At the same time, a strong endeavour has been made to present an account of modern magnetic theory which, it is hoped, will be readily understood by students of physics and also followed without undue difficulty by more general readers. The opportunity has been taken of including a large body of experimental work, much of foreign origin, which is of importance to all interested in industrial applications of magnetic materials. It is obvious that, in order to keep the book to a reasonable size, I have been compelled to make an arbitrary selection of material to some extent, but care has been taken to include accounts of all essential matter which a student cannot readily find elsewhere. It has been thought best, however, to exclude galvanomagnetic and most magneto-optic phenomena.

I wish to acknowledge my indebtedness to Stoner's *Magnetism and Matter* and his many research publications, to Van Vleck's *Electric and Magnetic Susceptibilities* and to Klemm's *Magnetochemie*. I have also made considerable use of the theses published by the University of Strasbourg and the technical publications of the Bell Telephone System. I desire to express my very warm thanks to all who have so kindly allowed me to use illustrations from their published papers, and it is hoped that proper acknowledgement is made in the text. My thanks are especially due to Dr W. Sucksmith with whom I had the pleasure of much happy discussion in the early stages of the work, to Mr B. Lloyd-Evans and Mr S. S. Watts who made many drawings for me, to Mr N. Davy and Mr J. C. Weston who assisted me in checking the manuscript and proofs, and, finally, to Mr J. A. Ratcliffe for his kind interest and advice on behalf of the Cambridge University Press. In addition, I have to record my debt to the late Dr A. P. Chattock, who first introduced me to the study of magnetism, and to whose memory this book is dedicated.

Nottingham   L. F. BATES
January, 1939

PREFACE TO THE SECOND EDITION

In preparing this edition I wished to incorporate much new material in the text, but the difficulties of the times made large-scale alterations of the existing pages impracticable. Fortunately, the first edition was still so modern that it was not necessary to remove much of its printed matter, and I have, in my opinion, been able to bring the new edition reasonably up-to-date by the addition of new chapters and a few supplementary notes. I am much indebted to many colleagues and students for helpful criticisms and suggestions in the preparation of these additions.

Nottingham   L. F. BATES
1947

CONTENTS

CONTENTS

CONTENTS

CONTENTS

CONTENTS

CHAPTER I

FUNDAMENTAL CONCEPTIONS IN MAGNETISM

Fundamental Definitions. It is well known that by artificial means we can cause certain substances, called *ferromagnetics*, to become more or less *permanently magnetised*,
i.e. to possess the power of attracting or repelling other
magnetised or magnetisable bodies, and we say that the region
around a magnetised body constitutes a *magnetic field*. If a
piece of unmagnetised iron be placed in such a region it becomes magnetised and is said to exhibit *induced magnetism*.
If we place a piece of permanently magnetised material in a
magnetic field, and find that it is not acted upon by translational forces but only by couples tending to set it in a particular direction, we say that it is in a *uniform field*; it is
assumed that the field is not disturbed by the introduction of
the material and that the latter's magnetism is not affected.
It can be shown that a uniform field exists in the middle of a
long solenoid or coil of wire wound uniformly upon a cylinder
and carrying a steady current, and for most purposes the
magnetic field due to the earth may be treated as uniform over
large regions.

Now when a long rod of permanently magnetised material,
e.g. a bar magnet, is suspended so that it may rotate freely in a
horizontal plane, it is acted upon by the magnetic field of the
earth, its axis comes to rest roughly along a north-south line,
and we speak of *north-seeking* and *south-seeking* ends of the
rod. However it is placed in a uniform field, two systems of
parallel forces act upon the rod, and the resultants of these
forces always appear to pass through two points or tiny
regions, one near each end of the rod. These points are called
poles and at them we may for purposes of calculation consider
north- and south-seeking magnetism to be located. The line
joining the poles is termed the *magnetic axis* of the magnet.

1

It is an experimental fact that like poles repel and unlike poles attract each other with forces which obey an inverse square law. This leads to the definition of a *unit pole* as one which will repel an equal and similar pole placed 1 cm. away *in vacuo* with a force of 1 dyne. The *pole strength* of the magnet is therefore measured by the number of unit poles to which each pole is equivalent. If we could place a unit pole at a fixed point in a magnetic field *in vacuo* it would be acted upon by a force which would be a measure of the *strength* or *magnetic intensity* of the field, and *unit magnetic intensity* exists at the point when the force on the unit pole is 1 dyne. The unit of magnetic intensity is now called the *oersted*, following the recommendation of the International Conference on Physics which met in London in 1934. It was previously termed the *gauss*, and although the reasons for the change are certainly not apparent, the term oersted will be used throughout this book in an attempt to carry out international obligations.

The work done in taking a unit north-seeking pole from a fixed point B to a fixed point A is known as the *potential difference* between the points A and B, and when B is located at infinity the work done in bringing the pole to A is known as the *potential* at the point A, for we may neglect magnetic forces at points an infinite distance from magnetised bodies. If A and B are separated by a small distance dx along the line AB, then the work done in taking unit pole from B to A, which must be independent of the path, is $-F_x dx = dV$, where F_x is the component of the magnetic intensity along AB at the point x; the minus sign is introduced because the field intensity decreases as the distance from a magnetic pole increases. Hence $F_x = -dV/dx$, or the field intensity is equal to the rate of decrease of potential with distance. The intensity at a point r from an isolated pole $+m$ is m/r^2, while the potential is m/r.

The direction in which a single north-seeking pole placed at a point in a magnetic field would be urged to move is known as the *direction of the line of force* at that point. In a uniform field such lines are parallel. By convention we may state that unit intensity exists at a point when one line of force passes

2

through unit area placed perpendicular to the lines of force at that point, and thus we may express the magnetic intensity at a point in any field in terms of lines of force per sq. cm. Since at a distance of 1 cm. from a unit pole the intensity is 1 oersted over a surface of area 4π sq. cm., we see that the convention requires that 4π lines of force leave unit north-seeking pole. A bundle of lines of force leaving unit north-seeking pole and ending on a unit south-seeking pole is known as a *unit tube of force*, and, on the above convention, consists of 4π lines.

We have seen that when a permanent magnet is placed in a uniform field it is acted upon by a couple. Referring to Fig. 1, suppose that the magnet NS of pole strength m unit poles is placed in a uniform field of strength $\mathscr{H}$ oersteds in the direction shown in the plane of the paper. If the poles are located at the points N and S, then the couple acting on the magnet is $2ml\mathscr{H}\sin\theta$ in the sense indicated, where $2l$ is the distance between N and S in cm. The product $2ml = M$ is termed the *magnetic moment* of the magnet, and is clearly equal to the maximum couple which can be exerted on the magnet

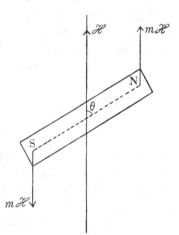

Fig. 1. Magnet in uniform field.

when it is placed in a uniform field of unit intensity; it is measured in dyne-cm. per oersted. It is perhaps unfortunate that no practical unit of magnetic moment has been named, for modern experiments show that there exists a fundamental unit of magnetic moment, the *Bohr magneton*, whose existence is just as real as that of the charge carried by an electron.

The extent to which a body is magnetised is measured by its magnetic moment per unit volume or *intensity of magnetisation*. This is a vector quantity, since magnetic moment has both magnitude and direction. If we could take a very long uniformly magnetised rod, we could imagine the cross-section at one end to be covered with m north-seeking or positive unit poles and at the other end with m south-seeking or negative unit poles. The intensity of magnetisation is, by definition, equal to $2ml$ divided by $2l\alpha$, where α is the area of cross-section of the rod; hence the intensity of magnetisation of a uniformly magnetised rod is equal to its pole strength per unit area of cross-section. In practice, however, the poles are never located at the ends of a rod, but always at a short distance from each end; in other words, the *magnetic length* or *equivalent length of a magnet* is always less than its geometrical length.

Magnetic Shell. For many purposes it is helpful to imagine a thin sheet of material uniformly magnetised in a direction normal to its surface, one side being north-seeking and the other south-seeking. Such an arrangement is known as a

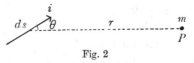

Fig. 2

magnetic shell, and the *strength of the shell* at any point on its surface is defined as the product of the intensity of magnetisation and the thickness of the shell at that point; a *uniform shell* is one in which this product is the same all over the shell. The conception is important in discussions of magnetic problems, as it may be shown that a closed circuit carrying a current of i e.m. units is equivalent to a uniform shell of strength i whose boundary is the circuit. It is sometimes convenient to picture an electron moving in a closed orbit or a spinning electron as a magnetic shell. Moreover, the potential at any point near a uniform shell is equal to the product of its strength and the solid angle subtended by the boundary of the shell at that point, the angle being taken as positive when lines radiating from the point fall first upon the north-seeking surface of the shell; the shape of the shell has no influence on the value for potential, the boundary remaining constant.

If we consider an element ds of a wire carrying a current i, it is deduced from experiment that the force on a pole of

4

strength m placed at a point P (Fig. 2) is proportional to $mi\dfrac{ds}{r^2}\sin\theta$, an expression due to Laplace. The force acts at right angles to the line joining the mid-point of the element to P and to the plane containing the element and this line. Hence if a wire is bent into a circle of radius r, the force on unit pole at the centre is proportional to $2\pi i/r$, and the *electromagnetic unit of current* is accordingly defined as that which, when flowing in a single complete turn of 1 cm. radius, causes a force of 2π dynes to act upon unit pole placed at the centre. When the current circulates in a clockwise direction, as viewed by an observer, the circuit behaves as a magnetic shell with its south-seeking face towards the observer.

It is easy to show, either by using Laplace's rule or the conception of a magnetic shell, that the work done in taking a unit pole once round a closed path surrounding a conductor carrying a current i e.m. units in opposition to the field is $4\pi i$ ergs. This result may be extended by taking a unit pole along the axis of a solenoid, completing the closed path by returning outside the windings, when the work done is $4\pi Ni$ ergs, where N is the total number of turns on the solenoid carrying the current i. If the solenoid is uniformly wound with n turns per cm. length, then it follows that the field along its axis is $4\pi ni$ oersteds.

Permeability. From the definition of unit pole it follows that the force between two poles of strength m_1 and m_2 a distance d apart *in vacuo* is $m_1 m_2/d^2$ dynes. If the poles are embedded in a medium, the force becomes equal to $m_1 m_2/\mu d^2$, where μ is a constant known as the *permeability* of the medium. We have at present no means of determining by direct experiment the value of the quantity μ used in this particular expression, and its dimensions are generally considered unknown. Let us now suppose that we have a long straight solenoid uniformly wound with n turns per cm. length and carrying a current i measured in electromagnetic units (1 ampere $= \frac{1}{10}$ e.m. unit). The magnetic field inside the middle portions of the solenoid and along its axis is uniform and equal to $4\pi ni$.

If now a small number of turns N' of wire are wound upon the solenoid and joined to a ballistic galvanometer, the quantity of electricity discharged through the latter when the current i is reversed is $2\dfrac{N'A . 4\pi ni}{R . 10^8}$ coulombs, where A is the mean area of cross-section of the solenoid and R is the total resistance of the galvanometer circuit in ohms. This quantity of electricity can be measured directly in terms of the deflection and known constants of the galvanometer. If, however, the solenoid were embedded in a medium of permeability μ, the quantity of electricity discharged through the galvanometer would be μ times the value observed when the solenoid was *in vacuo*, although the galvanometer deflection would be but imperceptibly changed unless the medium were ferromagnetic. Now, μ is here assumed to be the same quantity as that occurring in the expression for the force between two poles. We have, however, no means of proving this identity by experiment. Indeed, Abraham* has pointed out that μ may be defined in at least six different ways, and that it is impossible to prove experimentally that any two of them are the same.

A mathematical formula for μ may be obtained as follows. We consider such a long solenoid that when it is filled with a cylinder of magnetic material any free poles generated at the ends of the cylinder will produce no measurable change in the field acting at the middle portions of the cylinder. Let $\mathcal{H}$ be the field acting in the absence of magnetic material. The field inside a long narrow tunnel parallel to the axis of the cylinder will still be $\mathcal{H}$. If we cut a crevasse in the cylinder, or, rather, if we cut through the cylinder perpendicular to its axis and separate the two portions slightly, the field inside this gap will be augmented by the field arising from the magnetic poles, I per unit area, on either side of the gap. It is clear that the lines of force joining these poles must traverse the gap normally, and, hence, $4\pi I$ lines of force must pass through each sq. cm. of the gap cross-section. The total number of lines per unit area is therefore $\mathcal{H} + 4\pi I$, a quantity which we call the *magnetic induction* and denote by B. Hence we see that the ratio of

* H. Abraham, *Bull. Nat. Res. Coun., Washington*, No. 93.

the quantity of electricity discharged through the galvanometer is μ to 1 or B to $\mathscr{H}$. Hence

$$\mu = 1 + 4\pi I/\mathscr{H} = 1 + 4\pi k. \qquad \ldots\ldots(1)$$

We shall assume this expression to be correct for all isotropic substances. It is incorrect, in general, for crystalline substances, since in the latter cases I and $\mathscr{H}$ do not coincide in direction; we shall return to this matter in Chapter IV.

We may further imagine that *lines of induction* analogous to lines of force pass through a magnetised substance. If we consider lines of induction passing through an area perpendicular to the lines, then we define the product of the induction and the area as the *normal flux*, usually abbreviated to flux, so that the induction may be termed the *flux density* and measured in lines per sq. cm. In air or in vacuum the flux density is equal to the field strength (or intensity).

The ratio $I/\mathscr{H}$ or k is termed the *magnetic susceptibility*, and according to their susceptibilities substances may be roughly divided into three classes—*diamagnetic, paramagnetic* and *ferromagnetic*. In general, the susceptibilities of substances in the first two classes are small, diamagnetic susceptibilities being negative and paramagnetic susceptibilities positive. The susceptibility of a ferromagnetic is usually very large, unless the material is *saturated*—i.e. magnetised as strongly as possible in a powerful field—and depends upon the applied field and the previous magnetic history of the material. The magnetisation of diamagnetics and paramagnetics disappears when the exciting field is removed, but ferromagnetics then exhibit residual or permanent magnetisation.

In the English and foreign literatures certain symbols are commonly used in the description of susceptibility values. First, *the susceptibility per unit volume*, or the intensity of magnetisation per unit field, which is the quantity most commonly measured directly by experiment, is usually denoted by k or κ or x. Secondly, the *susceptibility per unit mass*, a quantity frequently required in theoretical discussions, is denoted by χ, and is equal to k/ρ, where ρ is the density of the substance. Thirdly, the *atomic or molar susceptibility* χ_A or χ_M

7

is respectively equal to the product of χ and the atomic or molecular weight, or to the product of k and the atomic or molecular volume. Thus, in the case of mercury, we have $k = -2 \cdot 285 \times 10^{-6}$ e.m.u. per c.c., $\rho = 13 \cdot 6$ gm. per c.c., and the atomic weight $= 200 \cdot 6$, so that, numerically,

$$\chi = 2 \cdot 285 \times 10^{-6} \div 13 \cdot 6 = 0 \cdot 168 \times 10^{-6} \text{ e.m.u. per gm.}$$

and

$$\chi_A = 0 \cdot 168 \times 10^{-6} \times 200 \cdot 6 = 33 \cdot 7 \times 10^{-6} \text{ e.m.u. per gm. atom.}$$

Energy of Magnetisation. There are several ways in which we may derive an expression for the energy stored in a magnetised medium, and one of the most satisfactory is the following. A thin ribbon of metal is wound as a uniform solenoid upon a very long cylinder of the magnetic material. Let n be the number of turns per cm. and l the total length of the cylinder, whose area of cross-section is α. Let R be the resistance of the winding. Then, the self-inductance L of the winding, i.e. the total number of lines of force linked with it when unit current flows, is $4\pi n(n\alpha l)\mu$. Hence, on applying an electromotive force E to the ends of the solenoid, we have, if i is the current at any instant,

$$E = Ri + L \cdot di/dt,$$

so that $$\int E i \, dt = \int R i^2 dt + \int L i \, di.$$

The second term on the right-hand side of the last equation is the work done in establishing the magnetic energy associated with the current i in the solenoid; on integration it gives

$$\tfrac{1}{2} L i^2 = \frac{1}{2} \frac{(4\pi n i)^2}{4\pi} \alpha l \mu = \tfrac{1}{2} \mu \frac{\mathscr{H}^2}{4\pi} \alpha l.$$

But αl is the total volume of the material upon which the solenoid is wound. Therefore the work done or energy stored per unit volume is $\mu \mathscr{H}^2 / 8\pi$. In the case of diamagnetics and paramagnetics this expression has a definite meaning, but in the case of ferromagnetics it is practically indeterminate as μ is almost an unknown function of $\mathscr{H}$.

It should be realised that the quantity $\mu \mathscr{H}^2 / 8\pi$ also represents the magnetic potential energy of unit volume of the material with respect to the applied magnetic field $\mathscr{H}$. It should therefore *in this connection* be preceded by a minus sign to indicate that it is negative, for we should have either to do work to turn the material, without change in magnetisation, through 180 degrees with respect to the lines of force of $\mathscr{H}$, or to supply thermal energy in order sufficiently to agitate the magnetic particles and so demagnetise the material. We shall see in Chapter III how this enables us to determine the direction in which diamagnetic and paramagnetic bodies tend to move in a non-uniform field.

Classical Theory of Diamagnetism. In 1854 Weber showed that all matter should exhibit diamagnetism, on the assumption that Ampèrian currents circulated within its molecules, but it was not until 1905 that Langevin showed that the electron theory provided a more satisfactory picture. Langevin examined the behaviour of an electron moving in a circular orbit of radius r when a magnetic field $\mathscr{H}$ was slowly established perpendicular to the plane of the orbit. He found that, unless the electron moved under an inverse cube-law force, the radius of the orbit remained unchanged but the velocity of the electron increased or diminished in accordance with Lenz's law, giving a change in magnetic moment of the system equal to $\Delta M = -\dfrac{e^2}{4m}\mathscr{H}r^2$. If, therefore, an atom contains a large number of electrons with their orbits orientated in all possible directions the total contribution made to the induced magnetic moment is

$$M_a = -\frac{e^2\mathscr{H}}{4m}\Sigma\bar{r}^2, \qquad\qquad \ldots\ldots(2)$$

where $\bar{r}^2$ represents the mean of the squares of the radii of the projections of the orbits on a plane perpendicular to the field. If we write $\bar{r}_0^2 = \bar{x}^2 + \bar{y}^2 + \bar{z}^2$, then $\bar{r}^2 = \bar{x}^2 + \bar{y}^2 = \frac{2}{3}\bar{r}_0^2$, so that the molar susceptibility is given by

$$\chi_M = N\frac{M_a}{\mathscr{H}} = -\frac{Ne^2}{6m}\Sigma\bar{r}_0^2. \qquad\qquad \ldots\ldots(3)$$

The last expression may be obtained in a more instructive manner; for, let us consider the special case of a circular orbit in the plane of the paper with a field $\mathcal{H}$ applied in the direction shown in Fig. 3a. Then it follows that when the electron is to the right of the line AB it experiences a force pushing it into the plane of the paper, and when to the left of AB a force pushing it out. Hence we may suppose that a

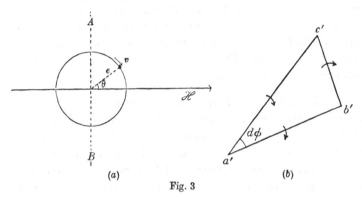

(a)   (b)

Fig. 3

torque acts on the system about the axis AB. The orbit accordingly precesses about $\mathcal{H}$. The angular momentum associated with the circular orbit is $U = mvr = 2m\alpha$, where α is the areal velocity, and the magnetic moment M, equal to the product of the equivalent current and the area A of the orbit, is $A\dfrac{e}{t} = \dfrac{A}{t}e = \alpha e$, where t is the time of description of the orbit. Now the torque Γ acting on the orbit is

$$M\mathcal{H} = -e\alpha\mathcal{H},$$

and if in Fig. 3b, $a'b'$ represents U at a given instant and $a'c'$ represents U a short time interval dt later, when the orbit has precessed through an angle $d\phi$, then $b'c' = U\,d\phi = \Gamma\,dt$. Hence $\Gamma = U\,d\phi/dt = U\sigma$, where σ is the angular velocity of precession.

Thus $\quad \Gamma = -e\alpha\mathcal{H} = 2m\alpha\sigma \quad$ or $\quad \sigma = -\dfrac{e}{2m}\mathcal{H}. \quad$ ......(4)

The change in magnetic moment brought about by this precession is now equal to

$$\Delta M = \frac{er^2}{2}\sigma = -\frac{e^2\mathcal{H}}{4m}r^2. \qquad ......(5)$$

10

Taking the general case of an electron orbit inclined at an angle to the lines of force, we may imagine the whole to precess about the field with the angular velocity given by (4), so that the change in magnetic moment is given by (5), where $\bar{r}^2$ now represents the mean of the squares of the radius upon a plane perpendicular to the field; in this way expression (3) for the susceptibility is again obtained.

Larmor first predicted the precession described above as the result of a general theorem which we shall now consider. If we take a central force X, Y, Z acting on the electron along fixed x, y and z (left-handed) axes and apply a constant magnetic field $\mathcal{H}$ parallel to the z axis, the equations of motion of the electron are

$$\left.\begin{array}{l} m\ddot{x} = e\mathcal{H}\dot{y} + X \\ m\ddot{y} = -e\mathcal{H}\dot{x} + Y \\ m\ddot{z} = Z \end{array}\right\}; \qquad \ldots\ldots(6)$$

therefore

$$m\left(\ddot{x} - \frac{e\mathcal{H}}{m}\dot{y}\right) = X,$$

$$m\left(\ddot{y} + \frac{e\mathcal{H}}{m}\dot{x}\right) = Y,$$

$$m\ddot{z} = Z.$$

If now we treat the electron as free of the magnetic field but referred to axes rotating about z in a right-handed direction with a constant angular velocity Ω, its velocities can be written

$$u = \dot{x} - y\Omega, \quad v = \dot{y} + x\Omega, \quad w = \dot{z},$$

and its accelerations

$$\dot{u} - v\Omega, \quad \dot{v} + u\Omega, \quad \dot{w}$$

or

$$\ddot{x} - 2\dot{y}\Omega - x\Omega^2, \quad \ddot{y} + 2\dot{x}\Omega - y\Omega^2, \quad \ddot{z},$$

so that the equations of motion are

$$\left.\begin{array}{l} m(\ddot{x} - 2\Omega\dot{y} - \Omega^2 x) = X \\ m(\ddot{y} + 2\Omega\dot{x} - \Omega^2 y) = Y \\ m\ddot{z} = Z \end{array}\right\}. \qquad \ldots\ldots(7)$$

Now, if $\Omega = e\mathcal{H}/2m$, the terms in Ω^2 are negligible compared with those in Ω, and the equations (7) referred to the moving axes without the magnetic field are the same as those of (6) for fixed axes with the field. In other words, apart from second-order disturbances, the effect of the field is merely to impress a rotation of angular velocity $\Omega = e\mathscr{H}/2m$ about the field as axis. It should be noted that Ω represents a right-handed motion about $\mathscr{H}$, whether the undisturbed motion of the electron is right- or left-handed about $\mathscr{H}$. This gives in a more precise way expression (3) for the susceptibility.

As in the later expression $\bar{r}^2$ must always be positive, or as the precession is always right-handed about $\mathscr{H}$, all matter must be diamagnetic. The diamagnetism is often not apparent, since any paramagnetism present may overwhelm it. As the expression contains no term involving temperature, the diamagnetism should be temperature independent. This is the case for bromine, phosphorus and sulphur, but frequently temperature variations, possibly due to superimposed para-magnetism, are found; in any case, some slight variation of $\Sigma\bar{r}_0^2$ with temperature is to be expected. When the expression (3) is used for the evaluation of r_0 the value found is of the order 10^{-8} cm., which is in satisfactory agreement with X-ray measurements. The large temperature variation of the susceptibility of bismuth is discussed on p. 148.

From the above theory it would be expected that the molecular susceptibility of a diamagnetic compound would be equal to the sum of the atomic susceptibilities of the atoms of which it is constituted, or

$$\chi_M = \Sigma r_1 \chi_{A_1}, \qquad \ldots\ldots(8)$$

where χ_{A_1} is the atomic susceptibility of the n_1 atoms of the same kind denoted by the suffix 1. As a particular case we may consider the molecular susceptibility χ_M of a diamagnetic compound as made up of the ionic susceptibilities of the anion and kation which it forms on solution, when

$$\chi_M = \chi_{\text{anion}} + \chi_{\text{kation}}. \qquad \ldots\ldots(9)$$

Equation (8) is but rarely obeyed and an extensive series of experiments by Pascal has shown that it should in general be replaced by the equation

$$\chi_M = \Sigma n_1 \chi_{A_1} + \lambda, \qquad \ldots\ldots(10)$$

where λ is a constant which has been found in a large number of important cases; e.g. for the benzene ring $\lambda = -1\cdot45 \times 10^{-6}$.

Only in the case of the saturated hydrocarbons, i.e. of the type C_nH_{2n+2}, is λ found to be zero. It is thus clear that λ is a constant which expresses the effect of chemical combination. Equation (9) appears to be somewhat more rigorously obeyed. It should be explained, however, that it can only be tested by making an assumption, either concerning the value of the ionic susceptibility of H^+ or concerning the radii of the electron orbits in ions which have a rare gas configuration. By taking the ionic susceptibility of H^+ to be zero Frivold* has shown that equation (9) is obeyed with reasonable accuracy. For, by assuming this value for H^+, the values for Cl^- and Br^- were directly obtained from measurements on solutions of the halogen acids. These values were then used to calculate the values for metallic ions on the basis of equation (9), and the two values so obtained for a given ion were in fair agreement. Joos† assumed that the ions possessing electron configurations similar to rare gas atoms have electron orbits for which $(\bar{r}^2)^{\frac{1}{2}}$ is inversely proportional to the atomic number of the atom forming the ion. There is not very good agreement between Frivold's results and those calculated on the basis of the Joos assumption. In any case, all the calculations depend upon the correctness of *Wiedemann's law*, which states that the susceptibility χ_l of a solution containing m gm. of salt of susceptibility χ_s dissolved in M gm. of solvent of susceptibility χ_w is given by

$$(M+m)\,\chi_l = m\,.\chi_s + M\,.\chi_w.$$

Classical Theory of Paramagnetism. Langevin was the first to give an electron theory of paramagnetism which still forms a valuable introduction to its study. He considered a paramagnetic gas in which each particle was endowed with a permanent magnetic moment μ so that when a mass of the gas was placed in a field each particle tended to set with its magnetic axis parallel to it. Now, at ordinary temperatures the particles are subject to considerable thermal agitation and would be prevented from taking up anything like exact align-

---

* O. E. Frivold, *Avhandlinger ut. av det Norske Videnskaps-Akad. Oslo*, No. 9, 1933. Cf. S. V. Anantakrishnan, *Proc. Ind. Acad. Sci.* **21**, 120, 1945.

† J. Joos, *Zeit. für Phys.* **32**, 835, 1925.

ment, but a kind of statistical equilibrium would be set up with a majority of the particles contributing to a magnetisation parallel to the field.

If the applied field is $\mathscr{H}$, then the field which would act upon an isolated particle placed at the centre of a spherical hole made in the medium (instead of the crevasse of p. 6) is $F = \mathscr{H} + \dfrac{4\pi}{3} I$. Now, the quantity I is so small in the case of a gas that we may neglect $\dfrac{4\pi}{3} I$ in comparison with $\mathscr{H}$. Even if

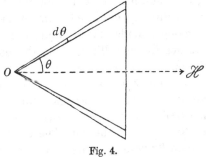

Fig. 4.

we took the paramagnetic gas oxygen and compressed it until its density was unity, as Debye* showed, $\dfrac{4\pi}{3} I$ would only be equal to $5 \times 10^{-4} \mathscr{H}$, approximately. However, we cannot always neglect this term in dealing with liquids and solids at very low temperatures, but in the case of a gas we can always suppose that its pressure and density is so low that $F = \mathscr{H}$ and there is no question of mutual action between the particles.

The potential energy of a particle whose magnetic axis makes an angle θ with the field is $-\mu F \cos\theta$, so that we may write that the number of particles whose axis make an angle θ with the field is proportional to $e^{-(-\mu F \cos\theta/kT)}$. This is merely an example of the application of the *Maxwell-Boltzmann law*, which states that when we deal with a large number of particles the number n_2 each possessing an energy E_2 is related to the number n_1 each possessing an energy E_1 by the equation

* P. Debye, *Handbuch der Radiologie*, Band **6**, 686, 1926.

THE LANGEVIN FUNCTION

$n_2 = n_1 e^{+(E_1-E_2)/kT}$. Let us now suppose that we can collect all
the particles in a given mass of gas and place them with their
centres at a point O, Fig. 4, and their axes pointing in the
directions in which they actually pointed before they were
collected. Then the number dn of particles whose axes fall
within the solid angle $d\omega$ between two hollow cones of semi-
angles θ and $\theta+d\theta$ is given by $dn = K \cdot e^{\mu F \cos\theta/kT} d\omega$, where
K is a constant, and $d\omega = 2\pi \sin\theta d\theta$. Since the total number
n of particles in the given mass of gas is equal to $\int dn$, then, if
we write $a = \mu F/kT$,

$$n = 2\pi K \int_0^\pi e^{a\cos\theta} \sin\theta d\theta.$$

Putting $\cos\theta = x$, we have

$$n = 2\pi K \int_{+1}^{-1} -e^{ax}dx = \frac{2\pi K}{a}(e^a - e^{-a}) = \frac{4\pi K}{a}\sinh a,$$

which gives K in terms of n.

Now the dn particles each contribute a component of
magnetic moment $\mu \cos\theta$ parallel to the field, while by
symmetry the components perpendicular to the field neutralise
one another. Hence the magnetisation of the given mass of
gas is equal to

$$M = \int_0^\pi \mu \cos\theta \, dn$$

$$= \int_0^\pi \mu \cos\theta \cdot K \cdot e^{a\cos\theta} 2\pi \sin\theta d\theta$$

$$= -2\pi\mu K \int_{+1}^{-1} x \cdot e^{ax}dx$$

$$= -2\pi\mu K \left[\frac{e^{ax}}{a^2}(1-ax)\right]_{+1}^{-1}$$

$$= 2\pi\mu K \left[\frac{e^a+e^{-a}}{a} - \frac{e^a-e^{-a}}{a^2}\right]$$

$$= \frac{2\pi\mu n}{4\pi\sinh a}\left[2\cosh a - \frac{2\sinh a}{a}\right]$$

$$= \mu n\left[\coth a - \frac{1}{a}\right]$$

$$= n\bar{\mu},$$

15

where $\bar{\mu}$ is the average magnetic moment each particle appears to possess under the experimental conditions.

Hence
$$\frac{\bar{\mu}}{\mu} = \coth a - \frac{1}{a}.$$

The expression $\coth a - 1/a$ is known as the *Langevin function* and is denoted by $L(a)$. When a is small

$$L(a) = \frac{a}{3} - \frac{a^3}{45} + \dots,$$

and vanishes with a. Hence, in this case, corresponding to weak fields and high temperatures,

$$\frac{\bar{\mu}}{\mu} = \frac{a}{3} = \frac{\mu F}{3kT} = \frac{\mu \mathscr{H}}{3kT} \text{ for a gas,}$$

or
$$N\bar{\mu} = \frac{N\mu^2}{3kT}\mathscr{H} = \frac{N^2\mu^2}{3RT}\mathscr{H},$$

where N is the number of particles in 1 gm. molecule of the gas. The last equation may be rewritten

$$\frac{N\bar{\mu}}{\mathscr{H}} = \chi_M = \frac{N^2\mu^2}{3RT} = \frac{C}{T}. \qquad \dots\dots(11)$$

Thus we see that the molar susceptibility of a paramagnetic substance should vary inversely as the absolute temperature, a statement which is known as the *Curie* or *Curie-Langevin law*. The law can be deduced by thermodynamics provided $\chi \neq f(\mathscr{H})$. The gas particles have here been treated as if they provided no diamagnetic effects. As we have seen, all matter is diamagnetic and an appropriate correction must be applied to experimental results before the paramagnetic susceptibility is calculated. There are only two common paramagnetic gases with which the law may be tested, namely O_2 and NO. In the case of O_2 the law holds exactly, but in the case of NO there are special circumstances which cause it to break down; these will be discussed on p. 47. It is remarkable that the Curie law holds so often in the case of solids and liquids, where a priori we would certainly not expect it to be valid. Thus it holds for many salts of the rare earth group. It holds for solutions of paramagnetic salts, where the ions may be looked upon as reasonably free. On the whole, however, paramagnetic susceptibilities tend to follow a modified Curie law, known as the *Curie-Weiss law*, viz. $\chi_M = C/T - \theta$, whose significance we will discuss later.

For large values of a, corresponding to intense fields and low temperatures

$$L(a) = 1 - \frac{1}{a} - 2e^{-2a} - \dots$$

and hence approaches unity. The complete curve of $L(a)$ plotted against a is shown in Fig. 67 on p. 234. We should therefore expect saturation effects to be apparent at low temperatures. Such effects have been found in the case of gadolinium sulphate, $Gd_2(SO_4)_3 . 8H_2O$; the results do not exactly fit the Langevin curve and the deviations will be discussed on p. 42.

Langevin's derivation of the Curie law is open to a serious criticism which was pointed out by Frl. van Leeuwen.* If the permanent magnetic moment is associated with and proportional to the angular momentum of moving electric charges, then the magnetic moment cannot have a fixed magnitude μ, but must take all values ranging from $-\infty$ to $+\infty$. When the Maxwell-Boltzmann statistics are applied to a system of such particles, the diamagnetic and paramagnetic contributions to the magnetisation annul one another. Hence we must regard the Langevin theory as incomplete, and we shall later see how the theory has been remodelled on quantum mechanics.

The Bohr Atom. Important changes in the theory of magnetism accompanied the publication of Bohr's theory of atomic spectra. In order to explain the emission of spectra lines by a hydrogen atom or a similar atomic system, Bohr treated the atom as a positively charged nucleus around which as centre an electron described a circular orbit, and he made three important postulates. First, the angular momentum associated with the orbit had to be an integral multiple of $h/2\pi$, where h is Planck's constant of action, $6 \cdot 610 \times 10^{-27}$ erg sec.; this effectively limited the radii of possible orbits to

---

* J. H. Van Vleck, *Electric and Magnetic Susceptibilities*, p. 94, 1932.

17

certain discrete values. Secondly, the electron while describing the orbit could not emit electromagnetic radiation; this meant the assumption of a number of stationary orbits or states in which the electron obeyed the classical laws of electrostatics, but not the requirements of electromagnetic theory. Thirdly, on passing from one stationary orbit to another nearer the nucleus, the electron emitted a quantum $h\nu$ of monochromatic radiation of frequency ν.

Bohr's first postulate is extremely important in magnetic theory; for we may write the angular momentum associated with the orbit

$$p_\phi = nh/2\pi = ma_n^2\omega, \qquad \ldots\ldots(12)$$

where n is an integer, a_n is the radius of the nth orbit and ω is the angular velocity. Now, the magnetic moment associated with the moving electron is equal to the area of the orbit multiplied by the current to which the moving electron is equivalent, i.e. to

$$\pi a_n^2 e \left/ \frac{2\pi}{\omega} \right. = \pi e \frac{a_n^2\omega}{2\pi} = \frac{nhe}{4\pi m}, \qquad \ldots\ldots(13)$$

where e is measured in electromagnetic units; if e is measured in electrostatic units the last expression must be divided by c, the velocity of light. We see, then, that the magnetic moment of the system is an integral multiple of $he/4\pi m$, a quantity which is therefore regarded as a fundamental unit of magnetic moment, the Bohr magneton. Its value, taking $h = 6 \cdot 610 \times 10^{-27}$ and $e/m = 1 \cdot 759 \times 10^7$, is $0 \cdot 925 \times 10^{-20}$ e.m. units and we shall denote it by μ_B. Let us note, in passing, that the ratio of the angular momentum of the system to its magnetic moment is $2m/e$.

If we further write

$$\frac{Ze^2}{a_n^2} = \frac{ma_n^2\omega^2}{a_n} \qquad \ldots\ldots(14)$$

for the condition that an electron must remain in a circular orbit under the electrostatic force between the nuclear charge Ze and the electronic charge e (both in e.s. units) and the centripetal force, we have

$$a_n = n^2h^2/4\pi^2 mZe^2, \qquad \ldots\ldots(15)$$

18

assuming that the nucleus is fixed in space. Since the sum W_n of the potential and kinetic energies of the moving electron is $-Ze^2/2a_n$, then

$$W_n = \frac{-2\pi^2 m Z^2 e^4}{n^2 h^2}. \qquad \ldots\ldots(16)$$

This is a minimum when $n = 1$, when we say that the system is in its fundamental, ground, or normal state.

According to Bohr's third postulate the emission of monochromatic radiation occurs when the electron passes from a stationary orbit n_2 to another n_1 when $n_2 > n_1$. The frequency is given by

$$\nu = \frac{2\pi^2 m Z^2 e^4}{h^3} \left(\frac{1}{n_1^2} - \frac{1}{n_2^2} \right). \qquad \ldots\ldots(17)$$

Expression (17) gives the frequency in vibrations per second, and in order to avoid the use of extremely large numbers it is usual to express the frequency in terms of *wave numbers*, or ν/c. The value of Z for hydrogen is 1, and, when n_1 is given the several values 1, 2, 3 and 4 while n_2 takes integral values in all cases greater than n_1, we find that (17) represents respectively the *Lyman, Balmer, Paschen* and *Brackett series* of lines emitted by the hydrogen atom. The factor $2\pi^2 m e^4/c h^3$ is known as the Rydberg constant for hydrogen, and its value is 109677·58.

The expression (17) is incomplete because the nucleus cannot be strictly regarded as a fixed point, and when account is taken of its rotation, the above value of ν/c must be multiplied by a factor $M/M + m$, where M is the mass of the nucleus. This fact has been used in calculating the ratio of the mass of the hydrogen atom to the mass of the electron from the experimental values of the Rydberg constants for hydrogen and ionised helium atoms.

The circular orbit is, however, only a special case for the motion of a point charge under an inverse square law, and Sommerfeld extended the Bohr theory to cover the general case of elliptical orbits. He separately quantised the angular momentum and the radial momentum of the electron, using the equation

$$\oint p_\phi d\phi = kh \quad \text{and} \quad \oint p_r dr = n_r h. \qquad \ldots\ldots(18)$$

19

We need not consider this work further as the net result is that the number n in equation (16) is merely replaced by $k + n_r$, so that the energy of the elliptical orbit is the same as that of the corresponding circular orbit, although the original single circular orbit for a total quantum number n is now replaced by one circular orbit of radius a_n and, in addition, $n - 1$ elliptical orbits. Sommerfeld also took account of another important property of a moving body, the variation of its mass with velocity. The mass m at a velocity v is equal to $m_0/\sqrt{1 - v^2/c^2}$, where m_0 is the mass at velocities small compared with that of light. On account of the variation of velocity in different parts of the orbit the path of the electron, instead of being a simple ellipse, becomes complicated and may best be likened to a rosette. There now result differences in the energies of orbits with the same total quantum number n, and the energy is the less the greater the eccentricity of the simple ellipse which would be described if m were independent of v. W_n, which now has n distinct values differing slightly from each other, may be written

$$W_n = -\frac{2\pi^2 m Z^2 e^4}{h^2} \left\{ \frac{1}{n^2} + \frac{4\pi^2 Z^2 e^4}{c^2 h^2} \left(\frac{n}{k} - \frac{3}{4} \right) \frac{1}{n^4} \right\} \quad \ldots\ldots(19)$$

$$= -\frac{2\pi^2 m Z^2 e^4}{h^2} \left\{ \frac{1}{n^2} + Z^2 \alpha^2 \left(\frac{n}{k} - \frac{3}{4} \right) \frac{1}{n^4} \right\},$$

where α is known as the *fine structure constant*.

The passage of an electron from a state with total quantum number n_2 to another with n_1 would at first sight appear to be possible in $n_2 n_1$ ways, thus giving a fine structure of $n_1 n_2$ lines very close together in place of a single spectra line, but many of these transitions are not found by experiment and a *Selection Rule* is used to account for the observed fine structure. Briefly, the rule is $\Delta k = \pm 1$, or only those transitions are possible in which the angular momentum increases or decreases by one unit, $h/2\pi$.

Spatial Quantisation. In discussing the behaviour of a paramagnetic gas it was assumed that the magnetic particles

20

could set with their axes at all possible orientations with respect to an applied magnetic field. Hence, if we could examine a gas composed of paramagnetic hydrogen atoms, their electron orbits could, on this view, set with their planes making all possible angles with the lines of force. It would then be difficult, however, to account for the sharpness of the lines emitted by a source of monochromatic light placed in a magnetic field, and Pauli* concluded that the view was incorrect and that an electron orbit could only set in certain discrete positions, viz. with the component of the angular momentum about the lines of force equal to an integral multiple of $h/2\pi$. As a matter of fact, a similar idea, but of a

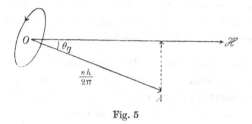

Fig. 5

more restricted nature, was put forward earlier by Lenz† to account for the behaviour of ferromagnetic crystals. So, if we represent the angular momentum of an orbit of quantum number n by a vector line OA perpendicular to the plane of the orbit and $nh/2\pi$ units long, then the orbit may set only in such positions that the projection of this vector upon the direction of the field is an integral multiple of $\pm h/2\pi$. This conception is of profound significance in the theory of magnetic susceptibility.

Referring to Fig. 5 we see that $\cos\theta_q = q/n$, where q takes the values $n, n-1, n-2, \ldots, -n$, and, hence, the component of magnetic moment along the field is $n\mu_B \cos\theta_q$, since $n\mu_B$ is the maximum magnetic moment which the orbit may exhibit. Now, the number of electron orbits setting with their magnetic axes at an angle θ_q with a field $\mathscr{H}$ must be proportional to

* W. Pauli, *Phys. Zeit.* **21**, 615, 1920.
† V. Lenz, *Phys. Zeit.* **21**, 613, 1920.

$e^{+n\mu_B \mathscr{H} \cos\theta_q/kT}$, so that the average magnetic moment $\overline{m}$ of each particle is given by

$$\overline{m} = \frac{n\mu_B \Sigma \cos\theta_q\, e^{+n\mu_B \mathscr{H} \cos\theta_q/kT}}{\Sigma\, e^{+n\mu_B \mathscr{H} \cos\theta_q/kT}}, \qquad \ldots\ldots(20)$$

where $\qquad \cos\theta_q = \dfrac{n}{n}, \ \dfrac{n-1}{n}, \ \ldots, \ \dfrac{-n+1}{n}, \ \dfrac{-n}{n}.$

For the fields of the moderate intensity obtained in the laboratory this reduces to

$$\overline{m} = n^2\mu_B^2 \frac{\mathscr{H}}{kT} \overline{\cos^2\theta_q} = \mu_B^2 \frac{\mathscr{H}}{kT} \frac{n}{3}(n+1), \qquad \ldots\ldots(21)$$

and when $n \to \infty$ we have, as we would expect, $\overline{m} = n^2\mu_B^2\mathscr{H}/3kT$, the Langevin expression, since so many discrete positions of the magnetic axis are now permissible that a condition of all possible orientations is simulated.

The Vector Atom Model. We have already made use of a vector line to represent in magnitude and direction the angular momentum of an electron orbit. In addition to the orbital motion of the electron, it has been found necessary to endow the electron with a spin in order to explain the occurrence of multiplets—doublets, triplets, etc.—in atomic spectra. This spin may also be represented vectorially, as, too, may be the spin of the nucleus. The latter produces relatively minute magnetic effects and we shall neglect it for the present. It is then found that spectroscopic requirements are satisfied by attributing four quantum numbers to each electron in an atom.

(1) The *total quantum number n*, with values 1, 2, 3, 4, ..., equivalent to the total quantum number introduced in the Bohr and Sommerfeld theories, determines the size of the orbit and the energy level in the atom. Thus, in a heavy atom there are two electrons in the innermost or *K X-ray level* with $n = 1$, eight electrons in the *L X-ray levels* with $n = 2$, eighteen in the *M X-ray levels* with $n = 3$, and so on, the maximum number of electrons in any level being $2n^2$.

(2) The *orbital angular momentum quantum number l* is a measure of the angular momentum of the orbital motion and

may assume any one of the values 0, 1, 2, 3, ..., $(n-1)$. For example, if $n = 5$, l may have one of the five values 0, 1, 2, 3 and 4. We see that l replaces the older azimuthal quantum number k except that $l = k-1$, or, more accurately, $k^2 = l(l+1)$. By convention, electrons with $l = 0, 1, 2, 3, 4, ...$ are respectively known as $s, p, d, f, g, ...$ *electrons*.

(3) The *spin momentum quantum number* s, whose value is always $\frac{1}{2}$, shows that, in addition to the orbital angular momentum $lh/2\pi$ units, the electron also possesses $sh/2\pi = h/4\pi$ units of spin momentum, which we may look upon as due to rotation about its own axis.

(4) The *"inner"* or *total angular momentum quantum number* j is the vector sum of l and s and can have only half-integral values. Thus if we have an electron for which $l = l_1$, $s = \frac{1}{2}$, then j can only have the values $l_1 \pm \frac{1}{2}$.

Although the quantum numbers n, l, s and j are sufficient to explain the behaviour of an electron emitting light in the absence of a magnetic field, two additional quantum numbers are required to account for its behaviour in a very strong magnetic field. It is then necessary to suppose that the direction of the field forms an axis with respect to which the spin and orbital momenta are separately quantised. We have a *spin magnetic quantum number* m_s, representing the projection of s upon the direction of the field, which may have only the two values $+\frac{1}{2}$ and $-\frac{1}{2}$, corresponding to the magnetic axis of the spinning electron set parallel or antiparallel to the field. We also have an *orbital magnetic quantum number* m_l, representing the projection of l upon the direction of the field, which may take only the values $l, l-1, ..., 0, ..., -(l-1), -l$.

The Pauli Principle. We now come to a very important rule or principle which enables us to fix the maximum number of the electrons which in any atom may have the same total quantum number n. The principle lays down that only one electron in a given atom can possess a particular set of values for the four quantum numbers, n, l, m_l and m_s. For example, only one electron in the chromium atom can have these four numbers respectively equal to 2, 1, 1 and $-\frac{1}{2}$. We have no direct proof of the principle; so far, however, experiment has provided no case in which it breaks down, and, although the principle is stated with reference to an atom placed in a strong magnetic field, it must be of more general application, since there must be some permanent or intrinsic peculiarity of the electrons which prohibits any two of them from taking the same values of the four quantum numbers when the field is applied. Hence, a given number of electrons with fixed n and l can only assume a definite number of quantum states, independent of the applied magnetic field, as the change from a strong to a weak field must be continuous. The values which the four quantum numbers may take with $n = 1$ and $n = 2$ are given below:

TABLE I

Atomic number	X-ray level	n	l	m_l	m_s
1	K	1	0	0	$+\frac{1}{2}$
2		1	0	0	$-\frac{1}{2}$
3	L	2	0	0	$-\frac{1}{2}$
4		2	0	0	$+\frac{1}{2}$
5		2	1	-1	$-\frac{1}{2}$
6		2	1	-1	$+\frac{1}{2}$
7		2	1	0	$-\frac{1}{2}$
8		2	1	0	$+\frac{1}{2}$
9		2	1	$+1$	$-\frac{1}{2}$
10		2	1	$+1$	$+\frac{1}{2}$

In drawing up this table it is assumed that the electrons are added with quantum numbers arranged in the order shown to give the lowest energy state of the atom. We see that for $n = 1$, we can have a K group of $2 = 2 \times 1^2$ electrons only, while for $n = 2$ a further L group of $8 = 2 \times 2^2$ electrons is added, while it can further be shown that for $n = 3$ an M group of $18 = 2 \times 3^2$ electrons is added, the possible values of l now being 0, 1 and 2; in general the maximum number of electrons in any X-ray level is $2 \times n^2$. In this way we may follow the construction of the electron systems of the various atoms and so obtain the classification of the periodic table reproduced on p. 30. We also see from Table I that the vector sums of m_l and m_s are zero for a complete electron shell as found in the case of a rare gas atom. This means that the projection of the resultant momentum of the electron system upon the direction of the field must be zero, since this must be equal to the sum of Σm_l and Σm_s. It means, too, that the resultant angular momentum and, therefore, the resultant magnetic moment must be zero for a closed electron shell. The latter can, therefore, according to the theory given on p. 11, exhibit diamagnetism alone. In any atom, then, the magnetic properties are mainly due to electrons outside the closed electron shells.

Russell-Saunders Coupling. Since an atom consists of a number of electrons occupying various energy levels, it is necessary to know how the orbital and spin momenta of the several electrons are combined or *coupled* to form the resultant orbital and spin momenta of the atom as a whole. The most common type of combination is termed *Russell-Saunders* coupling,* in which the several l vectors combine to form a resultant L and the s vectors a separate resultant S. The resultants L and S then combine to form a resultant J which represents the total angular momentum of the whole electron system of the atom. It should be emphasised that the Russell-Saunders coupling holds for an atom in a weak or zero field, whereas the four quantum members n, l, m_l and m_s were discussed in terms of a very strong field acting upon the atom. Again, although the conception of L, S and J is extremely helpful, it is not entirely satisfactory, as these quantities may be represented by matrices. The Russell-Saunders coupling is the only one which need be considered in magnetism, as it applies to the iron and rare earth groups, inert gas atoms in their normal state, and atoms with a single valency electron.

In forming the resultant J, certain quantum rules must be obeyed: viz. L may only assume the values 0, 1, 2, 3, ... and S may only assume the values 0, $\frac{1}{2}$, 1, $\frac{3}{2}$, 2, ..., determined by the number of electrons in the atom and the directions of their orbital and spin vectors. The resultant J takes the values 0, 1, 2, 3, ... when S is an integer and $\frac{1}{2}$, $\frac{3}{2}$, $\frac{5}{2}$, ... when S is a half-integer. Thus, for an atom with only one electron, in the

* H. N. Russell and F. A. Saunders, *Astrophys. Journ.* **61**, 38, 1925.

normal state, L can only have the value 0, S the value $+\frac{1}{2}$ and J the values $\pm\frac{1}{2}$; note that in this case $2J+1 = 2$. Similarly, for an atom with two electrons, one with $l = 0$ and the other in an excited state with $l = 1$, we have $L = 1$, while $S = 1$ or 0 according as the electron spins are parallel or antiparallel, so that $J = 0$, 1 or 2. In Fig. 6a are shown the three ways in which two electrons each with $l = 1$ may be coupled to give a resultant L, while in Figs. 6b and 6c are respectively shown the ways in which the s vectors of four electrons and five electrons may combine.

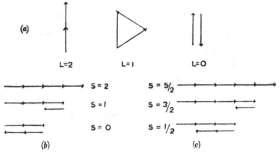

Fig. 6. Vector model of the atom.

In Figs. 7a and 7b are respectively shown the ways in which a vector $L = 2$ may combine with $S = \frac{3}{2}$ or with $S = 2$. There are clearly $2S+1$ ways in which L and S may combine if $L \geqslant S$, but $2L+1$ ways if $L < S$. Since the energy associated with a particular atom depends on the value of J, the value of either $2S+1$ or $2L+1$ gives the *maximum multiplicity* of energy levels and hence the types of spectra which may be emitted by the atom. For example, when $2S+1 = 3$, the atom may not emit spectra of higher multiplicity than triplets, but can, in addition, give singlets. The energy separation between the outermost components of a multiplet is known as the *overall width*.

In the presence of a moderately strong magnetic field we may picture the resultant L and S vectors precessing rapidly and independently about the lines of force to give projections M_L and M_S along the direction of the lines, while in a weak field the Russell-Saunders coupling obtains and the L and S

vectors form a resultant J about which they precess rapidly while J itself precesses slowly about the lines of force. In the case of a very intense field the separate L and S couplings would also be broken down and each l and s vector orientated independently with respect to the field, m_l and m_s giving the projection of $\sqrt{s(s+1)}$ and $\sqrt{l(l+1)}$ along the field direction. Fortunately, we know that the atomic susceptibility is unaffected by passage from a very strong to a weak magnetic field; cf. p. 34.

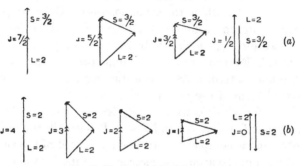

Fig. 7. Vector model of the atom.

Terminology. It is necessary to have a convenient shorthand notation to enable us to write down the electronic constitution of an atom in a simple manner. We have already seen that electrons in the K, L, M, N, ... X-ray levels may be respectively distinguished by the values of the quantum number $n = 1, 2, 3, 4, ...$, and we have denoted by the letters s, p, d, f, ... electrons for which $l = 0, 1, 2, 3, ...$, respectively. By adding numerical superscripts or indices we may indicate the number of electrons which have a particular value of l; thus, when we write $1s^2$ we mean that we are dealing with two electrons in the K level for which $n = 1$, and that both are s electrons with $l = 0$. The argon atom we may represent by $1s^2; 2s^2, 2p^6; 3s^2, 3p^6$, indicating that there are two K electrons with $n = 1$ and $l = 0$, two L electrons with $n = 2$ and $l = 0$, six L electrons with $n = 2$ and $l = 1$, two M electrons with $n = 3$ and $l = 0$ and six M electrons with $n = 3$ and $l = 1$. The potassium atom is given by the addition of one N electron

27

with $n = 4$ and $l = 0$, represented by $4s^1$ or $4s$, as the superscript in such cases is usually omitted.

The above statements, however, are incomplete, for they do not tell us the particular energy state in which the atom is found. This is indicated by the addition of a capital letter S, P, D, F, G, ..., which means that the resultant orbital momentum of the whole electron system is $0, 1, 2, 3, 4, ...$ units. A prefix to the capital letter represents the value of $2S + 1$; it gives the actual multiplicity of energy levels when $L \geqslant S$, but not when $L < S$, as shown on p. 26. A suffix represents the precise value of J which the atom possesses. For example, for the normal argon atom $L = 0$ and $S = 0$, and to the above formula we add $^1S_0$, which is to be read as "singlet S_0" and not as "one S_0". For the normal potassium atom $L = 0$ and $S = \frac{1}{2}$, and the lowest term is $^2S_{\frac{1}{2}}$. As a final example we may take the lowest term of the trebly ionised atom of chromium, the Cr^{+++} ion, for which is written $1s^2$; $2s^2$, $2p^6$; $3s^2$, $3p^6$, $3d^3$, $^4F_{\frac{3}{2}}$, showing that $L = 3$, $S = \frac{3}{2}$ and $J = \frac{3}{2}$. It is usually unnecessary to write out the complete shorthand statement of the electron constitution of the atom, as electrons forming complete shells make no contribution to the resultants L, S and J. In the case of the normal potassium atom all that is necessary is given by $4s$, $^2S_{\frac{1}{2}}$, telling us that the normal atom has a resultant angular momentum and therefore a resultant magnetic moment. The latter is due entirely to electron spin, since $L = 0$. The energy state of the potassium atom or the appropriate spectrum term is completely specified by the $^2S_{\frac{1}{2}}$.

Stability Rules. In finding the normal or "ground" state of an atom or ion use is made of a set of *stability rules* devised by Hund* from a study of spectra terms. In the case of S terms, for which $L = 0$, the maximum multiplicity of energy levels is unity, and, consequently, S terms are always singlets, P terms never more than triplets, D terms never more than quintets, and so on. Also, in an atom with an even number

---

* F. Hund, *Linienspektren und periodisches System der Elemente*, Springer, Berlin, 1927.

of electrons, S is always an integer, so that the systems of terms for such an atom will always have odd multiplicity, while in an atom with an odd number of electrons S has half-integer values and we get systems of terms of even multiplicity. Hence the multiplicities of systems of terms of successive atoms in the periodic table alternate between odd and even. In addition, Hund's rules tell us, first, that in the lowest term the spin momenta combine to give the maximum values of S consistent with the Pauli principle. Secondly, when the maximum value of S has been thus determined, the orbital momenta combine in a like manner to give the maximum values of L. Thirdly, for an incomplete shell of electrons the lowest term has $J = L - S$ when the shell is less than half occupied, and $J = L + S$ when the shell is more than half occupied.

As examples we quote first the normal Cr^{++} ion $1s^2$; $2s^2$, $2p^6$; $3s^2$, $3p^6$, $3d^3$; $4s$, with three d electrons and one s electron added to the electron configuration of the argon atom, so that $L = 2$, $S = 2$ and $J = 0$ giving a $^5D_0$ term. Secondly, we quote the Ce^{+++} ion, $1s^2$; $2s^2$, $2p^6$; $3s^2$, $3p^6$, $3d^{10}$; $4s^2$, $4p^6$, $4d^{10}$, $4f^1$; $5s^2$, $5p^6$, where the single electron in the incomplete 4f layer gives $S = \frac{1}{2}$, $L = 3$ and $J = L - S = \frac{5}{2}$, i.e. a $^2F_{\frac{5}{2}}$ term. Further examples are worked out by Gorter.*

The Periodic Table. In Table II are given the electron constitutions of the elements and the normal state of each atom as found by the rules discussed above. Those ground states which are not yet definitely proven by experiment are enclosed in brackets. Attention is particularly directed to the electron structures of the elements of the iron group, the elements of the rare earth group and the elements of the platinum group. In the elements of the iron group it is seen that a 3d set of electrons is built up after one or two 4s electrons have been added to the argon core. In the rare earth group a 4f set of electrons is completed after the addition of one 5d and two 6s electrons to the xenon core. Finally, in the platinum group, a set of 5d electrons is completed while in

* See C. J. Gorter, *Arch. du Musée Teyler*, **7**, 183, 1932.

	K	L		M			N				O			P			Q	Ground term	$\chi \times 10^6$
	1s	2s	2p	3s	3p	3d	4s	4p	4d	4f	5s	5p	5d	6s	6p	6d	7s		
H 1	1																	$^2S_{\frac{1}{2}}$	$-1{\cdot}97$
He 2	2																	$^1S_0$	$-0{\cdot}47$
Li 3	2	1																$^2S_{\frac{1}{2}}$	$+0{\cdot}50$
Be 4	2	2																$^1S_0$	$-1{\cdot}00$
B 5	2	2	1															$^2P_{\frac{1}{2}}$	$-0{\cdot}69$
C 6	2	2	2															$^3P_0$	$-0{\cdot}49$
N 7	2	2	3															$^4S_{1\frac{1}{2}}$	$-0{\cdot}8$
O 8	2	2	4															$^3P_2$	$+106{\cdot}2$
F 9	2	2	5															$^2P_{1\frac{1}{2}}$	—
Ne 10	2	2	6															$^1S_0$	$-0{\cdot}33$
Na 11		10 Ne core		1														$^2S_{\frac{1}{2}}$	$+0{\cdot}51$
Mg 12				2														$^1S_0$	$+0{\cdot}55$
Al 13				2	1													$^2P_{\frac{1}{2}}$	$+0{\cdot}65$
Si 14				2	2													$^3P_0$	$-0{\cdot}13$
P 15				2	3													$^4S_{1\frac{1}{2}}$	$-0{\cdot}90$
S 16				2	4													$^3P_2$	$-0{\cdot}49$
Cl 17				2	5													$^2P_{1\frac{1}{2}}$	$-0{\cdot}57?$
A 18	2	2	6	2	6													$^1S_0$	$-0{\cdot}48$
K 19							1											$^2S_{\frac{1}{2}}$	$+0{\cdot}52$
Ca 20							2											$^1S_0$	$+1{\cdot}10$
Sc 21						1	2											$^2D_{1\frac{1}{2}}$	—
Ti 22		18 A core				2	2											$^3F_2$	$+1{\cdot}25$
V 23						3	2											$^4F_{1\frac{1}{2}}$	$+1{\cdot}4$
Cr 24						5	1											$^7S_3$	$+3{\cdot}08$
Mn 25						5	2											$^6S_{2\frac{1}{2}}$	$+11{\cdot}8$
Fe 26						6	2											$^5D_4$	Ferro.
Co 27						7	2											$^4F_{4\frac{1}{2}}$	,,
Ni 28						8	2											$^3F_4$	,,
Cu$^+$	2	2	6	2	6	10												$^1S_0$	
Cu 29							1											$^2S_{\frac{1}{2}}$	$-0{\cdot}086$
Zn 30							2											$^1S_0$	$-0{\cdot}157$
Ga 31		28 Cu$^+$ core					2	1										$^2P_{\frac{1}{2}}$	$-0{\cdot}24$
Ge 32							2	2										$^3P_0$	$-0{\cdot}12$
As 33							2	3										$^4S_{1\frac{1}{2}}$	$-0{\cdot}31$
Se 34							2	4										$^3P_2$	$-0{\cdot}32$
Br 35							2	5										$^2P_{1\frac{1}{2}}$	$-0{\cdot}39$
Kr 36	2	2	6	2	6	10	2	6										$^1S_0$	$-0{\cdot}35$
Rb 37											1							$^2S_{\frac{1}{2}}$	$+0{\cdot}21$
Sr 38											2							$^1S_0$	$-0{\cdot}20$
Y 39									1		2							$^2D_{1\frac{1}{2}}$	$+5{\cdot}3$
Zr 40		36 Kr core							2		2							$^3F_2$	$-0{\cdot}45$
Cb 41									4		1							$^6D_{\frac{1}{2}}$	$+1{\cdot}5$
Mo 42									5		1							$^7S_3$	$+0{\cdot}04$
Ma 43									5		2							—	—
Ru 44									7		1							$^5F_5$	$+0{\cdot}50$
Rh 45									8		1							$^4F_{4\frac{1}{2}}$	$+1{\cdot}11$
Pd 46	2	2	6	2	6	10	2	6	10									$^1S_0$	$+5{\cdot}4$

TABLE II. Electronic configurations of the elements and mass susceptibilities of
with the exception

30

	K	L		M			N				O			P			Q	Ground term	$\chi \times 10^6$
	1s	2s	2p	3s	3p	3d	4s	4p	4d	4f	5s	5p	5d	6s	6p	6d	7s		
Ag 47											1							$^2S_{\frac12}$	−0·20
Cd 48											2							$^1S_0$	−0·18
In 49						46					2	1						$^2P_{\frac12}$	−0·11
Sn 50						Pd core					2	2						$^3P_0$	−0·25
Sb 51											2	3						$^4S_{1\frac12}$	−0·87
Te 52											2	4						$^3P_2$	−0·31
I 53											2	5						$^2P_{1\frac12}$	−0·36
Xe 54	2	2	6	2	6	10	2	6	10		2	6						$^1S_0$	−0·34
Cs 55														1				$^2S_{\frac12}$	−0·22
Ba 56						54								2				$^1S_0$	+0·9
La 57						Xe core					Xe		1	2				$^2D_{1\frac12}$	+1·04
Ce 58										1	core		1	2				$^3H_4$	+15·0
Pr 59										2			1	2				—	+25·0
Nd 60										3			1	2				—	+36·0
Il 61										4			1	2				—	—
Sm 62										6				2				$^7F_0$	—
Eu 63										7	Xe			2				$^8S_{3\frac12}$	+22·0
Gd 64										7	core		1	2				$^9D_2$	Ferro.
Tb 65						54				8			1	2				—	—
Dy 66						Xe core				9			1	2				—	—
Ho 67										10			1	2				—	—
Er 68										11			1	2				—	—
Tu 69										13				2				—	—
Yb 70										14				2				—	—
Lu+++	2	2	6	2	6	10	2	6	10	14	2	6							
Lu 71													1	2				$^2D_{1\frac12}$	—
Hf 72													2	2				$^3F_2$	—
Ta 73						68							3	2				$(^4F_{1\frac12})$	+0·87
W 74						Lu+++ core							4	2				$^5D_0$	+0·28
Re 75													5	2				$^6S_{2\frac12}$	—
Os 76													6	2				$(^5D_4)$	+0·05
Ir 77													9					$^2D_{2\frac12}?$	+0·15
Pt 78													9	1				$^3D_3$	+1·10
Au+	2	2	6	2	6	10	2	6	10	14	2	6	10					$^1S_0$	—
Au 79														1				$^2S_{\frac12}$	−0·15
Hg 80														2				$^1S_0$	−0·168
Tl 81						78								2	1			$^2P_{\frac12}$	−0·24
Pb 82						Au+ core								2	2			$^3P_0$	−0·12
Bi 83														2	3			$^4S_{1\frac12}$	−1·35
Po 84														2	4			$(^3P_2)$	—
— 85														2	5			—	—
Rn 86	2	2	6	2	6	10	2	6	10	14	2	6	10	2	6			$^1S_0$	—
— 87																	1	—	—
Ra 88																	2	$^1S_0$	—
Ac 89						86										1	2	$(^2D_{1\frac12})$	—
Th 90						Rn core										2	2	$(^3F_2)$	+0·11
Pa 91																3	2	$(^4F_{1\frac12})$	+2·6
U 92																4	2	$(^5D_0)$	—

bulk material. All the ground terms given have been derived from experiment of those in brackets.

most of these elements two 6s electrons are present. These details are of importance in a study of the electric and magnetic properties of the elements and will be discussed later. The mass susceptibilities are also given in the table.

The Splitting Factor. We have seen that an electron describing an orbit has associated with it an orbital angular momentum l units and therefore a magnetic moment $l\mu_B$, while to the electron spin must be apportioned an angular momentum of $\frac{1}{2}$ unit. Now, if the spinning electron also possessed a corresponding magnetic moment $\frac{1}{2}\mu_B$, we could at once write down the total magnetic moment of any atomic

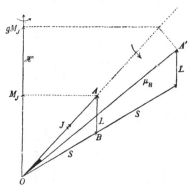

Fig. 8. The splitting factor.

state for which the value of J was known; it would simply be equal to $J\mu_B$. However, it is conclusively proved by experiments on the Zeeman effect, on the magnetic deflection of atomic beams and on the gyromagnetic effect with paramagnetic substances, that, in general, the magnetic moment is given by $gJ\mu_B$, where g is a constant known as the *Landé splitting factor*. It is found that the introduction of the factor is satisfactorily explained if we attribute a magnetic moment of $2 \times \frac{1}{2}\mu_B$ to the spinning electron; hence the splitting factor for the spinning electron is 2. Consequently, the ratio of the angular momentum of any electron system to its magnetic moment is equal to $\dfrac{2}{g}\dfrac{m}{e}$, instead of the simple value $2m/e$ noted on p. 18.

The vector model is particularly helpful in a discussion of the g factor. Referring to Fig. 8 let the L and S vectors combine to form the resultant J, represented by OA. Then J will precess slowly about an applied field $\mathscr{H}$ of ordinary intensity. If the corresponding magnetic moments are represented by L and $2S$ magnetons, their resultant magnetic moment μ_R may be represented by OA'. Since the L and S momentum vectors precess rapidly about OA, μ_R must also precess rapidly about OA, and thus give rise to a magnetic moment $\mu_R \cos A\hat{O}A'$ along OA, while, on account of the rapid precession there will, on the average, be no component perpendicular to OA. The atom thus behaves as if it had a magnetic moment $\mu_R \cos A\hat{O}A'$ with its axis along OA. Hence, we have

$$gJ = \mu_R \cos A\hat{O}A'$$

$$= J + S \cos A\hat{O}B.$$

But in the $\triangle AOB$,

$$L^2 = J^2 + S^2 - 2JS \cos A\hat{O}B;$$

therefore

$$gJ = \frac{J^2 + S^2 - L^2}{2J} + J$$

or

$$g = 1 + \frac{J^2 + S^2 - L^2}{2J^2}. \qquad \ldots\ldots(22)$$

Expression (22) for g does not, however, wholly explain the experimental results with the Zeeman effect, and Landé introduced the empirical expression for g, viz.

$$g = 1 + \frac{J(J+1) + S(S+1) - L(L+1)}{2J(J+1)}, \qquad \ldots\ldots(23)$$

which is known as the *Landé splitting factor*, and was given a theoretical basis in wave mechanics. In general, we may say that the wave mechanics treatment requires that the vectors L, S and J be replaced by $\sqrt{L(L+1)}$, $\sqrt{S(S+1)}$ and $\sqrt{J(J+1)}$ respectively. An interesting point arises in connection with these substitutions; we ought, strictly speaking, to attribute a spin of $\sqrt{S(S+1)} = \sqrt{\frac{1}{2}(\frac{1}{2}+1)} = \sqrt{3}/2$ units and a magnetic moment of $g\sqrt{S(S+1)} = 2\sqrt{3}/2 = \sqrt{3}$ magnetons to the spinning electron. It should be noted that, when the atomic system is exposed to an applied field, the component of the magnetic moment parallel to the field is gM_J and not $g\sqrt{M_J(M_J+1)}$, where $M_J = J, J-1, ..., -(J-1), -J$.

It might appear from earlier considerations that the observed susceptibility would depend upon whether a strong or a weak field is used in the measurements, i.e. whether or not the Russell-Saunders coupling persists. It can be shown that this is not so, for, in the newer quantum mechanics there is an important theorem, known as the *Principle of Spectroscopic Stability*,* which states that there is no change in the susceptibility as we pass from a strong to a weak field. Indeed, the vector model tends to give an unwarranted definiteness to spatial quantisation which the latter does not possess in the newer theory. The individual energy levels may, of course, be considerably altered by change in field strength, but in measuring the susceptibility we take a summation over all possible stationary states and this remains constant. We shall return to this point on p. 38.

The Zeeman Effect. When a source of monochromatic light is placed in a strong magnetic field the spectroscopic appearance of the light depends both on the intensity of the field and on the direction in which the source is viewed. For example, if a helium discharge tube is used as source the appearance of the yellow line in the absence of a magnetic field may be represented as in Fig. 9a. On exciting the field and looking along the lines of force towards the north seeking pole face of the electromagnet, the appearance is as shown in Fig. 9b. The singlet is now replaced by a doublet whose components lie on opposite sides of and separated from the original line by equal frequency differences $\Delta\nu$. The two components are circularly polarised in opposite directions, the component of higher frequency being polarised in the same direction as the current is required to flow in a coil exciting the magnetic field. Viewed perpendicularly to the lines of force the source gives the triplet of Fig. 9c, whose middle component coincides

* J. H. Van Vleck, *Electric and Magnetic Susceptibilities*, p. 137, 1932.

in position with the original line and whose outer components coincide with those of the doublet. All three components are plane polarised. The middle component is polarised with the electric vector parallel to the magnetic field and is known as a π component, while the others are polarised with their electric vectors perpendicular to the field and are known as σ components. We speak of the phenomena depicted in Fig. 9b, c as the *normal longitudinal* and *normal transverse Zeeman effects* respectively, and the triplet is called the *normal Lorentz or Zeeman triplet*, because in 1897 Lorentz gave an explanation of

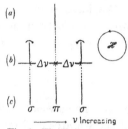

Fig. 9. The Zeeman effect.

its emission and the correct expression for the separation of its components on the basis of the classical electron theory. The effect is quite small, for, a field of 10,000 oersteds produces a separation $\Delta\lambda$ of the order of only 0·1Å., which may be contrasted with the 6 Å. separation of the D_1 and D_2 lines of sodium.

Normal triplets, however, occur only in special cases; e.g. in lines of singlet series, and, in general, the appearance of the transverse Zeeman pattern is much more complex, when it is termed *anomalous*. There are two main reasons why the transverse effect is always studied while the longitudinal effect is often neglected. First, it is impossible to excite a uniform field between the pierced pole-tips necessary for longitudinal observations, and, secondly, the transverse pattern gives more information, as it exhibits all possible components.

Two rules of great spectroscopic importance assist in the interpretation of Zeeman effects. *Preston's rule* states that lines of the same spectra series behave in the same manner in a magnetic field, and that corresponding lines of the same series in different elements behave in the same manner. Theoretically, this means that the Zeeman effect is determined solely by the L, S and J values of electron states and not by the total quantum number n. *Runge's rule*, which is generally but not always true, states that the wave-number separation of any

Zeeman component from the original line is a simple multiple of the normal triplet separation. Hence, in practice, it is usual to compare observed separations with those of a known singlet under the same conditions in order to compute the Runge factors or multiples.

We have seen that monochromatic radiation is emitted when an electron moves from one stationary state W_2 to another W_1 such that, in wave numbers,

$$\nu = \frac{1}{ch}(W_2 - W_1).$$

Now, when a weak magnetic field is applied the energy of each of these states is altered because of the Larmor precession of J about $\mathscr{H}$. Since the atom in the state W_2 may then have a resultant magnetic moment $M_2 g_2 \mu_B$ parallel to $\mathscr{H}$, the magnetic energy ΔW_2 associated with it is $-M_2 g_2 \mu_B \mathscr{H}$, where M_2 takes the values $J_2, J_2-1, \ldots, -(J_2-1), -J_2$. Hence, during an electron transition from W_2 to W_1 we may also have changes in magnetic energy $\Delta W_2 - \Delta W_1$, and, expressed in wave numbers, these correspond to frequency changes

$$\Delta\nu = \frac{1}{ch}(\Delta W_2 - \Delta W_1)$$

$$= (M_1 g_1 - M_2 g_2)\,\mu_B \mathscr{H}/ch$$

$$= (M_1 g_1 - M_2 g_2) \times 4 \cdot 66 \times 10^{-5}\mathscr{H}, \qquad \ldots\ldots(24)$$

which satisfactorily account for the Zeeman components.

It is found experimentally that in any electron transition the magnetic quantum number changes only by $\Delta M = 0$ or ± 1. Hence, in the particular case of a singlet source, for which $g_1 = g_2 = 1$, we get $\Delta\nu = (0 \text{ or } \pm 1)\, 4\cdot 66 \times 10^{-5}\mathscr{H}$, or the normal Lorentz triplet. We may take as a further example of the calculation of transverse Zeeman patterns those for the D_1 and D_2 lines of sodium, introducing first the *polarisation rules*, deduced from classical as well as quantum theory, that $\Delta M = \pm 1$ gives rise to σ components and $\Delta M = 0$ to π components; in the longitudinal direction $\Delta M = \pm 1$ gives circularly polarised components and $\Delta M = 0$ is forbidden. In Fig. 10a, b

are given the appropriate M_1g_1 and M_2g_2 values printed with
equal values of M directly below or above each other. The
vertical arrows indicate the emission of π components ($\Delta M = 0$)
and the diagonals the emission of σ components. Thus the D_1
line gives two π components with separations 1/3 that of the
normal triplet and four σ components with separations 3/3
and 5/3. The D_2 line gives two π components with separations
2/3 and two σ components with separations 4/3. The observed
separations agree exactly with calculation and are depicted in
Fig. 10c.

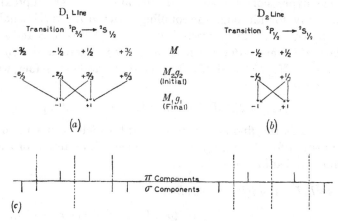

Fig. 10. Anomalous Zeeman effect.

The Paschen-Back Effect. It was observed by Paschen and
Back in 1913 that on the application of a sufficiently strong
magnetic field a complex Zeeman pattern might be replaced
by a triplet very similar to the Lorentz triplet. This Paschen-
Back effect arises when the field is able to break down the L, S
coupling, and, consequently, lines of narrow multiplet series
where this coupling is weak would be expected to show it best.
In a very strong field the L and S vectors set independently, as
described on p. 23, L with components $m_{l_1} = l_1$, $l_1 - 1$, ...,
$-(l-1)$, $-l_1$, giving a resultant $M_l = \Sigma m_{l_1}$, and S with
components $m_{s_1} = +\frac{1}{2}$ or $-\frac{1}{2}$, giving a resultant $M_s = \Sigma m_{s_1}$.
The quantity ΔW_2 of the previous section now becomes

FUNDAMENTAL CONCEPTIONS IN MAGNETISM

$-(M_{L_2}+2M_{S_2})\mu_B\mathscr{H}$, since the magnetic moment of a spinning electron is $2m_s\mu_B$. Hence, for an electron transition W_2 to W_1 we now have changes in magnetic energy giving rise to frequency differences, in wave numbers,

$$\Delta\nu = \frac{1}{ch}(\Delta W_2 - \Delta W_1)$$
$$= \{(M_{L_1}-M_{L_2})+2(M_{S_1}-M_{S_2})\}\frac{\mu_B\mathscr{H}}{ch}$$
$$= \{(M_{L_1}-M_{L_2})+2(M_{S_1}-M_{S_2})\}\times 4\cdot 66\times 10^{-5}\mathscr{H}. \quad\ldots(25)$$

The expressions for ΔW_2 and ΔW_1 are, however, approximate only, for, although the coupling between L and S is small compared with that between L and $\mathscr{H}$ and between S and $\mathscr{H}$ when $\mathscr{H}$ is very large, yet it is certainly not zero, and terms involving $M_{L_1}M_{S_1}$ and $M_{L_2}M_{S_2}$ should be added. In fact, we may write

$$\Delta\nu = \{(\Delta M_L)+2(\Delta M_S)\}\times 4\cdot 66\times 10^{-5}\mathscr{H}+A\,.\,\Delta(M_LM_S),$$

where A is the fine structure constant which occurs in the expression for the energy of the magnetic interaction of the orbital and spin momenta, viz.

$$A\sqrt{L(L+1)}\sqrt{S(S+1)}\cos(LS)$$
$$= \tfrac{1}{2}A\{J(J+1)-L(L+1)-S(S+1)\};$$

A is a constant for a given pair of values of L and S, and can therefore be obtained from the overall multiplet separation in the absence of a field. If now we assume that $\Delta M_L = 0$ or ± 1 and $\Delta M_S = 0$, that π components correspond to $\Delta M_L = 0$ and σ components to $\Delta M_L = \pm 1$, then

$$\Delta\nu = (0,\ \pm 1)\times 4\cdot 66\times 10^{-5}\mathscr{H}+A\,.\,\Delta(M_LM_S),$$

which accounts for the experimental facts so far obtained. Since the field must be very intense if the above magnetic energy changes are to be large compared with the differences in energy responsible for the ordinary multiplet structure, only very close doublets can be expected to show the Paschen-Back effect in the laboratory. Paschen and Back used a field of 43,000 oersteds to cause the principal series doublet of lithium, whose separation is only $0 \cdot 34$ cm. $^{-1}$, to give the normal triplet. We should need much stronger fields than can be produced at present in order to observe the Paschen-Back effect with the D_1 and D_2 lines. It has already been mentioned that the transition from the anomalous Zeeman to the Paschen-Back pattern is not accompanied by a change in magnetic susceptibility. See Note 1.

The Quantum Theory of Paramagnetism. We have seen that, when an atomic system with given L, S and J is placed in a weak magnetic field $\mathscr{H}$, J precesses slowly about $\mathscr{H}$ and gives a component of magnetic moment parallel to $\mathscr{H}$ equal to $M_J g \mu_B$, where $M_J = J, J-1, ..., -(J-1), -J$. The magnetic potential energy of the atom is therefore $-M_J g \mu_B \mathscr{H}$, and in a gas the number of atoms (or ions) with a particular value of M_J will be proportional to $e^{M_J g \mu_B \mathscr{H}/kT}$, assuming that Maxwell-Boltzmann law still holds. Hence, if 1 gm. atom containing N particles is placed in the field $\mathscr{H}$, the total magnetic moment $M_{\mathscr{H}}$ will be along the lines of force, and

$$M_{\mathscr{H}} = \frac{N \sum\limits_{M_J=-J}^{M_J=J} M_J g \mu_B e^{M_J g \mu_B \mathscr{H}/kT}}{\sum\limits_{M_J=-J}^{M_J=J} e^{M_J g \mu_B \mathscr{H}/kT}} \qquad \ldots\ldots(26)$$

$$= \frac{N g \mu_B \sum\limits_{M_J=-J}^{M_J=J} M_J (1 + M_J g \mu_B \mathscr{H}/kT)}{\sum\limits_{M_J=-J}^{M_J=J} (1 + M_J g \mu_B \mathscr{H}/kT)}$$

$$= N g \mu_B \frac{0 + \left(\dfrac{J^2 g \mu_B \mathscr{H}}{kT} + \dfrac{(J-1)^2 g \mu_B \mathscr{H}}{kT} + \ldots \right)}{2J+1}$$

$$= \frac{N g^2 J(J+1) \mu_B^2}{3kT} \mathscr{H}, \quad \text{since } \mathscr{H} \text{ is small,}$$

$$= g^2 J(J+1) \frac{N^2 \mu_B^2}{3RT} \mathscr{H},$$

whence
$$\chi_M = \frac{N^2 g^2 J(J+1)\mu_B^2}{3RT} = \frac{N^2 \mu_J^2}{3RT}. \qquad \ldots\ldots(27)$$

Thus by measurement of the atomic susceptibility we may obtain $g\sqrt{J(J+1)}\mu_B$, which we may also write equal to $p_{\text{eff}}\mu_B$, where p_{eff} represents the *effective magneton number* of the atomic system.

The Magnetic Significance of Multiplet Widths.

In deducing equation (27) we have definitely assumed that every particle is in the same L, S, J state. This may not be true, particularly if the particle requires but a small quantity of energy in order to excite it to a higher energy state. It is found convenient to distinguish between two cases, viz. when the energy of transition between two states is respectively very large and very small compared with kT, i.e. when the multiplets are wide or narrow respectively. Mathematically, we denote these cases by $h\nu(JJ') \gg kT$ and $h\nu(JJ') \ll kT$. We may further distinguish between the two cases by the largeness or smallness of the Zeeman separation with respect to kT.

In the first case, that of *wide multiplets*, $h\nu(JJ') \gg kT$, nearly all the particles must be in their lowest state, because so much energy is required to raise them to higher states. This is what is found with many ions of the rare earth group. Equation (27) for the magnetic susceptibility is then strictly accurate and the Curie law is obeyed.

In the second case, that of *narrow multiplets*, $h\nu(JJ') \ll kT$, the particles must exist in a variety of different energy states, and in the theoretical treatment it is reasonable to assume that the coupling of L and S is small compared with that of either L or S with $\mathscr{H}$. This is equivalent to the assumption that the atom or ion behaves as when in a strong field, so that the orbital and spin momenta are separately quantised, with

$$M_L = L, L-1, \ldots, -L; \quad g_L = 1,$$

$$M_S = S, S-1, \ldots, -S; \quad g_S = 2,$$

40

giving for the magnetic moment of 1 gm. atom,

$$M_{\mathscr{H}} = N\mu_B \left\{ \left(\sum_{M_L=-L}^{M_L=L} M_L e^{M_L \mu_B \mathscr{H}/kT} \middle/ \sum_{M_L=-L}^{M_L=L} e^{M_L \mu_B \mathscr{H}/kT} \right) \right.$$
$$\left. + \left(\sum_{M_S=-S}^{M_S=S} 2M_S e^{2M_S \mu_B \mathscr{H}/kT} \middle/ \sum_{M_S=-S}^{M_S=S} e^{M_S \mu_B \mathscr{H}/kT} \right) \right\},$$

so that, as the exponentials are still small,

$$M_{\mathscr{H}} = N\mu_B \{ \tfrac{1}{3} L(L+1) \mu_B \mathscr{H}/kT + \tfrac{4}{3} S(S+1) \mu_B \mathscr{H}/kT \},$$

whence $\qquad \chi_M = \dfrac{N^2 \mu_B^2}{3RT} \{ L(L+1) + 4S(S+1) \}$ $\qquad$ ......(28)

or $\qquad p_{\text{eff}} = \sqrt{L(L+1) + 4S(S+1)}.$ $\qquad$ ......(29)

We see that the Curie law is again deduced, but the value of p_{eff} is different from that in the case of wide multiplets. It is found that a modified form of equation (28) represents the magnetic behaviour of ions of elements of the iron group.

It should be noted that the difference in the expressions for χ_M for the two cases does not depend upon whether a weak or a strong field is used in the measurements. It arises because in the case of narrow multiplets all the particles are not in the same state irrespective of the field. As T falls the value of kT changes and $h\nu(JJ')$ may become comparable with it. Special calculations, taking into account peculiarities of the model, are then necessary; these are instanced by Van Vleck's* work on nitric oxide. When $h\nu(JJ')$ is comparable with kT, as in the cases of Sm^{+++} and Eu^{+++}, the calculations are very difficult,† the Curie law is not obeyed in such cases, and it also becomes necessary to take account of the temperature-independent paramagnetism contribution which a rigorous wave mechanics treatment shows to be present.

Saturation Phenomena. We have so far assumed that the field $\mathscr{H}$ and the absolute temperature T are such that $a = Jg\mu_B \mathscr{H}/kT \ll 1$. We must now consider the case when a is not small compared with 1, when, writing

$$x = g\mu_B \mathscr{H}/kT,$$

* J. H. Van Vleck, *Electric and Magnetic Susceptibilities*, p. 269, 1932.
† *Ibid.* p. 245.

equation (26) becomes

$$M_{\mathscr{H}} = Ng\mu_B \sum_{M_J=-J}^{M_J=J} M_J e^{M_J x} \Big/ \sum_{M_J=-J}^{M_J=J} e^{M_J x}$$

$$= Ng\mu_B \frac{d}{dx}\left(\log_e \sum_{M_J=-J}^{M_J=J} e^{M_J x}\right)$$

$$= Ng\mu_B \frac{d}{dx}\left(\log_e \frac{e^{Jx}(1 - e^{-(2J+1)x})}{1-e^{-x}}\right)$$

$$= Ng\mu_B \frac{d}{dx}\left\{\log_e\left(\frac{e^{(J+\frac{1}{2})x}-e^{-(J+\frac{1}{2})x}}{e^{x/2}-e^{-x/2}}\right)\right\}$$

$$= Ng\mu_B \frac{d}{dx}\log_e\left(\frac{\sinh(J+\frac{1}{2})x}{\sinh x/2}\right)$$

$$= NgJ\mu_B\left[\frac{2J+1}{2J}\coth\left(\frac{2J+1}{2J}\right)a - \frac{1}{2J}\coth\left(\frac{a}{2J}\right)\right]$$

or $\quad \dfrac{\bar{\mu}}{\mu} = M_{\mathscr{H}}/NgJ\mu_B$

$$= \frac{2J+1}{2J}\coth\left(\frac{2J+1}{2J}\right)a - \frac{1}{2J}\coth\left(\frac{a}{2J}\right). \quad \ldots\ldots(30)$$

The expression on the right side of equation (30) is known as the *Brillouin function*.

There are two cases of special interest, viz. when $J = \frac{1}{2}$ and when $J \to \infty$. When $J = \frac{1}{2}$ equation (30) reduces to

$$\bar{\mu}/\mu = \tanh a = a - \tfrac{1}{3}a^3 + \tfrac{2}{15}a^5 - \tfrac{17}{315}a^7 + \ldots \quad \ldots\ldots(31)$$

When $J \to \infty$ equation (30) gives, in the limit,

$$\bar{\mu}/\mu = \coth a - 1/a = L(a).$$

It is found that equation (30) expresses very satisfactorily the results obtained by Woltjer and Kamerlingh Onnes on the susceptibility of hydrated gadolinium sulphate, $Gd_2(SO_4)_3 \cdot 8H_2O$ at liquid helium temperatures. Here, with very high fields applied at very low temperatures we get the phenomenon of *paramagnetic saturation*.

Paramagnetism Independent of Temperature. It has already been mentioned on p. 41 that a rigorous wave mechanics

treatment shows that to the expression for magnetic susceptibility deduced with the vector model of the atom a constant
term, due to what is called a non-diagonal element in the
vector matrix, must be added. This term may be somewhat
inadequately pictured as arising from the different rates at
which L and S precess about J. For, if the coupling between
L and S is not infinitely great, irregularities in the precession
about J must set in when a magnetic field is applied, resulting
in a component of magnetisation parallel to and proportional
to $\mathscr{H}$. Referring to Fig. 8 we may suppose that μ_R precesses
regularly around J in the absence of a field, but when J is itself
caused to precess about a field $\mathscr{H}$, the precession of μ_R is no
longer symmetrical and a small increment in magnetisation
parallel to $\mathscr{H}$ is produced. Since this increase for a given
L, S, J state is independent of the temperature, we have an
example of paramagnetism independent of the temperature.
Its contribution to the atomic magnetisation is generally
negligible, but with rise in temperature, it becomes relatively
more important because the normal paramagnetism is decreasing, and it is not negligible when the L and S coupling is
very weak, as in the case of Eu^{+++}. Van Vleck* has shown that
it should be particularly important in the middle of a period,
where L and S are antiparallel, making the coupling between
successive energy levels very great. It is observed with strongly
coloured ions with $^1S_0$ terms, such as the permanganate ion,
and the ions of the yellow cobaltamines.

The Stark Effect. We shall see later that the effects of the
internal electric fields or crystalline fields in solids are very
important in magnetic theory, and we must now discuss the
effect of an applied electric field upon an atomic system. We
find that if we take an electric field in a given direction, the
orbital motion of each electron is affected while the spin is not.
In a weak field the L, S coupling is unaffected and the L and S
vectors precess about the resultant J; the latter then precesses
about the direction of the electric field as axis with $2J+1$
discrete settings such that the projection M_J along the field

* J. H. Van Vleck, *Phys. Rev.* **31**, 587, 1928.

43

is given by $M_J = J, J-1, ..., -(J-1), -J$. However, as the same energy is now associated with equal and opposite values of M_J we obtain only $J + \frac{1}{2}$ or $J + 1$ energy states instead of the $2J + 1$ associated with an applied magnetic field. Hence we get a splitting of the lines emitted by the atomic system, known as the *Stark effect*.

In an electric field sufficiently strong to break down the L, S coupling, only the L vector precesses about the field, giving a projection M_L upon it, where $M_L = L, L-1, ..., -L$. Again, the energy is the same for equal and opposite values of M_L. Now, while the spin is unaffected by the electric field, S must precess about the magnetic field associated with L, and so must precess about the field; hence we get a projection M_S along the field, where $M_S = S, S-1, ..., -S$.

The Magnetic Properties of Free Molecules. Attention has so far been confined to electron states in single atoms and ions, and we must now consider the electron levels in free molecules, because of their importance in connection with the magnetic properties of the diatomic gases oxygen and nitric oxide. Now, whereas the electric field in an atom may be assumed spherically symmetrical, the field in a non-rotating diatomic molecule is symmetrical about an axis joining the two nuclei, and to a first approximation may be treated as a strong axial field superimposed upon the spherical field which would result if the two nuclei coalesced to form the single core of an "united atom". Hence, the electron states of the molecule are similar to those of an atom in a strong electric field, described above.

It is sometimes helpful to compare the electron sub-groups in a molecule with those in an atom possessing an equal number and same type of valency electrons. Thus we may compare NO and Al as follows:

NO $(1s\sigma)^2$ $(2p\sigma)^2$ $(2s\sigma)^2$ $(3p\sigma)^2$ $(3d\sigma)^2$ $(2p\pi)^4$ $(3d\pi)$, $^2\Pi_{\frac{1}{2}}$

Al $\quad$ $1s^2$ $\quad\quad$ $2s^2$ $\quad\quad$ $2p^6$ $\quad\quad$ $3s^2$ $\quad$ $3p$, $\quad$ $^2P_{\frac{1}{2}}$

In such statements the symbols σ, π, δ, ... indicate that λ, the resolved angular momentum of the individual electron along the nuclear axis, is 0, 1, 2, ...

44

In the case of a molecule the symbols Λ, Σ and Ω are used to denote the values of M_L, M_S and their resultant projection M corresponding to a given multiplet level. Thus

$$\Lambda = M_L = 0, 1, 2, ..., L, \quad \Sigma = M_S = S, S-1, ..., -S,$$

and $\quad \Omega = M = \Lambda + \Sigma = \Lambda + S, \Lambda + S - 1, ..., \Lambda - S.$

The energy states in the molecule are accordingly characterised by the values of Λ and S. States with $\Lambda = 0, 1, 2, 3, ...$ are termed $\Sigma, \Pi, \Delta, \Phi, ...$, corresponding to atomic S, P, D, F, ... states; states with $S = 0, \frac{1}{2}, 1, ...$ are singlet, doublet, triplet, ... states, while $\Lambda + \Sigma$ corresponds to M_J in the atomic case. The energy state of a molecule is specified by $^{2S+1}\Lambda_{\Lambda+\Sigma}$. (Confusion between the general symbol Σ and the symbol Σ representing the value $S = 0$ must be avoided.)

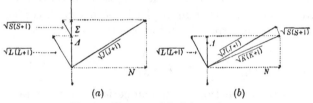

Fig. 11. Vector model of the molecule.

We must now consider what happens when the molecule is no longer fixed, but rotates about an axis perpendicular to the line of nuclei with a nuclear angular momentum specified by a quantum number N. A magnetic field then exists parallel to the axis of rotation, i.e. to the vector representing N, and an outer electron in the molecule is exposed to a resultant magnetic field due to L and to N, as well as to the electric field. Hence, the angular momentum of the pair of nuclei and that of the orbital and spin motions interact magnetically, and of the ways in which these angular momenta may be coupled, two, designated case (a) and case (b) by Hund, are of importance in magnetic theory.

In case (a) the coupling between S and Λ is much stronger than that between S and N and the resultant $\Omega = |\Lambda + \Sigma|$ compounds vectorially with N to form a resultant J, with $\sqrt{J(J+1)}\,h/2\pi$ units of angular momentum, about which the vectors N and Ω precess, as shown in Fig. 11a. J takes the values Ω, $\Omega+1$, $\Omega+2$, ..., and for each of the $2S+1$ values of Ω there is a set of rotational energy sub-levels denoted by a set of J values beginning with the lowest value of $J = \Omega$. From the figure it follows that $N^2 = J(J+1)-\Omega^2$.

In case (b) the coupling between S and Λ is small compared with the effects of molecular rotation and Σ disappears. Λ and N interact to form a resultant K vector, with values Λ, $\Lambda+1$, $\Lambda+2$, ... Now, the magnetic field parallel to K causes the K and S vectors to form a resultant J in a fixed direction about which they precess, as shown in Fig. 11b, and J takes the $2S+1$ values, $|K+S|$, $|K+S|-1$, ..., $|K-S|$, and we have, in general, $2S+1$ component rotational levels differing in J values for each value of K.

Van Vleck* has calculated the magnetic susceptibility for a diatomic molecule with multiplet intervals small compared with kT, and finds that, both for case (a) and case (b),

$$\chi_M = \frac{N^2\mu_B^2}{3RT}\{\overline{(2\Sigma-\Lambda)^2}\} = \frac{N^2\mu_B^2}{3RT}[4S(S+1)+\Lambda^2]$$

+ a small correction for diamagnetism and for paramagnetism independent of the temperature. ......(32)

When, however, the multiplet interval is large compared to kT, when case (a) generally obtains, the susceptibility is given by

$$\chi_M = \frac{N^2\mu_B^2}{3RT}(\Lambda+2\Sigma)^2, \qquad ......(33)$$

since Σ now takes only that value which gives the lowest energy, and the component of the magnetic moment along the nuclear axis takes the place of μ in equation (11) of p. 16.

Now, the oxygen molecule normally exists in a $^3\Sigma$ state (the subscript 0 is usually omitted), while most other diatomic gas molecules exist in a diamagnetic $^1\Sigma$ state. Hence, on substituting $S = 1$ and $\Lambda = 0$ in equation (32), we get

$$\chi_M = \frac{8N^2\mu_B^2}{3RT} = \frac{0\cdot993}{T}.$$

* J. H. Van Vleck, *Electric and Magnetic Susceptibilities*, p. 265, 1934.

This gives an effective magneton number of 2·83 and agrees with the experimental result that the susceptibility follows the Curie law, the calculated value of χ_M at 20° C. being 3390×10^{-6}, while the experimental value is about 3420×10^{-6} c.g.s. units per gm. mol.

The nitric oxide molecule normally exists in a $^2\Pi_{\frac{1}{2}}$ state at low temperatures and in a $^2\Pi_{\frac{3}{2}}$ state at high temperatures, the relevant multiplet width being 120·9 cm.$^{-1}$ and, therefore, comparable with kT at low values of T. The effective magneton number at high temperatures, obtained by substituting $S = \frac{1}{2}$ and $\varLambda = 1$ in equation (32), is 2, while that for very low temperatures obtained by substituting $\varSigma = -\frac{1}{2}$ and $\varLambda = 1$ in equation (33) would be zero. Hence, nitric oxide does not follow the Curie law. Van Vleck has deduced an expression for the molar susceptibility at intermediate temperatures which gives very satisfactory agreement with experiment from room temperature to that of the liquefied gas. See Note 2.

Magnetic Properties of Matter in Bulk. The calculation of susceptibility has hitherto been based upon the conception of a paramagnetic gas in which all the particles are uninfluenced by the electric or magnetic fields of their neighbours. The purely magnetic forces between neighbouring atoms are indeed very small and may be neglected, but the phenomenon of ferromagnetism, which is now attributed to an interchange interaction between electron spins, shows that in solids the orbital momenta may be quenched and only the spins left free. This is a state of affairs common to most salts of the iron group. Van Vleck has shown that if strong asymmetrical electrostatic fields acted inside these salts, they could quench the orbital momenta of the 3d electrons, even if they were too weak to change the Russell-Saunders coupling of the atom appreciably, as long as they were strong enough to change the spatial quantisation of the atoms. By *quenching of the orbital momenta*, we mean that on the application of an external field the orbital momenta remain fixed under the influence of the internal field and do not orientate with respect to the external one. The spin momenta are not affected by internal (electro-

47

static) fields, except those of the exchange interaction type between paramagnetic atoms or ions which give rise to ferromagnetism as described on p. 239. We should therefore expect equation (29) to be replaced by

$$p_{\text{eff}} = \sqrt{4S(S+1)} \qquad \ldots\ldots(34)$$

when the orbital moments are completely quenched. Hence, in the general case, where quenching is incomplete, the value of p_{eff} should lie between $\sqrt{4S(S+1)}$ and $\sqrt{L(L+1)+4S(S+1)}$.

This quenching does not occur in salts of the rare earths, because the $4f$ electrons, which determine their magnetic properties, are deep-seated within the atom and are practically unaffected by neighbouring atoms, a statement which is supported by the sharpness of the absorption lines in rare earth salts. On the whole, very little effect of concentration upon the molar susceptibilities of salts in the crystalline state or in solution in different solvents is found, although in the latter case electrostatic fields depending on the dipole constitution of the solvent are present. This is undoubtedly because the orbital momenta are completely quenched and not merely altered in some indeterminate manner.

In the case of diamagnetic salts Hoare and Brindley* have examined the experimental data and find that values of the molar susceptibility calculated from measurements with solutions differ from the values obtained for the same salts in the crystalline state. They find a definite effect of the ions upon the susceptibility of the surrounding water molecules, and consider that small univalent ions and all bivalent ions have smaller molar susceptibilities in solution than in crystalline form because of the high electric fields at the boundaries of the ions. Large univalent ions have larger molar susceptibilities in solution than in crystalline form, mainly owing to the change in the number of water molecules attached to each ion. As one would expect, the molar susceptibilities of hydrated salts are more nearly equal to the values obtained with solutions than with anhydrous salts.

* F. E. Hoare and G. W. Brindley, *Phys. Soc. Proc.* **49**, 619, 1937.

The theoretical treatment of the effects of crystalline fields is very difficult, and, as we shall see later, more experimental data for single crystals at different temperatures are desirable in order to assist the development of the theory; a summary of recent work has been given by Schlapp and Penney.* If it is supposed that an internal electric field acts upon an atomic system, then a kind of Stark effect takes place and the field causes a splitting of the atomic levels. Now, if in the absence of an electric field a large number of atoms are all in a particular J energy state and a magnetic field is applied to them, then $2J + 1$ energy states are established and the average magnetic susceptibility is determined by the distribution of the atoms in the several states. In the presence of an asymmetrical internal electric field strong enough to split the energy level, or remove the degeneracy of the level, the level splits into $2J + 1$ states. The separation of these states naturally depends on the magnitude of the internal field, and the latter may be considered strong or weak according as the splitting is large or small compared with the multiplet separation. There thus results a redistribution of the magnetic moment which may markedly affect the temperature variation of the susceptibility, especially at low temperatures.

The symmetry type of the internal field is of great importance, and, of course, it need not necessarily have the symmetry of the crystal. Excellent agreement between experiment and theory has been obtained in the case of the hydrated sulphates of Pr and Nd on the assumption that the crystal field has cubic symmetry and can be represented by a potential of the form $D(x^4 + y^4 + z^4)$. Fields of a monoclinic or triclinic character may be obtained by superimposing cubic and rhombic fields with different axes, and other fields may similarly be built up. The magnetic behaviour of Ni^{++} and Co^{++} in hydrated sulphates has satisfactorily been explained by a crystalline field of rhombic symmetry but with the departures from cubic symmetry relatively small.

The clearest indication that internal crystalline fields play a part in susceptibility values is given when the Curie law,

* R. Schlapp and W. G. Penney, *Rep. on Progress in Physics*, **2**, 60, 1935.

$\chi_M = C/T$, is not obeyed, and is replaced by the Curie-Weiss law, $\chi_M = C/T - \theta$. The latter is deduced on p. 234 for the special case of a ferromagnetic substance at high temperatures, and θ is there shown to be a parameter which is a direct measure of the intensity of the internal field, and, by analogy, we always attach this significance to θ whenever a Curie-Weiss law is found. In practice θ takes negative as well as positive values for different materials, so that it is presumed that internal fields may on occasion cause spins to set antiparallel and reduce the susceptibility. It is perhaps well to point out that the determination of θ is not always easy, since allowances have to be made for diamagnetism and temperature-constant paramagnetism which are important when θ is very large. See Chapter XIII.

Magnetic Properties of Free Electrons. It is clear that, in the case of a normal (non-ferromagnetic) metal, the atoms may be treated as diamagnetic cores while the valency electrons outside the cores may be treated as free or conduction electrons. In a way, then, the orbital momenta of the atoms are reduced to zero, and, apart from the diamagnetic cores, the properties of the metal are decided by the spin of the conduction electrons. The latter obey the Fermi-Dirac statistics, and when a magnetic field is applied can only set parallel or antiparallel to the field. Now, when a spin sets parallel to the field the energy is decreased, so that more electrons tend to set with parallel than with antiparallel spin, and the former must accordingly occupy higher quantum states, the energy being obtained from the field. In this way we get another type of temperature-constant paramagnetism, because the Fermi-Dirac distribution is so little affected by change in temperature.

At the same time, however, the free electrons exhibit a temperature-constant diamagnetism. The effect of a magnetic field upon moving electrons is of course merely to cause them to move in spiral paths, and on classical theory there is no change of energy except that already described in connection with the spin, and hence there is no diamagnetism. On the quantum theory, the projections of the spiral paths upon a plane perpendicular to the lines of force represent circular periodic motions which must satisfy quantum conditions. These result in the electron changing its energy when the field is excited. As the energy increases, a diamagnetic effect is produced, and it can be shown that this is almost exactly one-third the paramagnetic effect just described, provided that $\mu_B \mathscr{H} \ll kT$.

Stoner* has shown that for ordinary temperatures, where $\xi_0 \gg kT$, the respective paramagnetic and diamagnetic susceptibilities per electron, χ_P and χ_D, are given by

$$\chi_P = \frac{3}{2} \frac{\mu_B^2}{\xi_0} \left\{ 1 + \frac{\pi^2}{12} \left(\frac{kT}{\xi_0}\right)^2 - \frac{1}{6} \left(\frac{\mu_B \mathscr{H}}{\xi_0}\right)^2 \ldots \right\}, \qquad \ldots\ldots(35)$$

$$\chi_D = -\frac{1}{2} \frac{\mu_B^2}{\xi_0} \left\{ 1 - \frac{\pi^2}{12} \left(\frac{kT}{\xi_0}\right)^2 + \frac{1}{10} \left(\frac{\mu_B \mathscr{H}}{\xi_0}\right)^2 \ldots \right\}, \qquad \ldots\ldots(36)$$

where ξ_0 is the maximum electron energy at absolute zero and is equal to $\dfrac{h^2}{2m} \left(\dfrac{3N}{8\pi V}\right)^{\frac{2}{3}}$, N being the number of free electrons in a volume V. The resultant paramagnetism, obtained by adding (35) and (36), is practically independent of the temperature as the term in T^2 is so small. Now, Sucksmith† found that the alkali metals show an increase in paramagnetic susceptibility with rise in temperature. This Stoner attributes to a change in ξ_0 brought about by thermal expansion, an explanation which also appears to hold for the observed decrease of the diamagnetic susceptibility of mercury‡ with rise in temperature.

The Quantum Theory of Diamagnetism. On the Bohr theory the value of $\bar{r}^2$ to be inserted in the Langevin expression for the diamagnetic susceptibility of a hydrogen-like atom is

$$a_0^2 \frac{n^2}{Z^2} (\tfrac{5}{2}n^2 - \tfrac{3}{2}k^2),$$

which gives $\quad \chi_A = -0.79 \times 10^{-6} \dfrac{n^2}{Z^2} (\tfrac{5}{2}n^2 - \tfrac{3}{2}k^2), \qquad \ldots\ldots(37)$

* E. C. Stoner, *Roy. Soc. Proc.* A, **152**, 672, 1935.
† W. Sucksmith, *Phil. Mag.* **2**, 21, 1926.
‡ L. F. Bates and C. J. W. Baker, *Phys. Soc. Proc.* **50**, 409, 1938.

a_0, the radius of an hydrogen atom in its normal state, being 0.528×10^{-8} cm. The more exact theory of Van Vleck replaces this expression by

$$\chi_A = -0.79 \times 10^{-6} \frac{n^2}{Z^2} \left\{ \frac{5}{2} n^2 - \frac{3l(l+1)-1}{2} \right\}. \quad \ldots\ldots(38)$$

Now, we cannot expect to test this result by experiments on isolated atoms, since each hydrogen atom in a $^2S_{\frac{1}{2}}$ state possesses a permanent magnetic moment of $1\mu_B$ which completely swamps its diamagnetism. However, it is interesting to note that the value of χ_M for molecular hydrogen is -4.00×10^{-6}, while from (38) we have

$$\chi_M = 2\chi_A = 2[-0.79 \times 10^{-6}\{\tfrac{5}{2}+0+\tfrac{1}{2}\}] = -4.74 \times 10^{-6},$$

which is in satisfactory agreement with experiment, while the value
$$\chi_M = 2[-0.79 \times 10^{-6}\{\tfrac{5}{2}-\tfrac{3}{2}\}] = -1.6 \times 10^{-6}$$

from (37) is not.

Equation (38) may be extended to atoms with many electrons if we use appropriate *screening constants* to allow for the shielding of the nuclear charge by the electrons in the various rings or levels. Then, following Pauling,* the contribution of an individual electron in a more complicated atom to the atomic susceptibility may be written

$$\chi_{n,l} = -0.79 \times 10^{-6} \frac{n^2}{(Z-\sigma)^2} \left\{ \frac{5}{2} n^2 - \frac{3l(l+1)+1}{2} \right\}, \quad \ldots\ldots(39)$$

where the value of the screening constant σ is approximately calculated for the electrons in each sub-group. On substituting the appropriate values of σ in (39) we may calculate $\chi_A = \Sigma\chi_{n,l}$, but the method, although useful, is not very precise.

Stoner† has successfully applied the results of the *Hartree self-consistent field* calculations to the evaluation of $\Sigma\bar{r}^2$. The Hartree method consists in substituting a trial value of the electric field, based on an assumed electron density in the

---

* L. Pauling, *Roy. Soc. Proc.* A, **114**, 181, 1927.
† E. C. Stoner, *Magnetism and Matter*, p. 258, 1934.

atom, in the Schrödinger equation. From the solutions to this equation for the several electrons the field may be calculated; this is compared with the trial value previously assumed, and if the two agree, or are self-consistent, then the assumed value and electron density may be taken as correct. With much labour the electronic distributions have been found for a number of atoms, and Stoner uses the following expression:

$$\chi_A = -2{\cdot}83 \times 10^{-10} \int_0^\infty r^2\!\left(\frac{dN}{dr}\right) dr, \qquad \ldots\ldots(40)$$

where dN/dr is the charge in electron units in a spherical shell of unit thickness around the nucleus; the integral may be evaluated numerically or graphically. Thus, Stoner gives $\chi_A = -1{\cdot}86 \times 10^{-6}$ for helium, while the experimental value is $-1{\cdot}906 \times 10^{-6}$ (Havens).*

On the basis of approximate wave functions Slater,† Angus‡ and others have given the useful working expression

$$\chi_A = -0{\cdot}79 \times 10^{-6} \sum_z \frac{(n')^2\,(n'+\tfrac{1}{2})\,(n'+1)}{(Z-\sigma)^2}, \qquad \ldots\ldots(41)$$

where n' is a computed *effective quantum number* and $(Z-\sigma)$ the effective nuclear charge. It is arranged that n' and σ have values which give satisfactory agreement between calculated and experimental values of the energy levels, and the following n' replace the corresponding n values:

n	1	2	3	4	5	6
n'	1	2	3	3·7	4·0	4·2

Fundamental Atomic Constants. The values of the fundamental atomic constants have recently been subjected§ to much critical examination and it appears that the most acceptable values are those given below, where the ± terms are estimates of reasonable limits of error and not probable errors.

* G. Havens, *Phys. Rev.* **43**, 999, 1933.
† J. C. Slater, *Phys. Rev.* **36**, 57, 1930.
‡ W. R. Angus, *Roy. Soc. Proc.* A, **136**, 569, 1932.
§ S. von Friessen, *Roy. Soc. Proc.* A, **160**, 424, 1937; R. T. Birge, *Amer. Journ. Phys.* **13**, 71, 1945; and *Rep. on Progress in Physics*, **8**, 90, 1942.

Specific charge of the electron	$1{\cdot}7592 \pm 0{\cdot}0005 \times 10^7$ e.m.u./gm.
Electronic charge	$4{\cdot}8025 \pm 0{\cdot}0010 \times 10^{-10}$ e.s.u.
Planck's constant	$6{\cdot}624 \pm 0{\cdot}002 \times 10^{-27}$ erg. sec.
Avogadro's number	$6{\cdot}0228 \pm 0{\cdot}0011 \times 10^{23}$ mole^{-1}.
Mass of the hydrogen atom	$1{\cdot}673 \pm 0{\cdot}003 \times 10^{-24}$ gm.

From these data the value of the Bohr magneton is 5586 c.g.s. units per gm. atom. Now, most books of reference give values around 5563 and 5565. As the differences between the most recent and the older values are small, the value 5565 will be retained in the following chapters in order to facilitate comparison with other works.

CHAPTER II

THE PRODUCTION AND MEASUREMENT OF MAGNETIC FIELDS

The Solenoid. While a solenoid is extremely valuable for producing uniform fields it is seldom used for the production of fields greater than 1000 oersteds. In determinations[*] of the ratio e/m from measurements of the Zeeman effect a solenoid 80 cm. long wound with 18 layers of No. 4 square, cotton-covered, copper wire was used. The successive layers were separated by fibre spacers and cooled oil circulated between them. With a current of 200 amperes at 270 volts a field of 7300 oersteds was continuously maintained over a 6 cm. length along the axis occupied by a source of light. The field of this main solenoid was calibrated by placing inside it a standard solenoid carrying a current I_s, and mounting a search coil in the middle of both solenoids. When the two fields were in opposition and I_s suitably adjusted there was no deflection of a galvanometer in series with the search coil when the latter was rotated. The required field per unit current was then $\mathscr{H}_s I_s / I$, where I was the current in the main solenoid and $\mathscr{H}_s$ the field produced per unit current in the standard. See Note 3.

Gerloff and Löwe[†] have described the construction of solenoids for producing large uniform fields, and Künze[‡] in a study of cosmic ray tracks produced a field of 18,000 oersteds over the region occupied by a Wilson chamber, with a coil carrying a current of 1000 amperes at 500 volts. On account of the large heating effects the current could be maintained for about 50 sec. only, after which the coil required a day in which to cool. The stray field is considerable and loose pieces of iron can be sent hurtling through the air when the current is switched on.

[*] J. S. Campbell and W. V. Houston, *Phys. Rev.* **39**, 601, 1932; L. E. Kinsler and W. V. Houston, *Phys. Rev.* **45**, 104, 1934.
[†] G. Gerloff and E. Löwe, *Zeit. für Phys.* **98**, 559, 1936.
[‡] P. Künze, *Zeit. für Phys.* **79**, 203, 1932; **80**, 559, 1933

The Properties of Ferromagnetics: Ballistic Investigations. For the production of fields greater than 1000 oersteds ferromagnetic materials in the form of permanent or electromagnets are normally used, and we must now discuss the special magnetic properties of these materials. These are most satisfactorily examined when the material can be obtained in the form of an anchor ring of fairly large diameter and moderate area of cross-section α. A primary coil of N turns wound uniformly upon the whole ring and carrying a current i e.m. units produces a magnetic field $4\pi Ni/2\pi R = 2Ni/R$, which

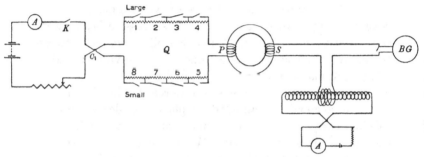

Fig. 12. Ballistic method for B, $\mathscr{H}$ curves.

may be regarded as uniform if R is sufficiently large. Any uncertainty due to lack of uniformity of the field over the cross-section is far outweighed by the absence of demagnetisation effects, since there are no free poles.

Referring to Fig. 12, the primary winding P is joined in series with a set of fixed resistances Q, an ammeter, adjustable resistance and commutator. Each of the resistances in Q is provided with a separate switch which allows it to be short-circuited; these resistances are previously chosen by trial and decrease in magnitude from numbers 1 to 8. On the ring is further wound a secondary coil S of n turns, which need not be uniformly spaced. S is permanently joined in series with the secondary coil of a standard mutual inductance and a sensitive ballistic galvanometer, preferably of the moving magnet type. The primary of the mutual inductance is connected to a battery, ammeter, rheostat and commutator as shown.

Let us suppose that we start with the specimen in the virgin or completely unmagnetised state, previously brought about by heating the specimen to a high temperature, or, more conveniently, by exposing it to an alternating field of slowly diminishing intensity, e.g. by gradually reducing an alternating current supplied to the coil P. Then, on closing the switch K a small current i_0 e.m. units flows in the coil P, and a ballistic deflection θ_0 is recorded by the galvanometer. This deflection corresponds to a change $n\alpha\,dB_0$ in the number of lines of induction linked with the secondary coil S. A small correction, due to the fact that the primary winding is not infinitely thin and therefore some additional lines of force are collected by S, is usually disregarded. The deflection θ_0 is later calibrated by finding the current I_0 whose reversal in the primary of the mutual inductance would produce the same deflection; then $2MI_0 = n\alpha\,dB_0$. We may therefore consider that each galvanometer deflection corresponds to a known change in the induction of the specimen.

On closing the switch 1, the largest of the resistances in Q is shortcircuited, a further increment of current i_1 passes through the primary and a galvanometer deflection θ_1 indicates a further increment of induction dB_1. Proceeding in this way we may successively close the switches 2, 3, ..., 8 and measure the corresponding increments of current and induction. Then,

writing $i = \Sigma i_0 + i_1 + ..., \mathcal{H} = \dfrac{2Ni}{R}$, and $B = \Sigma dB_0 + dB_1 + ...,$

we may tabulate a series of values of i and B and plot a graph of B against $\mathcal{H}$ as in Fig. 13, when the procedure so far described gives the initial portion Oa of the curve. The upper portion of Oa should, generally, run almost parallel to the $\mathcal{H}$ axis; the material is then said to be *saturated* (in a *technical*, as distinct from a theoretical sense). It is found that the purer the metal the higher is the value of B_{max}, the saturation induction. The

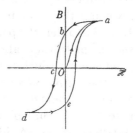

Fig. 13. B, $\mathcal{H}$ curve.

slope of the tangent to the curve at O gives the *initial per-*

meability, while the steepest slope of *Oa* gives the *maximum permeability*. The *initial susceptibility* for very weak fields obeys a law of the form $k = a + b\mathscr{H}$, known as the *Rayleigh law*.

On reopening the switches in the order 8, 7, ..., 1, and, finally, reopening K, the portion *ab* of the graph is obtained, the values of B being obtained by subtracting the measured changes from the value at *a*. The specimen is now no longer exposed to a magnetic field, but it still retains a considerable intensity of magnetisation, equal to $Ob/4\pi$, which is termed the *retentivity* of the material and is a measure of *the ability to retain magnetism when not subjected to adverse treatment*. Some authorities, e.g. Ewing, term the residual induction Ob the *retentiveness*.

The commutator C_1 is now changed over so that the magnetic field will be reversed and the key K and the switches 1, ..., 8 closed in order. In this way, data for the portion *bcd* of the graph are obtained, the successive values of B being obtained by subtraction of the changes in induction from the value at *b*. The graph cuts the $\mathscr{H}$ axis at the point *c*, which means that when a demagnetising field Oc is applied the specimen loses the magnetisation previously acquired in the field at the point *a*. We speak of Oc as the *coercivity*, and it measures *the ability to retain magnetism in spite of adverse treatment*; the force which would act on a unit pole placed in the field Oc is known as the *coercive force*.

The portion *de* of the graph may be obtained by reopening the switches 8, ..., 1 in order and reopening K; and, on further reversing the commutator C_1, closing K and the switches 1, ..., 8 in order, the portion *ea* may be found. It is found that the magnetisation always lags behind the applied field, a phenomenon which is termed *hysteresis*, the complete curve *abcdea* being a *hysteresis cycle*. The curve should be symmetrical about O, although to reach this state it may be necessary to take the ring through about a dozen complete cycles before starting to obtain readings from the point *a*, when the material is said to be in a *cyclic state*.

The area enclosed by the curve *abcdea* is a measure of the

energy wasted in taking unit volume of the material through a complete cycle. Following the procedure on p. 8 we see that the work done in establishing a current i in the primary P is

$$\int_{i_1}^{i_1} Li\,di = \int_{i_1}^{i_1} \frac{2N}{R}(\mu\alpha N)\,i\,di,$$

$$= \int_{i_1}^{i_1} \frac{2N^2}{R}\mu\alpha i\,di = \frac{R\alpha}{2}\int_{i_1}^{i_1} \frac{2Ni}{R}\left(\frac{2N\mu}{R}di\right),$$

$$= \frac{2\pi R\alpha}{4\pi}\int_{\mathscr{H}_1}^{\mathscr{H}_1} \mathscr{H}\,dB \text{ for a volume } 2\pi R\alpha,$$

$$= \frac{1}{4\pi}\int_{\mathscr{H}_1}^{\mathscr{H}_1} \mathscr{H}\,dB \text{ per unit volume,}$$

$$= \int_{\mathscr{H}_1}^{\mathscr{H}_1} \mathscr{H}\,dI \text{ per unit volume,}$$

as the limits of integration are the same.

Magnetometer Investigations. When the specimen cannot be obtained in the form of a solid anchor ring, it is sometimes possible to use a composite ring or frame of stampings or a core of metal tape closely wound to give a ring of rectangular section, with which ballistic measurements may be made; it is necessary, of course, to know the effective cross-section and volume of the metal in the ring. If, however, the material can be obtained only in the form of rods or bars, at least two other methods of investigation are possible.

The specimen may be placed inside a solenoid—preferably with its axis vertical—which is appreciably longer than the specimen, and the magnetic moment induced by the action of a series of applied fields may be measured with a mirror magnetometer. Two difficulties arise; first, free poles are set up at the ends of the specimen and produce demagnetising effects which must be taken into account, and secondly, there is much uncertainty as to the exact location of these poles. Further, elementary considerations show that the intensity of magnetisation is greatest at the middle of the rod. Consequently, if an induction helix were wound upon the middle portion of the rod and connected to a ballistic galvanometer,

59

we might measure by the ballistic method changes in induction, but they would be greater than the average over the whole rod.

Referring to Fig. 14a, it is advisable to place the specimen vertical and to raise or lower it until the pole at one end is level with the magnetometer needle M. Then, if the intensity of magnetisation is I and the horizontal component of the earth's field H, the magnetometer deflection θ is given by

$$I\alpha\left(\frac{1}{d_1^2} - \frac{1}{d_2^2}\frac{d_1}{d_2}\right) = I\alpha d_1\left(\frac{1}{d_1^3} - \frac{1}{d_2^3}\right) = H\tan\theta. \quad\ldots\ldots(42)$$

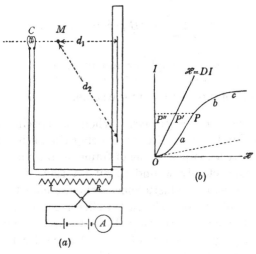

Fig. 14 a. Magnetometer method for I, $\mathscr{H}$ curve.
Fig. 14 b. Correction of I, $\mathscr{H}$ curve.

Hence, I is known and the applied field $\mathscr{H}$ is known from the solenoid current and constant. In this case an accurate knowledge of d_1 is much more important than an accurate knowledge of d_2. The coil C is a compensating coil to neutralise the effect of the solenoid upon the magnetometer. A coil wound upon the solenoid and supplied with current from a separate battery, not shown in the figure, is used to neutralise the vertical component of the earth's magnetic field; this is imperative when alloys of high initial permeability are being used. From equation (42) it follows that H must be accurately known. This, however, may be avoided by using a null method described by Bates,* or by passing a known current through the coil C or a similar coil and noting the deflection θ' of the magnetometer produced by the known field of the coil. The applied field may be varied continuously by the sliding rheostat R, which is often supplemented with an electrolytic resistance for fine, continuous adjustment.

The free poles at the ends produce a demagnetising effect which is assumed equal to DI, where D is a constant known as the *demagnetisation factor*. Unfortunately, we can calculate D with accuracy only in the case of an ellipsoidal specimen placed with its long axis parallel to the field; for an ellipsoid of revolution Maxwell found

$$D = 4\pi\left(\frac{1}{e^2}-1\right)\left(\frac{1}{2e}\log\frac{1+e}{1-e}-1\right),$$

where the semi-axes are $a = b = \sqrt{1-e^2}\,c$. Using this result the values of D corresponding to a series of values of c/a may be calculated. In the case of a long rod of narrow cross-section the ratio of the length to the diameter of the rod may be taken equal to the ratio c/a. A series of values of D for ellipsoids and rods are given in Table III.

Hence, if we plot the values of I found experimentally and using equation (42) against the corresponding values of $\mathcal{H}$ calculated from the solenoid current, we obtain a curve like $Oabc$ of Fig. 14b. Now, on account of the demagnetisation the plotted values of $\mathcal{H}$ are too large, and each must be decreased by an appropriate quantity DI. This may be done by drawing the line $\mathcal{H} = DI$. Considering the point P, the effective field producing the magnetisation OP'' is not $P''P$ but

$$P''P - P''P' = P'P.$$

We see, then, that the curve $Oabc$ is sheared with respect to the I axis, and should be corrected by plotting values of OP'' against the corresponding values of $P'P$. Now, the angle $P''OP'$ is clearly the greater the shorter and thicker the

* L. F. Bates, *Journ. Sci. Instr.* 8, 376, 1931.

specimen. When the specimen is very short and thick the magnetisation it may acquire is limited, for the demagnetising field balances the applied field over a wide range of values of the latter, and we then have $I = \mathcal{H}/D$ irrespective of the true value of the initial permeability of the material.

The magnetometer method has special advantages when we wish to measure the permeability under different physical conditions, e.g. under torsional or other stress, and when time effects are under investigation. It is, of course, possible to wind an induction helix upon a rod specimen and to measure the induction ballistically, but in this case the demagnetisation factor is slightly less than that appropriate to the magnetometer method, because the induction helix is wound in a region where the effect of the free poles is a minimum. It is found that the demagnetisation factor is constant only below $B = 10,000$ in the case of iron. A method of estimating the value of D for a very short specimen is described on p. 260.

TABLE III. VALUES OF D

Rates c/a or l/d	Ellipsoids (calculated)	Rod		
		Magnetometer* measurements	Ballistic† measurements	Ballistic‡ measurements
20	0·0848	0·0898	0·0672	0·0675
25	0·0587	0·0628	0·0464	0·0476
30	0·0432	0·0460	0·0344	0·0364
40	0·0266	0·0274	0·0211	0·0223
50	0·0181	0·0183	0·0144	0·0153
100	0·0054	—	—	—
200	0·0016	—	—	—
400	0·0004	—	—	—

* Mann's measurements quoted by J. Würschmidt, *Zeit. für Phys.* **19**, 388, 1923.

† C. L. B. Shuddemagen, *Phys. Rev.* **31**, 165, 1910, $d = 0·6$ to $2·0$ cm. The factors for $d = 0·36$ cm. are some 10 per cent higher.

‡ J. Würschmidt, *loc. cit.* The rod was systematically tapped during these measurements. The data in the last two columns refer to rods approximately 0·6 cm. in diameter. See Note 4.

Magnetic Viscosity. Ewing and Rayleigh showed that when long wires of soft iron are placed inside a solenoid in which a current is either gradually or suddenly changed there occurs a distinct, slow change of induction as manifested by a magnetometer deflection, for some time after the current has reached a steady value, especially when weak magnetising fields are used. Hence, it may happen that ballistic measurements of magnetic induction give results slightly lower than magnetometer measurements. The difference, which does not normally exceed 1 or 2 per cent, is attributed to *magnetic viscosity*. When a ferromagnetic is exposed to an alternating field, losses due to hysteresis, eddy currents and viscosity must arise, and, on account of their importance in telephony the latter have recently been studied. Carbonyl-iron in the form of tape exhibits magnetic viscosity to an extraordinary extent when recrystallised by heating for 2 hours in a stream of commercial hydrogen at 1000° C., and Richter* showed that the viscosity and ordinary hysteresis effects are not clearly related. Schulze† by a neat power loss method discovered that the magnetic viscosity of the material could be divided into two parts, one of which was a function of the frequency and varied with the temperature of the iron; he explained this phenomenon on the assumption that the annealed material contained parallel, threadlike Weiss domains, p. 238, magnetised in parallel and antiparallel directions only. As the threads increase in length the energy associated with the demagnetisation effects decreases while that associated with their surfaces of separation increases, so that at each temperature there must be a stable state with a definite degree of thread formation.

The Magnetic Potentiometer. To understand another method of investigating the ferromagnetic properties of a bar or sheet of material we must first discuss the use of the magnetic potentiometer, a device invented by Chattock‡ and reinvented by Rogowski, § while Wolman‖ designed the form described below. Suppose that a rod of insulating material of

* G. Richter, *Ann. der Phys.* 29, 605, 1937.
† H. Schulze, *Wissen. Veröff. aus den Siemens-Werken*, 17, 151, 1938.
‡ A. P. Chattock, *Phil. Mag.* 24, 94, 1887.
§ W. Rogowski and W. Steinhaus, *Arch. für Elektrotechnik*, 1, 141, 1912.
‖ W. Wolman, *Arch. für Elektrotechnik*, 19, 385, 1928.

circular or rectangular cross-section is uniformly wound with a very large number of turns of thin insulated wire. Further, let it now be bent into a semicircle so that the two ends of the rod will now rest upon the same plane with the ends of the winding flush with the ends of the rod. Consider it placed in a magnetic field as shown in Fig. 15a, then the total number of lines of force linked with the winding will be

$$\int_B^A n\mathscr{H}_s \alpha \, ds = \alpha n \int_B^A \mathscr{H}_s ds = \alpha n V_{AB}$$

$$= \alpha n \mathscr{H} \, l, \qquad \qquad \ldots\ldots(43)$$

when AB is parallel to a uniform field. In the above expressions α is the mean area of cross-section of the potentiometer, n the

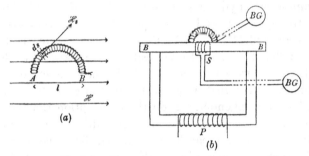

Fig. 15 a. Magnetic potentiometer.
Fig. 15 b. Use of magnetic potentiometer.

number of turns per cm., $\mathscr{H}_s$ the field normal to ds and V_{AB} is the difference in magnetic potential between the points A and B, which are at the mid-points of the potentiometer ends and a distance l apart. Hence, on joining the potentiometer to a ballistic galvanometer and suddenly removing the potentiometer from the field we should get a deflection proportional to $\mathscr{H}$. The instrument may therefore be calibrated by placing it inside a standard solenoid and reversing a known current i, when the galvanometer deflection should be proportional to $8\pi n i$.

Turning now to Fig. 15b, the bar specimen BB is clamped to a yoke which may be excited by a current in the coil P. A winding S upon the bar can be connected to a ballistic galvanometer, which permits the measurement of changes in the induction in the bar. The potentiometer is placed upon the bar, with its ends outside S, and may also be connected to a ballistic galvanometer to measure the effective field acting upon the bar in the region where the induction is measured. Hence the actual field producing a measured induction is found. The main difficulty lies in winding the potentiometer, with, say 3000 turns, sufficiently uniform to enable it to be used in non-uniform fields. Small potentiometers with $AB = 3$ cm. approximately and about 500 turns have been used by Bates* and Lloyd-Evans to investigate the field in the neighbourhood of the pole tips of electromagnets.

Measurements with Short Specimens. The investigation of the magnetic properties of very short specimens of ferromagnetics is described in Chapter IV. Commercial specimens in the form of short bars or stampings are investigated by means of a testing instrument called a permeameter. Thus, if the bar BB of Fig. 15b were replaced by a standard bar of the same dimensions, we could compare the inductions for known currents in the coil P by the corresponding deflections of the galvanometer when joined to the coils S. We might then refer to the apparatus as a permeameter. The reader will find a full account of such instruments elsewhere.†

Reversible and Incremental Permeability. Let us suppose that we are taking a specimen of material through a hysteresis cycle, and that we reach a point P on Oa or ea of Fig. 13. Then, if instead of proceeding to higher field values we reduce the field by a small quantity $\Delta\mathcal{H}$, the magnetisation falls by a small quantity ΔI and the induction by a small quantity ΔB. If these changes are sufficiently small they are reversible, i.e. unaccompanied by hysteresis. The ratios $\Delta I/\Delta\mathcal{H}$ and $\Delta B/\Delta\mathcal{H}$ are then known as the *reversible susceptibility* and the *reversible permeability* for the material in the state represented by the

---

* L. F. Bates and B. Lloyd-Evans, *Phys. Soc. Proc.* **45**, 425, 1933.
† T. Spooner, *Properties and Testing of Magnetic Materials*, London, 1927. See also R. L. Sanford, *Congr. Intern. d'Électricité*, **3**, 395, 1932, and *Bur. of Stds. Journ. of Research*, **16**, 563, 1936, and **23**, 415, 1939.

point P. Gans* has shown that if k_0 is the initial susceptibility at the origin, k_r the reversible susceptibility at P, I_s the saturation intensity of magnetisation and I the intensity at a point P on Oa, then $k_r/k_0 = f(I/I_s)$. Gans has given an expression for $f(I/I_s)$ but a theoretical basis for this expression has only recently† been found.

In modern alternating current practice the core of a transformer is often exposed to a steady field upon which an alternating field is superimposed, so that an unsymmetrical hysteresis cycle is continuously described. If the difference between the extreme values of the induction is divided by the difference between the extreme values of the field the quotient is termed the *incremental permeability*, and, in the limit, for very small differences reduces to the reversible permeability defined above. The symbols used by communications engineers are now being standardised.‡ Specimens on which alternating field tests are made must be laminated to reduce eddy current losses, and reference may be made to the work of Sims and Greig for details of modern technique.§

The Steinmetz Coefficient. A relation between the energy W_B lost per unit volume per cycle and the maximum induction B_{max} attained during the cycle was deduced empirically by Steinmetz.‖ It takes the form

$$W_B = \eta B_{max}^x, \qquad \ldots\ldots(44)$$

where η is a constant, depending on the material, known as the *Steinmetz coefficient*, and x is also a constant, depending on the material, approximately equal to 1·6. Equation (44) is not entirely reliable but may be used to compare the hysteresis properties of different materials.

Special Ferromagnetic Materials. Ferromagnetic materials may for practical purposes be divided roughly into three

* R. Gans, *Phys. Zeit.* 12, 1053, 1911, and *Ann. der Phys.* 61, 379, 1920.
† W. F. Brown, Jr, *Phys. Rev.* 54, 279, 1938.
‡ B. A. *Report*, Nottingham Meeting, 1937, Section G.
§ L. G. A. Sims, *Engineering*, 143, 23, 1937 and 144, 387, 1937, and J. Greig, *Engineering*, 144, 439, 1937.
‖ W. Steinmetz. Cf. *Bur. of Stds.* Circular No. 17, J. D. Ball, *Gen. Elect. Rev.* 19, 369, 1916, and E. V. Legg, *Bell Syst. Tech. Journ.* 18, 438, 1939.

FERROMAGNETIC ALLOYS

classes, viz. those with high values of the maximum induction in moderate fields, those with high initial and maximum permeabilities and those with pronounced hysteresis and high coercive forces. The first class is suitable for the construction of electromagnets, the second for the transformers, loading coils, etc. of the communications engineer, and the third for the manufacture of permanent magnets; the latter materials will be discussed in a subsequent section.

It has long been known that the purer the iron the more suitable it is for the construction of electromagnets. Weiss showed, however, that the addition of 50 per cent commercial cobalt to the iron gave an alloy with a maximum induction of over 23,000 instead of 20,000 for pure iron. This material is brittle and difficult to work, and it is usual to add about 2 per cent vanadium to make it easier to work; and, incidentally, the vanadium increases the electrical resistivity more than three-fold without affecting the magnetic properties to a marked extent. A useful alloy for pole tips, etc. is D.C.I. alloy,* consisting of 38 per cent cobalt and 62 per cent iron, which gives a maximum induction of over 24,000 and can be cast readily, forged and rolled into bars.

Pure iron has an initial permeability of about 250, which may be increased to about 400 by the addition of some 4 per cent silicon, which also raises the resistivity five-fold. The initial and maximum permeabilities of an iron wire may be enormously increased by heating it in an atmosphere of moist hydrogen† at 1400° to 1500° C. for about 18 hours and subsequently annealing it at a temperature just below 910° C., care being taken to let it cool slowly over the range 910° to 890° where the γ to α phase change occurs. See Note 5.

Nickel and iron combine to form alloys, called permalloys, with extremely high initial permeabilities, provided that the appropriate heat treatment is given. Elmen‡ kept his preparations at about 900° for an hour, cooled them slowly to room temperature, reheated them to 600° C. and subsequently

* Made by Darwins, Ltd., Sheffield.
† P. P. Coiffi, *Phys. Rev.* 45, 742, 1932, and 55, 673, 1939.
‡ G. W. Elmen, *Journ. Franklin Inst.* 297, 583, 1929. For other references see J. C. Chaston, *Elect. Communication*, 15, 38, 1936.

cooled them to room temperature on a copper plate in the open air. This process he calls "air quenching", and it is particularly effective in the case of a permalloy containing 78·5 per cent nickel and 21·5 per cent iron. The heat treatment is most important and its effects may be increased by the judicious application of a magnetic field during the treatment. For example, Bozorth and Dillinger* applied a field of only 10 oersteds to a 65 per cent nickel permalloy as it cooled some 200° from a temperature just above its Curie point, and found that the maximum permeability was thereby increased fifty-fold. It is supposed that strains set up by magnetisation are released by heat treatment, so that, after cooling, the material may be more easily magnetised.

It is an interesting fact that the addition of traces of certain metals, in particular, chromium, manganese, molybdenum, silicon, and, especially, copper to nickel iron alloys increases their initial permeability in an extraordinary manner. This is presumably the secret of the high initial permeability of *mu-metal*, which is 10,000 to 30,000, while its maximum permeability is 60,000 to 100,000 for fields of the order 0·025 to 0·04 oersted. The resistivity of these alloys is also high and this makes them particularly valuable for use in small transformers, etc.

A group of materials with strange magnetic properties are the *perminvars*, alloys containing 10 to 40 per cent iron, 10 to 80 per cent nickel and 10 to 80 per cent cobalt. They are characterised by low hysteresis losses, and specimens baked at 400° C. exhibit constant permeability in weak fields. Moreover, specimens which have been air quenched give hysteresis cycles of peculiar shape, each with a "waist" in the region of the origin, a phenomenon which Auwers and Kühlewein† have examined in detail.

The Magnetic Circuit. We now turn to a conception which is particularly useful in practical work, as it enables us to make speedy, if only approximate, calculations of the magnetising currents necessary to produce stated fields with particular

* R. Bozorth and J. Dillinger, *Physics*, **6**, 279, 1935.
† O. v. Auwers and H. Kühlewein, *Ann. der Phys.* **17**, 107, 1933.

arrangements of magnets, solenoids, etc. Suppose we have a coil of N turns carrying a current C amperes and that we take a unit pole once around a closed path threading through the coil. Then, if we move in opposition to the field we do a quantity of work $\frac{4\pi}{10} NC$ ergs. The work done in taking the unit pole over a short distance dl is $\mathscr{H} dl$ and the total work over the complete path, $\int \mathscr{H} dl$, is known as the *line integral of magnetic force* or the *magnetomotive force* (M.M.F.), and is expressed in gilberts.

Hence
$$\int \mathscr{H} dl = \frac{4\pi}{10} NC. \qquad \ldots\ldots(45)$$

Field intensities are sometimes defined in terms of gilberts per cm. The product NC is known as the *ampere-turns linked with the circuit*, and, in general, when several coils are used the complete statement is

$$\text{M.M.F.} = \int \mathscr{H} dl = \frac{4\pi}{10} \Sigma NC. \qquad \ldots\ldots(46)$$

In the same way that we picture the E.M.F. of a battery as responsible for the passage of electricity through an electric circuit, so we may picture the M.M.F. of the coil system of an electromagnet as responsible for the flux of magnetism through the magnetic circuit. Now, the quantity $\mathscr{H} dl$ may be re-written as follows:

$$\mathscr{H} dl = \frac{B}{\mu} dl = \frac{Ba}{\mu a/dl} = \frac{d\phi}{\mu a/dl},$$

where $d\phi$ is the flux through a small area a. Hence

$$d\phi = \mathscr{H} dl/dl/\mu a = \mathscr{H} dl/dZ,$$

where dZ is termed the *magnetic reluctance* of an element of the circuit of length dl and area of cross-section a, and is analogous to resistance $R = \rho . dl/a$ in the electrical case; the reciprocal, $1/Z$, is known as the *magnetic conductance*. The total flux ϕ through the circuit is given by

$$\phi = \int d\phi = \int \mathscr{H} dl \Big/ \int dZ = \frac{\text{M.M.F.}}{Z}. \qquad \ldots\ldots(47)$$

To take a straightforward example, consider an iron anchor ring of large mean radius R and radius of cross-section r, in which a small gap of thickness g is cut. Then the total magnetic reluctance Z is $\dfrac{2\pi R - g}{\mu \pi r^2} + \dfrac{g}{\pi r^2}$, since the permeability of air is unity; it is here assumed that there is no magnetic leakage, i.e. that lines of force are not found outside the gap. Hence the M.M.F. required to give a total flux ϕ across the gap is equal to $\phi \left\{ \dfrac{2\pi R - g}{\mu \pi r^2} + \dfrac{g}{\pi r^2} \right\}$ in these circumstances. Two points require comment; first, the value of μ depends upon the M.M.F. used, and, secondly, the term $g/\pi r^2$ may be extremely large compared with the other term when μ is large, showing that most of the reluctance is in the air gap.

If we consider an electromagnet of the form shown on p. 71, the total reluctance consists of the reluctance of the yoke plus the reluctances of the pole pieces and pole tips, plus the reluctance of the air gap, there being as many terms in the expression for the total reluctance as there are components in the magnetic circuit. In each portion of the circuit leakage occurs; e.g. all lines of force from the coils do not enter the iron. In general, we must write

$$\phi = (\text{M.M.F.}) \div q_1 \frac{l_1}{\mu_1 a_1} + q_2 \frac{l_2}{\mu_2 a_2} + q_3 \frac{l_3}{\mu_3 a_3}, \quad \ldots\ldots(48)$$

where q_1, q_2, ... are *leakage factors* which must be known, either from theoretical considerations or from direct experiment.

Electromagnet Design. For ordinary laboratory purposes the design of an electromagnet usually follows the lines shown in Fig. 16. The yoke BC is a forging of dead mild carbon steel which is magnetically as soft as possible. The pole pieces, cores or limbs AA' and $D'D$ are also made of soft iron or of dead mild steel. Pole tips of special shape and material may be affixed at A' and D' by steel rods which pass through the pole pieces, and their distance apart is adjusted by the handwheels at A and D. The pole tips are usually truncated cones with an angle of $120°$.

For, referring to Fig. 17, suppose we have a cone of angle 2θ with a uniform intensity of magnetisation I_0, then the whole of the free magnetism upon the slant surface P of a ring between sections of radii r and $r + dr$ is $M = I_0 2\pi r\, dr$. This magnetism produces a field at O parallel to the axis equal to

$$\frac{M}{OP^2}\frac{OQ}{OP} = \frac{Mx}{(r^2 + x^2)^{\frac{3}{2}}} = f. \qquad \ldots\ldots(49)$$

Now f is a maximum when

$$df/dx = \frac{M}{(r^2 + x^2)^{\frac{3}{2}}} - \frac{3Mx^2}{(r^2 + x^2)^{\frac{5}{2}}} = 0,$$

i.e. when $\qquad r^2 + x^2 - 3x^2 = 0 \quad$ or $\quad x = r/\sqrt{2},$

i.e. when $\qquad \tan\theta = \sqrt{2} \quad$ or $\quad \theta = 54°\,44'.$

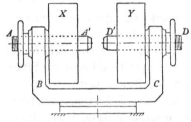

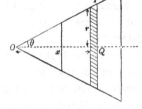

Fig. 16. Construction of an electromagnet.　　　Fig. 17. Design of a pole tip.

Substituting in (49) for M and putting $OP = r/\sin\theta$, the field at O due to the whole ring is

$$I_0 \sin\theta \left(\frac{2\pi r\, dr}{\sin\theta}\right)\left(\frac{\sin^2\theta}{r^2}\right)\cos\theta = 2\pi I_0 \sin^2\theta \cos\theta\, dr/r.$$

Hence for a truncated cone of base b and gap face a the total field F is given by

$$F = 2\pi I_0 \sin^2\theta \cos\theta \int_a^b \frac{dr}{r}.$$

Remembering that there are two pole tips, the gap field is $2F$ and equal to $4\pi I_0 \sin^2\theta \cos\theta \log_e b/a.$

Now, in practice the pole pieces are not completely saturated and it is necessary to make θ somewhat greater, and 60° is usually chosen, for which the value of the gap field for cobalt steel pole tips with $I_0 = 1900$, is

$$4\pi(1900)\tfrac{3}{8}\,2{\cdot}303\log_{10}b/a = 2{\cdot}07 \times 10^4 \log_{10}b/a.$$

If θ is made greater than 60° the field in the gap is reduced because the flux falls off too slowly; if much less than 60° the field is much too intense near the pole faces and the pole tips become saturated.

For the condition of most uniform field between the pole faces we have from (49) the condition

$$\frac{d^2f}{dx^2} = \frac{9Mx}{(r^2+x^2)^{\frac{5}{2}}} - \frac{15Mx^3}{(r^2+x^2)^{\frac{7}{2}}} = 0,$$

whence $x = r\sqrt{\tfrac{3}{2}}$ or $\tan\theta = \sqrt{\tfrac{2}{3}}$, i.e. $\theta = 39°\,14'$, which for cobalt steel pole tips gives a gap field of

$$4\pi(1900)\,(0{\cdot}6324)^2\,(0{\cdot}7745)\,2{\cdot}303\log_{10}b/a = 1{\cdot}71 \times 10^4 \log_{10}b/a.$$

Returning to Fig. 16 the magnet is excited by a current supplied to the coils X and Y, a good working value being 10 to 20 amperes from a 220 volt supply. The coils are wound with thick cotton-covered wire and may be oil cooled. When heavier currents are employed, the coils are wound with bare copper strip or tubing which must be cooled either by forcing air or oil through the windings or passing water through the tubing. It is impossible to construct a powerful electromagnet without using a substantial quantity of iron, although by increasing the weight of a magnet we do not very markedly increase the maximum field beyond a certain value, but we do increase the area of the gap over which that field may be maintained.

About 90 per cent of the magnetic reluctance is located in the air gap and pole tips, and, consequently, the field which acts upon the yoke is really quite small, being of the order of a few oersteds only, as is self-evident when the huge demagnetising effect of the strong poles of the magnet is remembered. Hence, the yoke must be magnetically as soft as possible to give a high magnetisation in a small applied field. To avoid leakage losses the cores should be as short as possible and they must not be saturated, and, as the number of stray lines increases as we move away from the gap, the cores of large magnets should be not cylindrical but conical to ensure that saturation is never attained.

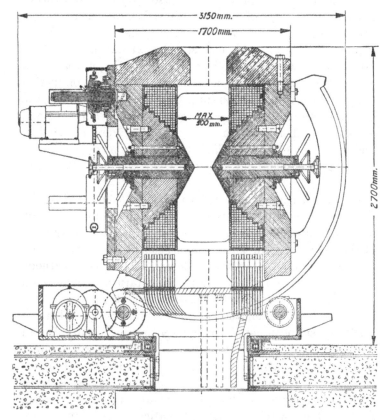

Fig. 18. Design of large magnet at Upsala. (Dreyfus.)

Comprehensive calculations for the design of large magnets have been given by Dreyfus,* and incorporated in the construction of the large magnet at Upsala University,† of which a diagram is given in Fig. 18. The magnet housing consists of a cylindrical body of cast steel to which is bolted a cast steel end block, while a similar block slides in an axial direction. The

* L. Dreyfus, *Arch. für Elektrotechnik*, **25**, 392, 1931.
† L. Dreyfus, *Asea. Journ.* **12**, 8, 1935.

main wrought-iron pole cores are mounted on these blocks; they are 34 cm. long and taper from a diameter of 112 cm. to one of 59 cm., and through each runs a 20 cm. hole carrying an inner core whose position is adjusted by the handwheels. A conical pole piece of wrought iron is fitted to the main core and the pole tips are of cobalt steel. The magnet coils are wound with rectangular copper tube, 21×23 mm.^{2}, with a 10 mm. diameter hole which carries the cooling water. With a current of a little over 4 amperes per sq. mm. of conducting surface of the tube and 150 turns on each of the two coils, the available ampere turns are about 500,000. This magnet, although smaller than that described by Häder* at Leyden, and still smaller than that described by Cotton and Mabboux† at the Academie des Sciences, is perhaps the most efficient in the world.

In using a magnet precautions must be taken against the danger of accidental failure of the current supply, when the large self-inductance of the magnet might cause a breakdown of the coil insulation. Thus, Dupouy‡ uses an electrolytic cell consisting of an aluminium anode and a gas carbon cathode immersed in a 10 per cent solution of tartaric acid, connected in parallel with the magnet coils, so that practically no current passes through the cell when the magnet is working. When the electricity supply is cut off the induced current in the coils can flow through the cell, as the carbon now becomes the positive pole, and the magnetic field falls off less suddenly. Copper-oxide rectifiers could be used for the same purpose. Even with smaller magnets when such devices are not employed, field-breaking switches, which automatically place a resistance in parallel with the coils before breaking the circuit, should be used. It is also advisable after using a magnet to demagnetise the pole pieces by reversing a safe current through the coils several times while gradually reducing the current to zero.

Magnets for Special Purposes. Special magnets have recently been built for investigators in nuclear physics. Thus,

* G. Häder, *Siemens Zeit.* **10**, 481, 1930.
† A. Cotton and G. Mabboux, *Recherches et Inventions*, Dec. 1929.
‡ M. G. Dupouy, *Ann. de Phys.* **15**, 495, 1931.

while Rosenblum* used the large magnet at the Academie des Sciences with pole faces 35 cm. in diameter for the examination of α-particles of long range emitted by radioactive elements, a special magnet for the purpose was constructed by Cockcroft.† It consists of a main cylindrical casting A, Fig. 19, of high permeability steel, over whose central core the exciting coils C are placed. The upper cylindrical pole piece A' is then slipped over the central core and mounted accurately in position by means of a carefully ground ring R. The annular gap across which the field is maintained is seen at E. It is 80 cm. in mean

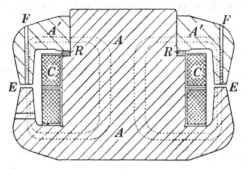

Fig. 19. Electromagnet for α-ray investigations. (Cockcroft.)

diameter, 5 cm. in radial width and 1 cm. across. Fields of 18,000 oersteds, constant to 1 part in 1000 over the gap, are obtained when 2·2 kilowatts are supplied to the coils, but as most of the fields required can be produced with 0·3 kilowatt or less, a battery can be used as a constant source of current. A group of α-rays of a fixed velocity, emitted from a source inserted through one of the holes F, can be bent by an appropriate field and brought to a focus on a detector inserted through the other hole F, not shown in the above figure. By adjustment of the field groups of rays of different velocity can in turn be focused on the detector; a high vacuum must be maintained in the gap or the α-rays will be absorbed before reaching the detector.

* S. Rosenblum, *Compt. Rend.* **188**, 1401, 1929.
† J. D. Cockcroft, *Journ. Sci. Instr.* **10**, 71, 1933. See also L. H. Martin and A. A. Townsend, *Roy. Soc. Proc.* A, **170**, 192, 1939.

Lawrence and Livingston* used a magnet with pole faces 27·5 in. in diameter in the more recent form of their cyclotron apparatus for the multiple acceleration of ions to very high speeds. The ions, formed near the centre of the diametrical region between semicircular, hollow, accelerating electrodes, are caused to describe ever-widening but approximately semicircular paths under the action of a powerful high-frequency oscillating field between the electrodes and the magnetic field perpendicular to the plane in which the ions move. The frequency of the oscillations is so adjusted that the ions cross the gap between the electrodes in synchronism with the oscillations, and thus gain an increase of energy at each passage, finally emerging with the very high velocity corresponding to about 5 million volts or ionre. The total weight of iron in the magnetic circuit is about 65 tons and the weight of the exciting coils about 9 tons.

Permanent Magnets. When constant fields of a few thousand oersteds are required in air gaps of large volume for long periods of time, permanent magnets of cobalt steel may be used with advantage. The theory of their construction is essentially that given by Evershed.† Let us suppose that we have a simple form of permanent magnet consisting of a cobalt steel yoke with an air gap between plane faces—the split anchor ring of p. 70 may be pictured. Then, because the M.M.F. in the circuit is zero,

$$\int \mathcal{H}_a dl_a + \int \mathcal{H}_s dl_s = 0, \qquad \ldots\ldots(50)$$

where $\mathcal{H}_a$ and $\mathcal{H}_s$ represent the fields within the air gap and steel and l_a and l_s denote the magnetic paths in air and steel respectively. For a stable state of the magnet we may assume that $\mathcal{H}_a$ and $\mathcal{H}_s$ are constant, and, integrating (50), we have

$$\mathcal{H}_a l_a = - \mathcal{H}_s l_s. \qquad \ldots\ldots(51)$$

---

* E. O. Lawrence and M. S. Livingston, *Phys. Rev.* **45**, 608, 1934. See also M. C. Henderson and M. G. White, *Rev. Sci. Instr.* **9**, 19, 1938.
† S. Evershed, *Journ. Inst. Elect. Eng.* **58**, 780, 1920; **63**, 725, 1925.

Now, the flux through the magnet must be constant, so that

$$B_s A_s = q \mathcal{H}_a A_a, \qquad \ldots\ldots(52)$$

where B_s is the induction in the steel and A_a and A_s are the respective areas of cross-section of the gap faces and steel, while q is a leakage factor, since lines of force are found outside the gap. The two relations (51) and (52) enable us to calculate the minimum volume of steel required to produce a stated field over a particular gap, for, the volume of steel is

$$A_s l_s = \frac{q \mathcal{H}_a^2 A_a l_a}{B_s \mathcal{H}_s}, \qquad \ldots\ldots(53)$$

and, when q, $\mathcal{H}_a$, A_a and l_a are specified, the expression is a minimum when $B_s \mathcal{H}_s$ is a maximum.

It is useful to have a *criterion of the magnetic quality* of permanent magnet steel. Suppose that, after being strongly magnetised, a steel has a *residual induction* or *remanence* B_r and we proceed to demagnetise it by applying magnetic fields of increasing intensity. Then we may record the values of the induction B and the applied demagnetising field $\mathcal{H}$, and obtain values of the product $B\mathcal{H}$. The latter starts from zero, passes through a maximum, $(B\mathcal{H})_{\text{max}}$, and ends at zero when the coercive field $\mathcal{H}_c$ is reached. As a criterion of magnetic quality Evershed takes $(B\mathcal{H})_{\text{max}}$, since this is the maximum amount of the external magnetic energy per unit volume of the material that can be supported by a given steel. Scott* has shown by direct experiment on several steels that the product $B_r \mathcal{H}_c$ is directly proportional to $(B\mathcal{H})_{\text{max}}$. This means that in order to assess the magnetic quality of a series of magnet steels we need only find their values of $(B_r \mathcal{H}_c)$ instead of measuring the complete demagnetisation curve to get $(B\mathcal{H})_{\text{max}}$.

Permanent magnets with large pole faces are used for α- and β-ray spectroscopy. Thus, Briggs† uses a magnet giving a field of 5300 oersteds between poles $26 \times 17 \cdot 8$ cm. and $1 \cdot 19$ cm. apart, and Surugue‡ uses one with pole faces 33×45 cm. giving fields up to 1430 oersteds for β-ray work. The magnet

* K. L. Scott, *Bell Sys. Tech. Journ.* **11**, 383, 1932.
† G. H. Briggs, *Journ. Sci. Instr.* **9**, 5, 1932.
‡ J. Surugue, *Journ. de Phys.* **6**, 94, 1935.

designed by Cockcroft, Ellis and Kershaw* consists of a rigid U-shaped yoke of soft iron from which are supported two horizontal arms, each consisting of twenty-three laminated plates of cobalt steel bolted together and fitted with pole pieces of high permeability steel. The cobalt steel contains 35 to 36 per cent cobalt. The pole pieces have plane faces 500 sq. cm. in area and when 5·5 cm. apart can give a maximum field of 2300 oersteds. Field coils are placed on the horizontal arms, and the field can easily be changed and set to new "permanent" values by the passage for a few moments of suitable exciting currents. The field is but little affected by small changes in temperature.

Materials for Permanent Magnets.† Nearly all commonly used permanent magnet steels are practically carbon steels to which quantities of other elements have been added. Such steels must be heat-treated; they are first hardened by heating and quenching from some critical temperature, so that they may have the most suitable crystalline structure. Quenching is usually done in oil or water, and, unfortunately, $\mathscr{H}_c$ and B_r vary with the quenching temperature in different ways. Thus for a high content chromium steel the higher the quenching temperature the greater $\mathscr{H}_c$ and the lower B_r, so that some compromise in choosing the quenching temperature is necessary. The effect of adding chromium *to* a steel is to raise $\mathscr{H}_c$ and also to permit it to be quenched in oil instead of water, thus avoiding quenching cracks; $\mathscr{H}_c$ is raised from 40 to 70 oersteds by the addition of 1 to 4 per cent chromium.

An early tungsten steel consisting of 6 to 8 per cent tungsten, 0·7 to 0·9 per cent carbon and the remainder iron had a coercive force of 60 to 80 oersteds; the *dimension ratio*, i.e. length divided by diameter, of the specimen used, was 15. K.S. magnet steel, discovered by Honda, is composed of 0·7 to 1·0 per cent carbon, 6 to 8 per cent tungsten, 1 to 4 per cent

* J. D. Cockcroft, C. D. Ellis and H. Kershaw, *Roy. Soc. Proc.* A, **135**, 628, 1932.

† A valuable review and record of magnetic materials is given by C. Webb, *Journ. Inst. Elect. Eng.* **82**, 303, 1938. See also D. A. Oliver, *Magnetism*, p. 71, Inst. Physics, 1938.

chromium, 15 to 36 per cent cobalt and the remainder iron, and has a coercive force of 200 to 250 oersteds, while its residual induction for small dimension ratios is about $2\frac{1}{2}$ times as great as that of the earlier tungsten steels.

Körster in 1931 produced an alloy containing iron, cobalt and molybdenum with a coercive force of 100 to 350 oersteds and a residual induction of 12,000 to 6700 for very high dimension ratios. About the same time Mishima invented M.K. steel, consisting of about 30 per cent nickel, 12 per cent aluminium, somewhat less than 20 per cent cobalt and the remainder iron, whose coercive force is between 200 and 700 oersteds with residual induction between 11,000 and 7500 for very high dimension ratios, but its general properties cause its use to be restricted. The new K.S. magnetic alloy discovered by Honda, Masumoto and Shirakawa* contains 15 to 36 per cent cobalt, 10 to 25 per cent nickel, 8 to 25 per cent titanium and the remainder iron. When cast into a metal mould and annealed at about 660° C. it gives a coercive force of about 800 to 900 with a residual induction of about 7500 for very large dimension ratios, and is specially suitable for short magnets. This alloy is difficult to forge and a magnet must first be cast and afterwards shaped by grinding. Jellinghaus† finds that an alloy of 50 per cent cobalt and platinum possesses a value of $\mathscr{H}_c = 2650$ oersteds and $B_r = 4530$ gauss.

Now, whereas substances of high initial permeability must be as free as possible from internal strains, permanent magnet materials must be heavily strained. This is borne out by Bozorth's experiments on a *precipitation hardened* iron-nickel-aluminium alloy. Precipitation hardening can be applied when the alloy can exist at room temperature in a stable condition consisting of two phases and when at a higher temperature one phase dissolves completely in the other to form a solid solution. In making a precipitation-hardened magnet, the material is quenched rapidly from a high temperature, and then reheated to an intermediate temperature at which the

---

* K. Honda, H. Masumoto and Y. Shirakawa, *Sci. Rep. Tok. Univ.* 23, 365, 1934.

† W. Jellinghaus, *Zeit. tech. Phys.* 17, 33, 1936.

second phase precipitates slowly in a very finely divided form. When the optimum amount has precipitated the material is cooled to room temperature and no more marked changes occur, but each submicroscopic precipitated particle forms a centre of strain. The strains produced by quenching are ordinarily regarded as sufficient, and further heat treatment is avoided except for "ageing" purposes. Bradley and Taylor* have shown by X-ray examination that in the Fe_2NiAl lattice there are tiny regions of iron rich β-phase on the point of separating out. In their equilibrium state these should have lattice dimensions 0·3 per cent greater than the parent lattice, but as this is not permitted, they exist in a state of enormous strain.

Extremely Intense Fields. The most intense magnetic fields so far recorded have been produced by means of coils carrying an enormous current for such a short interval of time that the coil is not unduly heated. Thus, in shortcircuiting what was virtually an accumulator pile through a coil of very low resistance Kapitza† produced fields of the order of 100,000 oersteds; a similar method was used by Wall. Kapitza obtained still more intense fields by shortcircuiting a special generator similar to a single phase alternating turbogenerator. The armature of the generator is sufficiently robust to permit it to be shortcircuited for one-hundredth of a second, even when running at a very high speed. Moreover, as only a half-wave of the current is used, the current may be broken by means of a synchronously adjusted break. Employing impulses of 50,000 kilowatts, Kapitza‡ obtained momentary fields of 300,000 oersteds over a small region.

As such fields cannot be maintained for more than a few thousandths of a second, a special switch is used to keep the current on for about one-hundredth of a second, and is so designed that the time required to make or break the current is only a few ten-thousandths of a second. The coil used to

---

* A. J. Bradley and A. Taylor, *Nature*, **140**, 1012, 1937, and "Magnetism," *Inst. Phys.*, 1938.

† P. Kapitza, *Roy. Soc. Proc.* A, **105**, 693, 1924.

‡ P. Kapitza, *Phys. Soc. Proc.* **42**, 425, 1930.

produce the field is subjected to huge electrodynamical forces and is of special design. The generator is placed at such a distance from the coil that the mechanical disturbances caused by shortcircuiting the machine only reach the coil after the field has vanished and any experiment is over.

Measurement of Strong Fields. The most direct laboratory method of measuring the intensity of the field between the poles of an electromagnet is to use a search coil in series with the secondary coil of a standard mutual inductance and a ballistic galvanometer. The search coil should consist of a number of turns of fine insulated wire wound upon a non-conducting frame, whose area of cross-section is accurately known. The search coil is placed with its plane perpendicular to the field, and on suddenly removing it from the field the galvanometer is deflected. An equal deflection of the galvano-meter is now obtained by the reversal of a steady current in the primary coil of the inductance. If the mutual inductance is M henries, I the steady current in amperes, n the number of turns each of *effective area* a on the search coil and $\mathcal{H}$ the field in oersteds, then

$$2MI = na\mathcal{H}/10^8. \qquad \ldots\ldots(54)$$

By effective area is meant the area which actually embraces lines of force; it is therefore for a coil of one layer equal to the area of a circle whose radius is that of the former + the radius of the wire + one thickness of the insulation layer on the wire. The effective area of a coil of several layers may be found experimentally by placing it inside a standard solenoid and measuring the mutual inductance between the coil and the solenoid by balancing it against an inductometer.

If instead of removing the coil from the magnetic field, we turn it through 180° about an axis perpendicular to the lines of force, equation (54) becomes $2MI = 2na\mathcal{H}/10^8$. It is also possible to remove the coil and to reverse the current in the primary of the mutual inductance at the same time, and by proper adjustment of the current to obtain zero deflection of the galvanometer, when equation (54) is satisfied.

The Fluxmeter. In practice the ballistic galvanometer is replaced by a fluxmeter which is virtually a moving coil ballistic galvanometer in which torsional control is extremely small and heavy damping is produced almost entirely by the electromagnetic couples due to currents induced in the coil by its motion. It consists of a rectangular coil of wire C, Fig. 20, wound on a non-conducting former and suspended by a fine silk fibre F—often attached to a flat spiral spring to avoid damage from mechanical shocks—between the poles of a strong permanent magnet. The ends of the coil are connected through spirals of thin silver strip to the terminals TT, to which the ends of the search coil are joined.

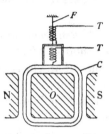

Fig. 20. Construction of fluxmeter.

On removing the search coil from the magnetic field a quantity of electricity Q coulombs flows through the fluxmeter, where

$$Q = na\mathscr{H}/10^8 R, \qquad \ldots\ldots(55)$$

when the total resistance in the fluxmeter circuit is R ohms. The coil receives an impulse and moves through a definite angle θ, which is practically independent of the rate at which the search coil is removed from the field. If the total effective area of the coil is NA and it moves in a radial field H, then a rotation through an angle θ would cause a quantity of electricity $Q' = NAH\theta/10^8 R$ to flow through the circuit. Now, if electromagnetic damping alone occurs, then Q must equal Q', i.e.

$$na\mathscr{H} = NAH\theta, \qquad \ldots\ldots(56)$$

or

$$\mathscr{H} = \frac{NAH}{na}\theta = \text{constant} \times \theta.$$

The above theory is adequate provided that negligible torsion control and no air damping exist; the following theory covers the general case.

The equation of motion of the fluxmeter coil is

$$I\frac{d^2\theta}{dt^2} + k\frac{d\theta}{dt} + C\theta = NAHi, \qquad \ldots\ldots(57)$$

where I is the moment of inertia of the coil system, k the coefficient of mechanical damping, C the torsion constant of the fibre, and i is the momentary current in the coil during the removal of the search coil. Now, assuming that the self-inductances of the search coil and fluxmeter coil are L_1 and L_2 respectively,

$$i = \frac{1}{R}\left(\frac{d}{dt}na\mathcal{H} - NAH\frac{d\theta}{dt} - L_1\frac{di}{dt} - L_2\frac{di}{dt}\right), \quad \ldots(58)$$

whence

$$I\frac{d^2\theta}{dt^2} + \left(k + \frac{(NAH)^2}{R}\right)\frac{d\theta}{dt} + C\theta = \frac{NAH}{R}\left[\frac{d}{dt}(na\mathcal{H}) - L_1\frac{di}{dt} - L_2\frac{di}{dt}\right]$$

$$\ldots\ldots(59)$$

Integrating the last equation,

$$I\int_0^t\frac{d^2\theta}{dt^2}dt + \left(k + \frac{(NAH)^2}{R}\right)\int_0^\theta d\theta + C\int_0^t\theta\,dt$$

$$= \frac{NAH}{R}\left[\int d(na\mathcal{H}) - \int_0^0 L_1\,di - \int_0^0 L_2\,di\right]. \quad \ldots(60)$$

If the search coil is withdrawn suddenly, i.e. t is very short, then θ remains very small and the second and third terms on the left side of equation (60) can be neglected, the last two terms on the right side disappear, and

$$I\omega_0 = \frac{NAH}{R}(na\mathcal{H}), \qquad \ldots\ldots(61)$$

where $I\omega_0$ is the total angular momentum imparted to the fluxmeter coil. If the torsion constant of the fluxmeter is zero the equation of the motion of the coil which ensues is given by

$$I\frac{d^2\theta}{dt^2} + \left(k + \frac{(NAH)^2}{R}\right)\frac{d\theta}{dt} = 0. \qquad \ldots\ldots(62)$$

Solving this equation and writing $\left(k + \dfrac{(NAH)^2}{R}\right) = K$, we have

$$\theta = C_0 + C_1 e^{-Kt/I}.$$

Now, when $t = 0$, $\theta = 0$ and therefore $C_0 + C_1 = 0$; so that

$$\theta = C_0(1 - e^{-Kt/I}). \qquad \ldots\ldots(63)$$

Differentiating (63),

$$\frac{d\theta}{dt} = C_0 \frac{K}{I} e^{-Kt/I}. \qquad \ldots\ldots(64)$$

To find C_0 we have $d\theta/dt = \omega_0$ when $t = 0$; therefore $\omega_0 = C_0 K/I$.
Hence from (63)

$$\theta = \frac{(NAH)(na\mathscr{H})}{RK}(1 - e^{-Kt/I}). \qquad \ldots\ldots(65)$$

Now, θ is a maximum when t is very great; therefore

$$\theta_{\max} = \frac{(NAH)(na\mathscr{H})}{R\left(k + \dfrac{(NAH)^2}{R}\right)}. \qquad \ldots\ldots(66)$$

When k is negligible the last equation reduces to that already obtained, viz.

$$\theta_{\max} = na\mathscr{H}/NAH. \qquad \ldots\ldots(67)$$

From equation (66) it follows that if we plot $1/\theta_{\max}$ against R we should get a straight line whose slope is a measure of the coefficient of mechanical damping, k.

When the torsion constant cannot be neglected, the equation of motion following the removal of the search coil is

$$I\frac{d^2\theta}{dt^2} + K\frac{d\theta}{dt} + C\theta = 0. \qquad \ldots\ldots(68)$$

The roots of the auxiliary equation are

$$-\frac{K}{2I}\left[1 \pm \sqrt{1 - \frac{4IC}{K^2}}\right],$$

and, since C is bound to be small, these reduce to

$$-\frac{K}{2I}\left[1 \pm \left(1 - \frac{2IC}{K^2}\right)\right] \text{ or to } \left(-\frac{K}{I} + \frac{C}{K}\right) \text{ and } -\frac{C}{K}.$$

Hence, we may write

$$\theta = Ae^{-Ct/K} + Be^{(C/K - K/I)t}.$$

When $t = 0$, $\theta = 0$ and $A + B = 0$; therefore $A = -B$.

When $t = 0$, $\dfrac{d\theta}{dt} = \omega_0$ and $\omega_0 = A\left(\dfrac{K}{I} - \dfrac{2C}{K}\right)$.

Therefore $\theta = \dfrac{K(NAH)(na\mathscr{H})}{R(K^2-2IC)}[e^{-Ct/K} - e^{(C/K-K/I)t}]$, ...(69)

which reduces to (65) when $C = 0$.

Again, since C must be small, $K^2 \gg IC$; then

$$\theta = \frac{(NAH)(na\mathscr{H})}{RK}[e^{-Ct/K} - e^{(C/K-K/I)t}]. \quad ......(70)$$

θ is therefore a maximum when

$$\frac{d}{dt}[e^{-Ct/K} - e^{(C/K-K/I)t}] = 0,$$

i.e. when $\qquad t = t_{\max} = \dfrac{KI}{K^2-2IC}\log_e\dfrac{K^2-IC}{IC}.$

Writing $\alpha = IC/K^2$ and substituting for $t_{\max}$ in (70), we have, following Surugue,*

$$\theta_{\max} = \frac{(NAH)(na\mathscr{H})}{R\left(k+\dfrac{(NAH)^2}{R}\right)}\left[\left(\frac{1-\alpha}{\alpha}\right)^{-\alpha/1-2\alpha} - \left(\frac{\alpha}{1-\alpha}\right)^{(1-\alpha)/1-2\alpha}\right].$$
$$......(71)$$

The last expression for $\theta_{\max}$ is important only when R is so great that the term $\left(k+\dfrac{(NAH)^2}{R}\right)$ does not explain the smallness of the deflection which is observed. In practice conditions are nearly always such that equation (67) adequately represents the behaviour of the instrument and the makers generally specify an upper limit to the value of R, beyond which equation (66) must be used; equation (71) need be invoked only when this limiting resistance is enormously exceeded.

The fluxmeter can be constructed so that a pointer moves over a circular scale on which each division corresponds to a known change of linkages, or known number of *maxwell-turns*. Instead of a pointer and scale, a lamp and mirror system may be used. Provided that the motion is heavily damped electromagnetically the resistance in the circuit may vary, as we have seen above, over fairly wide limits, and a number of search coils of different effective areas may be used to measure

* J. Surugue, *Journ. de Phys.* 6, 486, 1935.

a wide range of fields. The accuracy of the instrument is, of course, limited by the accuracy with which the deflection may be read, and, in general, there is a zero drift of which account must be taken. In fact, it takes a little time to acquire skill in using a fluxmeter for accurate work. The scale can be calibrated by measuring a known change of linkages by means of a standard mutual inductance. See Note 6.

The Bismuth Spiral. When the use of a search coil is not convenient, a bismuth spiral may be used for field measurement. The resistance of a metal is increased when it is placed in a strong magnetic field, as described on p. 292, and in the case of bismuth the change is pronounced. The bismuth is mounted as a flat non-inductively wound spiral between thin insulators and provided with flat copper leads. The resistance R_0 in zero field and the resistance $R_{\mathscr{H}}$ in a series of known fields $\mathscr{H}$ are measured and a calibration curve of $R_{\mathscr{H}}/R_0$ against $\mathscr{H}$ obtained, the plane of the spiral being perpendicular to the lines of force. This method is very suitable for the measurement of strong fields, but precautions must be taken to avoid sudden changes in the temperature of the spiral.

The Electromagnetic Balance. When extensive uniform fields are to be measured an electromagnetic balance, devised by Cotton,* is useful. In its simplest form it consists of a long narrow rectangle of copper wire wound on a non-conducting former and suspended from one arm of a balance. The rectangle hangs with the long sides vertical with one short side in and perpendicular to the lines of the field to be measured, while the other is in a region of no appreciable field. On passing a current i through the rectangle a vertical force $\mathscr{H}lni$ acts upon the short side of n turns each of length l, and can be measured with a sensitive balance. Horizontal forces acting on the vertical sides sometimes produce disturbances, but they may be avoided by winding the wire upon a former such that the long sides are arcs of circles and the short sides are directed along radii of circles with a common centre on the axis about which the

* A. Cotton, *Journ. de Phys.* 9, 383, 1900 and A. Cotton and G. Dupouy, *Congr. Intern. d'Électricité*, 3, 207, 1932.

balance turns. An accuracy of about 0·5 to 1 part in 1000 is claimed.

Briggs and Harper* have described a simple form of electromagnetic balance for use with permanent magnets of the type mentioned on p. 77. In general, with such balances, it is essential to make measurements with the current first in one direction and then in the other through the coil, in order to eliminate the effects of convection currents, magnetic impurities in the coil, distortion of the former due to lateral forces, and the induction or image effect between the coil and magnet.

Dupouy† has designed a direct-reading instrument, called a *gaussmeter*, which depends for its action upon the large couples which act upon crystals like siderose and oligiste when placed in a uniform magnetic field. In the case of siderose the two principal susceptibilities differ by $47·9 \times 10^{-6}$ e.m. units per c.c. at room temperature, so that an appreciable couple directly proportional to the field tends to set the crystal section with the axis of greater susceptibility parallel to the field, as explained on p. 138. The crystal is mounted on one end of a light tube supported on horizontal pivots. At the other end of the tube is a light spiral spring and pointer. The crystal is placed in the magnetic field with the lines of force perpendicular to the tube. The rotation of the tube is resisted by the spiral spring and the pointer takes up an equilibrium position on a circular scale, the field being read directly in oersteds. The range of the instrument is determined by the constant of the spring. For accuracy, the temperature of the crystal must be kept constant, as the susceptibility of a paramagnetic crystal varies considerably with temperature.

Field Control and Stabilisation. In many experiments it is necessary to maintain a magnetic field strictly constant over long periods of time. When the temperature of the apparatus in the gap can be kept constant to 0·01 to 0·1° C. the field may be controlled by a bismuth spiral. A bismuth spiral with a

---

* G. H. Briggs and A. F. A. Harper, *Journ. Sci. Instr.* **13**, 119, 1936.
† G. Dupouy, *Ann. de Phys.* **14**, 549, 1931.

resistance of the order of 10 ohms has its resistance increased by about 0·5 ohm in a field of 1000 oersteds. This large change can be measured accurately by a bridge and sensitive galvanometer. Now a change in temperature of the spiral by $\pm 0\cdot025°$ C. would give a change of only 0·000075 ohm, which is less than the change for a field change of ± 1 oersted. Consequently, this method is very suitable under the above conditions; and there seems to be no reason why the light reflected from the galvanometer lamp should not be used to work an automatic temperature control somewhat on the lines described below.

A fluxmeter may be used to control and measure accurately changes of magnetic field, as described by Rutherford, Wynn-Williams, Lewis and Bowden* in connection with the annular magnet described on p. 75. Two large search coils are fixed in the field gap and joined in series with the fluxmeter and a magnetic comparator. The latter consists of a permanent magnet which can be rotated in the region between the two Helmholtz coils connected in the fluxmeter circuit, so that a motion of the magnet will produce a change of flux through the circuit and bring the fluxmeter coil to any desired position. A vernier scale attached to the magnet permits small changes to be made. A spot of light reflected from the fluxmeter is kept as nearly as possible in a zero position, and a change of flux of 1 part in 100,000 in the annular magnet field is said to be detectable. A spiral groove cut in a wooden disc and filled with mercury forms a small resistance for fine adjustments in the field circuit. Small changes in the annular field can be measured by the magnetic comparator, which can be calibrated in terms of the annular field. Large controlled changes can be measured by joining the secondary of a mutual inductance in series opposition to the search coils in the fluxmeter circuit, and allowing a known fraction of the magnetising current to flow through the primary, so that there occurs only a small deflection of the fluxmeter and this can be measured by the comparator.

* Lord Rutherford, C. E. Wynn-Williams, W. B. Lewis and B. V. Bowden, *Roy. Soc. Proc.* A, **139**, 617, 1933.

The above fluxmeter method of control suffers from the disadvantage that it cannot deal with a slow continuous change of the magnetic field such as may be produced by a gradual change in temperature of the iron core and yoke. It has, however, been used by Wynn-Williams* in the construction of an automatic field stabiliser of great sensitivity designed to work with the system just described. A narrow rectangular spot of light reflected from the fluxmeter mirror falls upon a double cathode photoelectric cell, so that both cathodes are partly covered when the spot is in its zero position. Each cathode is connected through a simple valve and thyratron

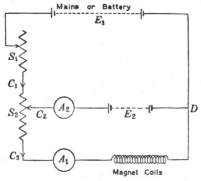

Fig. 21. Circuit for maintaining a steady current.

circuit, entirely worked from the A.C. mains, to a system of relays. Motion of the spot of light causes one relay system or the other to function and to actuate a series motor which operates a movable contact in a mercury resistance and thus alters the magnetic field current by the amount necessary to restore the fluxmeter spot to its equilibrium position. It is arranged that when the field has the required magnitude both thyratrons are alight, one going out when the field is too low and the other when the field is too high, and both are out when the field changes by such a large amount that the fluxmeter spot moves completely off the photoelectric cell, the direction of the change being indicated by the order in which the thyratrons extinguish. Arrangements are made to prevent

* C. E. Wynn-Williams, *Roy. Soc. Proc.* A, **145**, 250, 1934.

over-correction and to provide protection against failure of the current supplies or of the lamp.

For many purposes it is sufficient to maintain the magnet current constant and this may be accomplished by a simple battery circuit described by Potter.* The magnet current drawn from the town supply or a main laboratory battery is stabilised by a small compensating current from a bank of accumulators of small capacity, connected as in Fig. 21. The slider S_1 regulates the magnet current C_3, while S_2 is so adjusted that practically no current flows through the heavily shunted microammeter A_2. If the mains voltage changes by ΔE_1, then the magnet current changes by the very small amount

$$\Delta C_3 = \frac{R_2}{R_1 R_2 + R_2 R_3 + R_3 R_1} \Delta E_1, \quad \ldots\ldots(72)$$

where R_2 is the very small resistance of the arm $S_2 D$ via A_2, R_1 the resistance of $S_2 D$ via E_1, and R_3 the resistance of $S_2 D$ via the magnet coils. Hence, if R_2 is made very small, we have

$$\Delta C_3 = \frac{R_2}{R_3 R_1} \Delta E_1. \quad \ldots\ldots(73)$$

The control is automatic, and will easily eliminate 90 per cent of the mains fluctuations if R_2 is made sufficiently small. The device would also safeguard against the dangers of a sudden breakdown in the mains supply. More elaborate control systems, suitable for the electromagnet of a cyclotron, have been described by Lawson and Tyler.†

* H. H. Potter, *Journ. Sci. Instr.* 11, 95, 1934.
† J. L. Lawson and A. W. Tyler, *Rev. Sci. Instr.* 10, 304, 1939.

CHAPTER III

SUSCEPTIBILITY MEASUREMENTS WITH ISOTROPIC SUBSTANCES

Introduction. In this chapter will be discussed the experimental methods and technique for the determination of the magnetic susceptibilities of isotropic substances. Now, while gases and liquids have susceptibilities independent of the direction of the applied field in which they are measured, crystalline materials in general do not. Hence the measurements described in this chapter will be confined to polycrystalline solids or to finely powdered specimens. The information obtained in the case of a solid is of restricted use when such materials are used, although it becomes much more important when supplemented by the knowledge that the susceptibility follows the Curie law, for then it may safely be deduced that the internal crystalline fields are symmetrical and that the mass susceptibility of a powdered specimen is the same as that in any direction in a single crystal. Consequently, in the case of solids temperature investigations are usually much more important than in the cases of gases and liquids.

It is helpful to survey very briefly the magnetic behaviour of the materials most frequently handled. The magnetic susceptibilities of the elements have already been given in Table II, pp. 30–31, and will be further discussed on p. 124. All common gases, with the outstanding exceptions of oxygen, nitric oxide and nitrogen peroxide, are diamagnetic. The atoms of the rare gases with their closed electron shells clearly have no permanent magnetic moments, and it follows that all atoms and ions whose electron configurations are of the rare gas type are diamagnetic. Now, from Table II it is observed that many metals, e.g. copper and silver, which certainly possess electrons outside the closed electron shells or cores, are diamagnetic. It is therefore concluded that the individual atoms in the lattices of these metals are doubly ionised. Indeed, magnetic measurements on matter in bulk may help very much in elucidating problems in valency.

It is to be expected that all ions with incomplete electron shells possess permanent magnetic moments, and conspicuous examples are provided by many atoms and ions of the transition and rare earth series. In general, molecules are diamagnetic, although certain exceptions have been noted above. Thus, nearly all organic substances are diamagnetic, but there are a few rare exceptions to this rule, viz. where the total number of electrons in the molecule is odd and we have a free radicle; some of these special cases have been examined by Sugden* and Klemm.† Meta-phosphoric acid provides an interesting example. If its chemical formula is H_2PO_3 the molecule has an odd number of electrons and would presumably be paramagnetic, whereas if written $H_4P_2O_6$ the number of electrons is even and the molecule ought to be diamagnetic; as experiment shows the substance to be diamagnetic it is concluded that the latter formula is correct.

It is important to emphasise at this point that it is exceedingly difficult to obtain metals free of traces of ferromagnetic impurity. For example, pure copper is undoubtedly diamagnetic, but the author doubts strongly whether anyone has yet possessed a specimen which would be repelled from a strong to a weak magnetic field. Most of the common materials used as containers in experimental work are diamagnetic, but their diamagnetism cannot be taken for granted and they must be assumed to contain ferromagnetic impurity unless proved to be free therefrom. All glasses, in particular, must be tested; pyrex is diamagnetic but many glasses are paramagnetic.

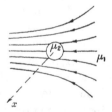

Fig. 22. Forces on a body in a non-uniform field.

Forces acting on a Body in a Magnetic Field. Before we pass to the experimental measurement of the susceptibilities of weakly magnetic substances, we must first examine the

* S. Sugden, *Trans. Farad. Soc.* **30**. 18, 1936.
† W. Klemm, *Magnetochemie*, p. 167, 1936.

forces which act upon such bodies when they are placed in a magnetic field. In Fig. 22 let the unbroken lines represent the lines of force of a magnetic field and the broken line a direction along which the small body O, of permeability μ_2 and volume v, is free to move. If the region around the body is evacuated, i.e. it is in a region of permeability unity, then its potential energy, see p. 8, is $-\dfrac{(\mu_2-1)\mathcal{H}^2}{8\pi}v$. If, however, the body is situated in a medium of permeability μ_1, it displaces a volume of the medium whose potential energy was $-\dfrac{(\mu_1-1)\mathcal{H}^2}{8\pi}v$. The change in potential energy which results from the introduction of the body is therefore $-\dfrac{\mu_2-\mu_1}{8\pi}\mathcal{H}^2v$, and the force which tends to move the body along the direction x is given by

$$F_x = -\frac{d}{dx}\left(\frac{-\mu_2+\mu_1}{8\pi}\right)\mathcal{H}^2v$$

$$= \frac{\mu_2-\mu_1}{8\pi}\frac{d\mathcal{H}^2}{dx}v$$

$$= (k_2-k_1)\left(\mathcal{H}_x\frac{d\mathcal{H}_x}{dx}+\mathcal{H}_y\frac{d\mathcal{H}_y}{dx}+\mathcal{H}_z\frac{d\mathcal{H}_z}{dx}\right)v, \quad \ldots\ldots(74)$$

since $\mu_1 = 1+4\pi k_1$, $\mu_2 = 1+4\pi k_2$ and $\mathcal{H}^2 = \mathcal{H}_x^2+\mathcal{H}_y^2+\mathcal{H}_z^2$, where $\mathcal{H}_x$, $\mathcal{H}_y$ and $\mathcal{H}_z$ are the components of the field along x and any two mutually perpendicular axes y and z.

There are, of course, similar expressions to (74) for the forces respectively parallel to y and z. These may be written

$$F_y = (k_2-k_1)\left(\mathcal{H}_y\frac{d\mathcal{H}_y}{dy}+\mathcal{H}_z\frac{d\mathcal{H}_z}{dy}+\mathcal{H}_x\frac{d\mathcal{H}_x}{dy}\right)v \quad \ldots\ldots(75)$$

and $$F_z = (k_2-k_1)\left(\mathcal{H}_z\frac{d\mathcal{H}_z}{dz}+\mathcal{H}_x\frac{d\mathcal{H}_x}{dz}+\mathcal{H}_y\frac{d\mathcal{H}_y}{dz}\right)v. \quad \ldots\ldots(76)$$

The expression (74) for F_x is correct only when the field $\mathcal{H}$ is not appreciably distorted by the introduction of the body. This means that we must be able to neglect all surface forces and any changes in the field, which we may do because of the low values of k_2 and k_1 with which we deal. The surface forces are proportional to $(k_2 - k_1)^2$ and are therefore extremely small. It follows that the direction in which a body tends to move along x, i.e. in the direction of increasing or decreasing $\mathscr{H}$, will depend on the quantity $(k_2 - k_1)$. If this is positive then the body will move in the direction of increasing $\mathscr{H}$ and so behave as a paramagnetic body. On the other hand, if k_1 is greater than k_2 the quantity $(k_2 - k_1)$ will be negative and the body will behave as a diamagnetic body although it may actually be paramagnetic.

This effect of the surrounding medium may be shown by the following lecture experiment. The ends of a horizontal fibre of soft glass are bent downwards and provided with small spheres of the glass A and B. The fibre is suspended by a thread of unspun silk with the sphere A between the poles of an electromagnet, one pole tip being pointed or wedge-shaped and the other flat. Soft glass is usually much more paramagnetic than air, and, on exciting the field, the sphere A moves away from the flat pole tip. A beaker containing a strongly paramagnetic solution of a salt of the iron group is then placed round A. The latter will now move towards the flat pole tip. On replacing the soft glass apparatus by one of pyrex (diamagnetic) the sphere A will move towards the flat pole tip in both cases.

Returning to equation (74), it is clear that the experimental arrangements are much simplified and the measurements more satisfactory if two of the three terms $\dfrac{d\mathscr{H}_x^2}{dx}$, $\dfrac{d\mathscr{H}_y^2}{dx}$ and $\dfrac{d\mathscr{H}_z^2}{dx}$ can be neglected. This is invariably done in practice, but it must be emphasised that the small components are always present to some extent, and, though they need not be considered in any calculations, they may cause slight disturbances which make the experimental procedure more difficult.

A point which has seldom been discussed since Kelvin* first raised it concerns the behaviour of freely suspended diamagnetic or paramagnetic cylinders when placed in an ideally uniform field. Surprising as it may seem a priori, we

* Lord Kelvin, *Papers on Electrostatics and Magnetism*, Section 691. See also E. C. Stoner, *Phil. Mag.* 23, 854, 1937.

may prove that both types set with their long axes parallel to the lines of force. For, on p. 93 we saw that the potential energy of a small body in a field $\mathscr{H}$ is given by $-\dfrac{(\mu-1)\mathscr{H}^2}{8\pi}\,v$. This expression assumes that the fields inside and outside the body are identical. For the present purpose, however, the last expression must be replaced by

$$\left\{-\frac{\mu-1}{8\pi}(\mathscr{H}-DI)^2\right\}v = -\frac{k}{2}(1-kD)^2\mathscr{H}^2v,$$

where D is the demagnetising factor of the cylinder which has its minimum value when the axis of the cylinder is parallel to the lines of force. Hence, as in the final expression for the energy we must write k positive and k negative for paramagnetic and diamagnetic cylinders respectively, the potential energy is least in both cases when D is a minimum, i.e. when the cylinders set with their long axes parallel to the lines of force.

Survey of Experimental Methods. So many different experimental arrangements have been used for the measurement of the susceptibilities of liquids and solids that one might almost think that variety had been the main object in the minds of many of the workers in this field. However, when we analyse the work we find that for ordinary purposes we need consider only two main methods for the measurement of these susceptibilities. Provided that an adequate quantity, say 10 c.c., of the material under investigation is available, the Gouy method described in the following section is, in the writer's view, always to be preferred, since it permits absolute measurements of a high order of accuracy to be made with the ordinary equipment of a university laboratory, and permits comparison measurements to be made with ease. Moreover, with special apparatus the method can be used for measurements on gases. Absolute and comparison determinations of the susceptibilities of liquids are often made by a method, devised by Quincke, of which the theory is essentially the same as that of the Gouy method. Both require the provision of magnetic fields from 5 to 20,000 oersteds sensibly constant over a reasonable volume between flat pole faces of an electromagnet.

When only small quantities of a substance are available or measurements have to be made at very high temperatures the Curie method is always to be recommended. It requires the provision of a strong field which varies rapidly over a short distance, so that special precautions have usually to be taken to ensure that the specimen under investigation is always placed in a standard position, although in a modification of the method, described on p. 104, such precautions are not necessary. Absolute measurements are impossible by this method, but it has proved particularly valuable in conjunction with the Sucksmith ring balance for measurements at very low temperatures. It has been used with sensitive torsion or electrodynamometer balances in several important surveys of the magnetic properties of the elements.

In the methods so far mentioned the specimen can only be placed in a suitable mount; it cannot be exposed to stresses, etc. while the magnetic observations are being made. If we wish to make experiments on the magnetic behaviour of a substance under tension or pressure, etc., we must use one of the methods designed by Kapitza and Webster and by Rankine, which are described on pp. 108 and 110 respectively. For extending susceptibility measurements to fields much more powerful than those produced by electromagnets, Kapitza has designed a special technique for which the appropriate balance is described on p. 112. Finally, modern developments require a knowledge of the magnetic behaviour of substances when exposed to fields of radio frequencies, and the special methods used for obtaining this information will be briefly described. In nearly all measurements with solids the effects of ferromagnetic impurities can be very large, and in a special section we shall discuss how these may be eliminated.

On account of the special difficulties peculiar to measurements with gases and vapours the experiments on them will be described separately. We need only mention here that the Gouy method can be used for the absolute determination of the susceptibility of a gas like oxygen and a variety of methods used for comparing the susceptibilities of other gases with that of oxygen. The experimental data in the case of vapours is very scanty.

The Gouy Method. The Gouy* method was first used extensively by Pascal.† The substance to be investigated must be provided in the form of a rod of uniform cross-section, or as a solution or very finely ground powder which can be placed inside a glass tube of uniform cross-section. The specimen or its container is suspended with its axis vertical from one arm of a sensitive balance, so that its lower end is near the mid-point of the field between two flat pole tips, while its upper end is in a region well outside the gap, as shown in Fig. 23a. The lower end is thus in a strong uniform field $\mathscr{H}$ while the upper end is in a very much weaker field $\mathscr{H}_0$. Hence the specimen must be some 10 to 15 cm. long. Now, the magnetic force acting on an element of a solid specimen of length dx and volume dv is equal to

$$\frac{k_2 - k_1}{2}\left(\frac{d\mathscr{H}_x^2}{dx} + \frac{d\mathscr{H}_y^2}{dx} + \frac{d\mathscr{H}_z^2}{dx}\right)dv, \quad \ldots\ldots(77)$$

where k_2 and k_1 are the respective susceptibilities of the specimen and the surrounding medium, usually air, and $\mathscr{H}_x$, etc. are the components of the field in the region lying between x and $x+dx$. From considerations of symmetry it is seen that $\mathscr{H}_x$ and $\mathscr{H}_z$ and $\frac{d\mathscr{H}_x^2}{dx}$ and $\frac{d\mathscr{H}_z^2}{dx}$ must be very small, hence the force acting on the element dx is equal to

$$\frac{k_2 - k_1}{2}\frac{d\mathscr{H}_y^2}{dx}dv = \frac{k_2 - k_1}{2}\frac{d\mathscr{H}_y^2}{dx}\alpha\,dx,$$

where α is the area of cross-section of the specimen. Therefore the total downward magnetic force acting on the whole specimen is given by

$$\frac{k_2 - k_1}{2}\alpha\int_{\mathscr{H}_y=\mathscr{H}_0}^{\mathscr{H}_y=\mathscr{H}}\frac{d\mathscr{H}_y^2}{dx}dx = \frac{k_2 - k_1}{2}\alpha\,[\mathscr{H}^2 - \mathscr{H}_0^2], \quad \ldots\ldots(78)$$

---

* L. G. Gouy, *Compt. Rend.* **109**, 935, 1889.
† P. Pascal, *Compt. Rend.* **150**, 1054, 1910.

In practice $\mathscr{H}$, the field in the gap, is usually of the order of 10,000 oersteds while $\mathscr{H}_0$, the field at the end of the specimen, is usually less than 100 oersteds, so that $\mathscr{H}_0^2$ may be neglected.

On making an experiment the specimen is weighed when the electromagnet is not excited and again when the current is switched on. The difference in weight, m gm., represents the magnetic pull on the specimen. Hence

$$\frac{k_2 - k_1}{2} \alpha \mathscr{H}^2 = mg \text{ dynes.} \qquad \ldots\ldots(79)$$

As an example, suppose that a rod of sulphur 1 sq. cm. in cross-section is suspended with its lower end in a field of 10,000 oersteds and its upper end in a field of 100 oersteds. Then the values of k_2 and k_1 being respectively $-1\cdot00$ and $+0\cdot03 \times 10^{-6}$ e.m. units per c.c., the magnetic force on the cylinder is

$$\tfrac{1}{2}(-1\cdot00 - 0\cdot03) \times 10^{-6} \times (10^8 - 10^4) = -51\cdot5 \text{ dynes.}$$

Hence there is an apparent decrease in weight of $0\cdot0525$ gm.

Now, the field between two flat pole faces is not uniform over an appreciable portion of the gap unless the latter is small and the pole faces large, and, accordingly, forces given by the expressions (75) and (76) of p. 93 are present. These cannot contribute to the downward pull on the specimen but they may cause a paramagnetic specimen to be displaced sideways. The amount of lateral displacement will, to some extent, depend on the length of the suspension connecting the specimen to the balance, and it must be avoided, particularly when the specimen is mounted inside a narrow furnace or a draught-proof case. It may be reduced by suspending the specimen with its upper end in the strong field

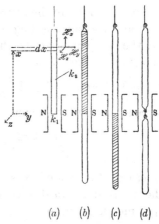

(a)   (b)   (c)   (d)

Fig. 23. The Gouy method of measuring susceptibilities.

and its lower end in the region of weak field. It is, of course, enormously increased by the presence of ferromagnetic impurities in the specimen.

In the measurement of the susceptibilities of liquids and of powders the containing vessel may be a simple tube (c), or a tube divided as shown in Fig. $23b$, into two compartments, the upper containing the specimen. Pyrex glass should be used for temperatures up to $500°$ C. and quartz up to $1200°$ C., provided that the material does not attack the quartz tube, which must in general be evacuated and sealed. Under these conditions the quantity m in (79) is the difference in the apparent mass of the container and specimen and that of the empty container when the field is on in both cases. The tube may also be provided with hooks, as shown in Fig. $23d$, so that a similar but empty tube may be attached by means of a thin copper wire; when both tubes are evacuated no correction should be necessary for the presence of the surrounding medium, i.e. k_1 disappears from the expression for the force acting on the specimen. Further, no allowance should be necessary for pull on the container, since the upward force on the diamagnetic container should be balanced by an equal downward pull on the attached tube. In any case, the specimen can be removed and the container again evacuated and sealed, and any necessary correction obtained by again finding the apparent change in weight when the electromagnet is excited. The susceptibility of a liquid may also be found* by noting the magnetic pull exerted upon a quartz cylinder immersed in it.

Provided that the field $\mathscr{H}$ can be measured accurately, the Gouy method gives the value of the susceptibility in absolute measure. It is, however, always advisable to check the field measurement at intervals by using a standard substance as specimen, e.g. water, or a solution of nickel chloride of known concentration or powdered crystals of manganese sulphate, $MnSO_4 . 4H_2O$. The method can be used at very low temperatures, but at temperatures above $400°$ to $500°$ C., depending on the apparatus, convection currents produce serious disturbances and the method is not then reliable. The order of

* G. Dupouy and C. Haenny, *Compt. Rend.* **199**, 781 and 843, 1934.

accuracy is limited by the correctness of the value of α and the value of k_2, but relative measurements with an accuracy of 1 part in 1000 are usually attained. In the case of powders the order of accuracy of an absolute determination depends upon the uniformity of packing in the container and is usually of the order of 1 part in 100; relative measurements with the same specimen are, of course, of higher order of accuracy. The method requires that at least some 10 c.c. of material is available in the case of substances of small susceptibility.

The Quincke Method. The Quincke method for the measurement of the susceptibility of a liquid has been extensively used in many forms. The liquid under investigation is placed in a vessel which is practically a vertical U-tube of which one limb is of narrow but very uniform cross-section, while the other is very wide. The narrow limb is placed between the poles of an electromagnet so that the surface of the liquid lies close to the line of centres of the pole pieces when the field is off. On exciting the magnet, a strong uniform field is established at the upper surface of the narrow column of liquid, while the lower portion of the column is in a region of comparatively weak field. As in the Gouy method, a force will act upon the column, and, if paramagnetic, the liquid will rise. Owing to the disparity in the cross-sections of the two limbs there will be no appreciable change in the level of the liquid in the wider limb.

The total upward magnetic force on the surface of the liquid is

$$\frac{k_2 - k_1}{2} \alpha \mathscr{H}^2 = (\rho - \sigma) gh\alpha = p\alpha, \qquad \ldots \ldots (80)$$

where k_2 and k_1 are respectively the susceptibilities of the liquid and the air above its surface, and ρ and σ are respectively the densities of the liquid and air; h is the observed rise in level of the liquid, and $(\rho - \sigma) gh$ is consequently the change in hydrostatic pressure to which the rise corresponds. Thus, by observing h the value of $k_2 - k_1$ may be found. It is clear that the air above the liquid may with advantage be replaced by a weakly diamagnetic gas such as hydrogen or carbon dioxide;

allowance may be made for the susceptibility of the vapour of the liquid. Alternatively, if $k_2 - k_1$ is known, $\mathscr{H}$ may be calculated; the method thus provides a satisfactory means of finding the intensity of a strong magnetic field in a narrow gap, when a standard liquid is available. In this method, of course, lateral forces can play no part.

The bore of the narrow limb must be strictly uniform if surface tension changes are to be avoided, although such effects are greatly diminished by the use of alcoholic or other solutions of low surface tension. If the bore is very narrow the surface of the liquid may be deformed when the field is applied and the column may not rise or fall. Accurately bored tubes are now readily obtainable, and it would appear that measurements could be made with containers in which the two limbs have identical cross-sections. The rise in level in one limb would be accompanied by a corresponding fall in the other. The surface remote from the magnetic field might therefore be observed by means of a reading microscope and the change in level noted when the field is applied, the quantity h in equation (80) then being twice the magnitude of the observed change. This method does not appear to have been used in research, although it can easily be adapted as a routine laboratory experiment.

The prevailing practice is to restore the liquid to its original level in the tube between the pole pieces either by raising or lowering the remote limb, or by adjusting the level of the liquid therein. The latter method has been extensively used by Piccard.* The method of raising the remote limb has been used by Fahlenbrach,† who used a flexible glass connection between the two limbs, taking care that the volume of the flexible portion was small compared with the non-flexible portion of his apparatus, in order to avoid distortion effects.

Auer's Experiments. An excellent example of the use of the Quincke method is provided by the work of Auer,‡ whose

---

* A. Piccard, *Arch. des Sci. Phys. et Nat.* **13**, 343, 1913; **2**, 455, 1920; see also S. Koch, *Avhandlinger ut. av det Norske Videnskaps. Akad. Oslo*, **7**, 1, 1932.

† H. Fahlenbrach, *Ann. der Phys.* **13**, 265, 1932.

‡ H. Auer, *Ann. der Phys.* **18**, 593, 1933.

apparatus, primarily designed for the absolute measurement of the susceptibility of water, is shown in Fig. 24. T represents a stout rectangular brass tube which can be rotated about a horizontal axis passing through two agate supports L, by turning the screw A. The liquid under examination is contained in the tripartite Jena glass tube GKM, which is rigidly clamped to T; there is therefore no question of serious deformation when T is rotated about L. The liquid surface in M lies between the pole pieces, while the remote surface is in

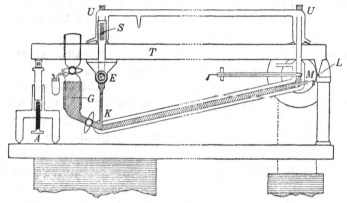

Fig. 24. Auer's apparatus for the susceptibility of water.

the tube K to which the microscope E is rigidly attached. On exciting the magnet the liquid (water) level in M falls and that in K rises. The level is restored to its original position in M by turning the screw A until the meniscus viewed in E returns to its former position on the cross-wires; this meniscus is illuminated by a point source of light and the microscope setting is exact. The distance through which the tube K is thus raised is read by means of a microscope, not shown, focused upon the glass scale S, whose divisions are the same distance from the axis of rotation as is the meniscus in K. The glass system UU is provided to enable the air above the liquid to be replaced by a diamagnetic gas. The tube G is used as a reservoir and is shut off from the rest of the apparatus during the measurements.

Certain portions of the apparatus are water jacketed as shown, in order that the temperature of the liquid in the field-column and in the inclined tube shall be at a constant temperature. This is most important. For, in order to restore the level in M the inclination of the slant tube KM to the horizontal is altered and the change in hydrostatic pressure thus brought about is equal to the product of the change in height measured on the scale S and the density of the liquid in KM. Let σ be the density of the gas, and ρ_1 and ρ_2 the densities of the liquids in the slant column KM and the vertical column M respectively. Then if h is the measured change in level

$$k_2 - k_1 = \frac{2gh(\rho_1 - \sigma)}{\mathscr{H}^2 - \mathscr{H}_0^2},$$

whence $\quad \chi_2 = \dfrac{k_2}{\rho_2} = \dfrac{2gh}{\mathscr{H}^2 - \mathscr{H}_0^2}\left(\dfrac{\rho_1}{\rho_2} - \dfrac{\sigma}{\rho_2}\right) + \dfrac{k_1}{\rho_2}.$

It is clearly very desirable that ρ_1 should be equal to ρ_2, and this is obtained by the use of the water jacket. The main difference between the methods adopted by Piccard and others and that adopted by Auer is that in the former the density to be used in the calculations is that of the liquid in the reservoir, while in the latter the density is that of the liquid actually in the field.

Auer used relatively wide tubes, 10 mm. in internal diameter, in order to work quickly and to avoid "creeping" errors, i.e. errors brought about by an asymptotic approach to the required surface level in M. With this apparatus he found the diamagnetic susceptibility of water at $16\cdot3°$ C. to be $0\cdot72145 \pm 0\cdot00048 \times 10^{-6}$ e.m. units per gm., with a temperature coefficient between $15°$ and $20°$ C. of $1\cdot45 \times 10^{-4}$ deg.$^{-1}$.

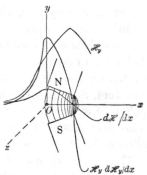

The Curie Method. Suppose that the two pole pieces of an electromagnet have flat surfaces and are set in

Fig. 25. The Curie method.

the somewhat exaggerated positions shown in Fig. 25, so that the lines of force are represented by the curved lines. Then a small spherical body placed at O will tend to move along the direction x under a force

$$F_x = \frac{k_2 - k_1}{2} v \left(\frac{d\,\mathscr{H}_x^2}{dx} + \frac{d\,\mathscr{H}_y^2}{dx} + \frac{d\,\mathscr{H}_z^2}{dx} \right)$$

$$= (k_2 - k_1) v\,\mathscr{H}_y \frac{d\,\mathscr{H}_y}{dx},$$

since, on account of symmetry, the quantities $d\,\mathscr{H}_x^2/dx$ and $d\,\mathscr{H}_z^2/dx$ are very small, and can be neglected. In the figure the quantities $\mathscr{H}_y$, $d\,\mathscr{H}_y/dx$ and $\mathscr{H}_y\,d\,\mathscr{H}_y/dx$ are all plotted against x. Usually $\mathscr{H}_y$ and $d\,\mathscr{H}_y/dx$ are of the order of 10^4 oersteds and 5×10^3 oersteds per cm. respectively. At first sight it might appear that the method would be less sensitive than the Gouy method, but its advantage lies in the fact that F_x may be measured with a delicate torsion balance, e.g. with a torsion constant of less than 1 dyne cm. per radian. The use of a sensitive torsion balance does not, of course, necessarily mean increased accuracy, for the three-dimensional freedom of the specimen makes precise replacement exceedingly difficult, since there is usually no strong constraint to avoid the effects of lateral forces.

The method requires that the curve of $d\,\mathscr{H}_y^2/dx$ against x shall exhibit a reasonably flat or extended maximum, so that the value is fairly constant over a short distance; otherwise, in addition to difficulties of stability, serious errors may arise through failure to maintain the body in a standard position. This defect may be avoided by using pole tips of special design so that the force is fairly constant over distances of as much as 1 to 2 cm. Such pole tips have been designed by Fereday;[*] one pole face is convex and the other is bigger and concave with a smaller radius of curvature, and constant values of $\mathscr{H}\,d\mathscr{H}/dx$ of the order of 10^6 over a distance of about 1·5 cm. are obtained.

The extensive use of the Curie method for comparison measurements has resulted in the construction of a large number of torsion balances or translation balances of varied design, and some of the more important will now be described.

* R. A. Fereday, *Phys. Soc. Proc.* **43**, 383, 1931.

The Sucksmith Ring Balance. For rapid determinations of susceptibilities which are not small the neat ring balance designed by Sucksmith* may be used. It consists of a ring, shown in Fig. 26, made of phosphor bronze strip, whose width and thickness depend on the sensitivity required, contained in a thick brass box to maintain a uniform temperature. It is rigidly fixed at O with its plane vertical and carries a light scale pan P, mica damping vane E and hook F all attached to a copper wire soldered to the lowest portion of the ring. Light from a distant straight filament lamp passes through the lens L to be reflected from the plane mirrors B and C as shown, to form an image upon a scale S. The specimen is placed in a small spherical glass phial suspended in a nonuniform magnetic field, e.g. as indicated in the figure.† When the field is excited the specimen moves through a very small distance in which

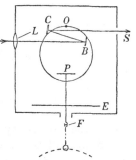

$$\mathscr{H}_y \, d\mathscr{H}_y / dx$$

is constant to a high degree of approximation, while the deformation of the ring produced by the displacement is sufficient to give an appreciable vertical displacement of the image.

Fig. 26. Sucksmith ring balance.

For, when a force F_x is applied vertically, causing the phial to move up or down, the radial displacement u at any point on the ring is given by

$$u = \frac{F_x r^3}{4EI} \left\{ \theta \sin\theta + \cos\theta - \frac{4}{\pi} \right\},$$

where r is the radius of the ring, I the moment of inertia of the cross-section, E Young's modulus, and θ the angle between the radius through the point and the horizontal. When $\theta = 90°$, then the displacement of the lowest point of the ring relative to O is

$$u_1 = \frac{F_x r^3}{2EI} \left\{ \frac{\pi}{4} - \frac{4}{\pi} \right\} = \frac{0 \cdot 298 \, F_x r^3}{2EI}.$$

* W. Sucksmith, *Phil. Mag.* **8**, 158, 1929.
† Cf. also D. Shoenberg, *Roy. Soc. Proc.* A, **155**, 712, 1936.

The angle turned through by the tangent to the ring is, to a first approximation, given by

$$\phi = \frac{1}{r}\frac{du}{d\theta} = \frac{F_x r^2}{4EI}(\theta \cos \theta),$$

which is a maximum at about 49°, when

$$\phi_{49} = \frac{0\cdot561\, F_x r^2}{4EI}.$$

Now the displacement v_1 on a vertical scale D cm. away from the mirror C is given by

$$v_1 = \phi_{49}(4D + 2d),$$

where $2d$ is the distance between the mirrors, so that finally

$$v_1 = \frac{(4D + 2d)\,0\cdot561 F_x r^2}{4EI},$$

and
$$v_1/u_1 = \frac{0\cdot943\,(4D + 2d)}{r},$$

so that the sensitivity is inversely proportional to r. With $r = 2\cdot5$ cm. and $D = 100$ cm. the ratio v_1/u_1 is about 150, or a 1 mm. motion of the body results in a 15 cm. displacement of the image, which can be read to 0·01 mm. when the scale is replaced by a microscope; this corresponds to a 0·000007 cm. displacement of the specimen. Of course, the mirrors B and C need not be mounted accurately at $\theta = 49°$. The displacement, then, is strictly proportional to the load, and may be calibrated by placing known weights on the pan. It is so small, less than 0·1 mm., that the force must be uniform over it, and on account of the linear relation between force and displacement there is no necessity to restore the specimen to some standard position.

By changing the weights on the pan the specimen may be set in a suitable position in the field. Lateral motion may be prevented either by the use of a single horizontal fibre of unspun silk mounted inside the brass box parallel to the line of centres of the pole pieces, or by using two weak metal spirals attached to the copper wire, as described on p. 263 and shown in Fig. 76. The use of these spirals also permits the whole arrangement to be inverted, and so to be used at high temperatures without the serious consequences of convection currents which would affect the form of apparatus shown in Fig. 26. The accuracy is about 0·5 per cent.

The Föex and Forrer Balance. In Fig. 27 is given a diagram of a translation balance designed by Föex and Forrer* which is probably the best of its type hitherto designed. At one end of a light metal rod the specimen S is attached and a coil C is attached to the other. The rod is suspended by five threads, of which four are shown in the diagram, so that it can move only

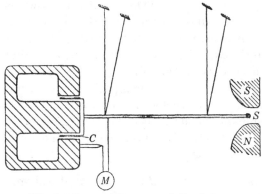

Fig. 27. Föex and Forrer translation balance.

in one direction parallel to the rod. The specimen is placed as shown between shaped pole pieces, and when the field is excited the specimen moves in the direction perpendicular to the lines of force. It can be restored to its former position by passing a current through the coil C, which lies in the gap of a permanent magnet system, like the coil of a loud-speaker. The restoring force is therefore approximately directly proportional to the current in the coil. A special form of Kelvin double suspension mirror M is used to indicate the position of the coil. An accuracy of about 1 part in 750 is claimed for a measurement in which only 0·35 gm. of a substance of susceptibility $0·55 \times 10^{-6}$ e.m. units per c.c. was used, the measured force being less than 3 dynes. The apparatus is,

* G. Föex et R. Forrer, *Journ. de Phys.* **7**, 180, 1926.

however, rather elaborate and bulky, and the stray field of the electromagnet may cause errors. These could no doubt be avoided by replacing the permanent magnet by a fixed coil and supplying alternating current to the coil C and the fixed coil in series, a method used by Gans in measurements in which such high sensitivity was not claimed.

The Curie-Chéneveau Balance. Rapid measurements at fixed temperatures may be made by a form of torsion balance designed by Chéneveau* and improved by Gray and Farquharson.† A vertical tube is suitably counterpoised and suspended from a long torsion fibre so that the lower end of the tube lies in a non-uniform portion of the field of a permanent horse-shoe magnet, mounted with the longer edge of the gap parallel to the axis of the tube. On placing a specimen in the tube a deflection occurs and the angle of re-torsion necessary to restore the tube to its original position is determined. The instrument is calibrated by placing water in the tube. It is clearly unreliable when the specimen contains a ferromagnetic impurity, for practically no variation of the experimental conditions is possible.

The Method of Kapitza and Webster. As already mentioned, a method of measuring magnetic susceptibilities, which is particularly valuable when the specimen is subjected to deformation or other interference while the magnetic measurements are in progress, has been devised by Kapitza and Webster.‡ It requires the production of a strong, constant, magnetic field, $\mathscr{H}$, uniform to 1 part in 1000 over an extended region. Such a field may be obtained with some difficulty between two large flat and accurately parallel pole pieces. A small spherical body placed in this field will not tend to move perpendicularly to the lines of force unless the field is artificially distorted. Distortion may be produced by placing a short iron rod with its axis parallel to the lines of force, as shown in Fig. 28.

* C. Chéneveau, *Phil. Mag.* **20**, 357, 1910.
† F. W. Gray and J. Farquharson, *Journ. Sci. Instr.* **9**, 1, 1932.
‡ P. Kapitza and W. L. Webster, *Roy. Soc. Proc.* A, **132**, 442, 1931.

It is readily seen that the magnetisation of the rod produces a disturbing field H with a gradient dH/dx in the neighbourhood of the body B, and a force $k(\mathscr{H}+H)\dfrac{dH}{dx}v$ will act on the latter, tending to make it move towards or away from the rod, and may be measured if B is suitably mounted on a torsion balance. The force acting on the iron rod due to the presence of B must be equal and opposite to that acting on B, and, therefore, instead of mounting the latter on the torsion balance it is fixed in position while the iron rod is free to move. Now, it is ob-

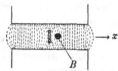

Fig. 28. Kapitza and Webster's method of measuring susceptibilities.

viously impossible to produce such a strictly uniform field that the force $M\dfrac{d\mathscr{H}}{dx}$, due to the action of the field $\mathscr{H}$ on the magnetic moment M of the rod, is zero, but Kapitza and Webster state that this force can adequately be compensated by gravitational or other controls. It is essential that $d\mathscr{H}/dx$ shall not vary appreciably during the period in which the experiments are made with the iron attached to the balance. Kapitza and Webster have designed a special form of balance in which the iron rod is suspended from a horizontal torsion fibre, and this has been used by Webster in an investigation of the magnetic properties of bismuth exposed to stress. The disturbances produced by ferromagnetic impurity in the metal system used to clamp B must, of course, be taken into account.

Difficulties occurred because of unavoidable changes of zero due to fluctuations in $\mathscr{H}$. The latter were reduced to a minimum, by compensating for high frequency eddy current effects, by using an automatic field current regulator to prevent sudden changes, and by bracing the pole tips to prevent mechanical distortion when the temperature of the magnet rose.

Potter* has used the method in his measurements of the saturation intensity of magnetisation possessed by an iron

* H. H. Potter, *Roy. Soc. Proc.* A, **146**, 362, 1934.

sphere at high temperatures, as described on p. 262; in this case the iron sphere is fixed and the body B replaced by a movable coil of wire carrying a current.

The Rankine Method. A simple, but extraordinarily sensitive method of dealing with the properties of weakly magnetic substances has been devised by Rankine.*

Instead of fixing the magnet and allowing the specimen to move, Rankine fixes the specimen and allows the magnet to move. The apparatus, shown diagrammatically in Fig. 29, consists of a light horizontal cross-bar suspended from a torsion head by a vertical metal strip of very low torsion constant. From one end of the cross-bar a vertical cobalt-steel magnet SN is suspended by a fine silk fibre, while a brass rod B of equal mass is similarly suspended from the other end. The magnet is about 5 cm. long and weighs about 8 gm. Such a system takes up an equilibrium position which is practically independent of the earth's magnetic field, although no special mounting devices appear necessary to ensure this most desirable independence. On bringing a cylinder of a weakly magnetic substance parallel to and close to the magnet, it is attracted or repelled by both poles, and the magnet moves. In practice the cylinder is so placed that only the upper pole of the magnet is really effective. In the figure C, C represent

Fig. 29. Rankine's apparatus.

glass containers in which specimens may be placed. A hollow copper cylinder, not shown, is placed round the lower pole to provide electromagnetic damping and thus render the motion of the magnet almost dead-beat.

Suppose that the couple acting on the magnet is Γ, then the angle of restitution, θ, through which the torsion head K must be turned to restore the magnet to its original position, is given by $\Gamma/\tau + \tau'$, where τ and τ' are respectively the torsion

* A. O. Rankine, *Phys. Soc. Proc.* **46**, 391, 1934.

constants of the metal and silk suspensions; τ is usually much greater than τ'. It is, however, not necessary to turn the torsion head, for a suitable restoring couple may be provided by passing a current of the order of a few milliamperes through a single long straight wire running horizontally, directly below the axis of the magnet and parallel to the cross-bar. It is not possible mathematically to calculate the couple produced between the magnet and specimen. Temple* has calculated the couple in the case of a horizontal magnet and a spherical specimen. He has also shown that a small error arises because of the magnetisation of the specimen by the earth's field. In general, then, the apparatus must be calibrated by placing a liquid of known susceptibility in one of the containers.

A serious difficulty is encountered in the construction of the instrument, namely, the elimination of iron from all portions of the apparatus. If a little free iron is present the spurious couples acting on the magnet may exert much greater control upon the motion of the magnet than the torsion of the metal suspension. Consequently, all metal parts should be iron free, and, if possible, made of aluminium; copper should always be substituted for brass when aluminium cannot be used. Special care is also necessary in the choice of specimens, although it would appear possible to devise a form of correction similar to that used by Owen and Honda, described on p. 115. The method should, however, be particularly valuable in the study of gases and vapours, changes of state and changes accompanying strain. The efficient use of the apparatus really necessitates the provision of a special "magnetic" laboratory in which the vertical gradient of the earth's field is small. A satisfactory form of the apparatus† has been used for a comparison of the susceptibilities of H_2O, HDO and D_2O.

Measurements in Very Intense Fields. A special balance is obviously necessary for the measurement of the considerable forces which are exerted upon a body when placed in a very intense field. In these circumstances $\mathscr{H} \, d\mathscr{H}/dx$ may be as

* G. Temple, *Phys. Soc. Proc.* 48, 393, 1936.
† H. P. Iskenderian, *Phys. Rev.* 51, 1092, 1937.

large as 10^{10}, when the corresponding force on unit volume of a weakly magnetic substance is of the order of 10 gm. As such a force persists only for about 0·01 sec., it must be measured ballistically by a balance with a very short period of oscillation, high sensitivity, and fairly low inertia. Moreover, the balance must be critically damped, itself be unaffected by the field, and provide a large linear magnification of any displacement of the magnetised body.

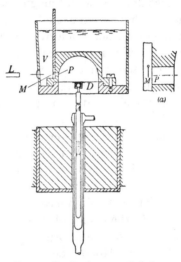

Fig. 30. Kapitza's magnetic balance.

An instrument designed by Kapitza* is represented diagrammatically in Fig. 30. The specimen is placed in the container A attached to a glass rod, and so placed in the magnetising coil, mentioned on p. 81, that it is pulled downwards when the field is established and depresses the constantan diaphragm D. The latter forms the floor of an oil-filled chamber communicating through a small outlet P with an outer vessel V, also oil-filled. The exterior end of P is partially closed by a small mirror M, only 0·9 by 0·7 mm., cemented to a horizontal glass fibre supported between two vertical constantan strips. The mirror could thus swing, as

* P. Kapitza, *Roy. Soc. Proc.* A **131**, 224, 1931.

depicted in the enlargement, Fig. 30a, to open and close the hole. When the diaphragm D is suddenly depressed, oil must find its way into the inner chamber, and, in so doing, cause the mirror to tilt. In this way an achromatic spot of light may be deflected on a horizontally moving photographic plate. The linear magnification so obtained depends on the ratio of the area of the constantan diaphragm to that of the outlet, and on the distance between the mirror and the plate; the final magnification is arranged to be of the order 10^5. Measurements are conducted in such a manner that an oscillograph gives a record of the value of the field current on the same photographic plate as the record of the tilted mirror. Satisfactory determinations of susceptibility can be made in fields between 85,000 and 200,000 oersteds. The balance can be calibrated by applying weights suddenly by an electromagnetic device; so far it has been used for comparison measurements only.

The balance is free from thermal disturbances, since these can produce merely slow motion of the oil in the outlet. A surprisingly steady zero is obtained, although the instrument is found to act as a delicate seismograph when a weight is attached to D. Effects due to static friction and to gravitational control do not appear to arise. Certain errors are incidental to the use of the very intense fields, namely, those produced by eddy currents and by adiabatic magnetisation. The first produce an apparent diamagnetism of the specimen when the field is increasing, and an apparent paramagnetism when decreasing, so that some idea of their relative importance can always be obtained from the magnetisation record. The second produce thermal changes which may alter the temperature of the specimen, particularly during low-temperature determinations. When a ferromagnetic or paramagnetic substance is suddenly placed in a magnetic field its temperature rises, while the temperature of a diamagnetic substance does not change, unless its diamagnetism is a function of the temperature. Fortunately, the maximum changes of temperature due to these magnetocaloric effects, which are discussed in detail in Chapter IX, are only about 1° or 2° C. and can usually be neglected in susceptibility determinations, although these changes are much greater than those observed in measurements with electromagnets.

Radiofrequency Measurements. If we place a paramagnetic within a coil forming part of an oscillatory circuit, there results a frequency change which can be measured by the heterodyne beat method. The change in beat note is directly proportional to the susceptibility, when due account is taken of diamagnetism corrections and capacity changes caused by placing the substance inside the coil; such allowances may be found by repeating the measurements with a diamagnetic of similar chemical constitution. Gorter* made experiments on paramagnetic alums at liquid hydrogen and nitrogen temperatures with frequencies of about 2 megacycles per sec. The susceptibility was found to decrease by about 75 per cent with iron and chrome alums when a steady field of about 2000 oersteds was applied parallel to the axis of the coil and by 20 per cent in the case of vanadium alum.

When an alternating field is applied to a paramagnetic the carriers follow it completely when the frequency is low, but if the frequency is too high they are unable to do so. With intermediate frequencies there is a phase difference between the applied field and the magnetisation, so that energy is lost and heat is dissipated in the paramagnetic. This heat is found to be markedly reduced in the above cases when the strong steady field is maintained, and the following explanation fits the facts. The steady field produces Zeeman separations much larger than the interaction energy between the spins so that changes in the distribution of the carriers in the various Zeeman levels cannot be supplied at the expense of the coupling between the carriers. Moreover, the weak coupling between the carriers and the crystal lattice cannot provide the necessary energy unless the frequency is sufficiently low, and apparently this is not the case. See Chapter XVI.

Correction for Ferromagnetic Impurity. In measurements described above the possibility of the introduction of ferro-

* C. J. Gorter, *Physica*, **3**, 503, 998 and 1006, 1936; **4**, 579 and 667, 1937. Cf. I. Waller, *Zeit. für Phys.* **79**, 370, 1932 for theory. See also A. H. Cooke and R. A. Hull, *Proc. Roy. Soc.* A, **162**, 404, 1937.

magnetic impurity into the specimen or its container has frequently been emphasised. Such impurity is often very difficult to remove and the elimination of the effects to which it gives rise must now be considered. Little is known of the precise way in which ferromagnetic impurities behave when they are finely divided and spread throughout the body of a specimen. It is often stated that a small amount of iron introduced into copper or even a large amount into aluminium will not exhibit ferromagnetic character. In the case of liquid mercury iron amalgams, the iron appears to exist in a colloidal form and rises to the top of the fluid. A drop containing nearly all the iron can be removed from the surface in a magnetic field. In frozen mercury-iron amalgams the ferromagnetism of the iron is definite enough even in concentrations of 10^{-4} gm. iron in 100 gm. mercury.

If the quantity of ferromagnetic impurity is not too large the magnetic effects can be eliminated by a method due to Honda* and Owen. It is assumed that in strong non-uniform fields while the diamagnetic or paramagnetic constituents of a specimen experience forces proportional to $\mathscr{H} \, d\mathscr{H}/dx$, the ferromagnetic impurity is saturated and experiences a force proportional to $d\mathscr{H}/dx$. Thus, when measurements are made by the Curie or Faraday methods the apparent susceptibility $k_{\mathscr{H}}$ of a substance in a field $\mathscr{H}$ is connected with the true susceptibility k_∞ in an extremely high field by the equation

$$k_\infty = k_{\mathscr{H}} - c'\sigma_s/\mathscr{H}, \qquad \ldots\ldots(81)$$

where σ_s is the saturation specific magnetisation of the impurity of concentration c' gm. per c.c. of specimen. The value of k_∞ may be found by plotting values of $k_{\mathscr{H}}$ against the corresponding values of $1/\mathscr{H}$, and finding the intercept at $1/\mathscr{H} = 0$.

Equation (81) does not correctly express the results obtained by the Gouy method. As Vogt† showed, in this case the expression on p. 97 may be written

$$mg = \alpha \int_0^x I \frac{d\mathscr{H}_y}{dx} dx, \qquad \ldots\ldots(82)$$

* K. Honda, *Ann. der Phys.* **32**, 1048, 1910; M. Owen, *Ann. der Phys.* **37**, 657, 1912.　　　† E. Vogt, *Ann. der Phys.* **14**, 1, 1932.

where I is the intensity of magnetisation of the material lying between x and $x + dx$ of Fig. 23a. The quantity I contains a paramagnetic or diamagnetic contribution which we may separate from the ferromagnetic, thus

$$mg = \alpha \frac{k_\infty}{2}\left[\mathscr{H}_y^2\right]_{\mathscr{H}_0}^{\mathscr{H}} + \alpha \int_{\mathscr{H}_0}^{\mathscr{H}} c'\sigma d\mathscr{H}_y, \quad \ldots\ldots(83)$$

where $c'\sigma$ is the intensity of magnetisation of the impurity in the field $\mathscr{H}_y$.

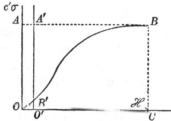

Fig. 31. Behaviour of ferromagnetic impurity.

Referring now to Fig. 31, where the curve OB is supposed to represent the $c'\sigma$, $\mathscr{H}$ curve for the ferromagnetic impurity, it is seen that the last equation may be rewritten

$$mg = \alpha \left\{ \frac{k_\infty}{2}\left[\mathscr{H}_y^2\right]_{\mathscr{H}_0}^{\mathscr{H}} + \text{area } O'B'BC \right\}.$$

But, since the area $O'B'BC = $ area $O'A'BC - B'A'B$

$$= c'\sigma_s(\mathscr{H} - \mathscr{H}_0) - Q,$$

and $2mg/\alpha(\mathscr{H}^2 - \mathscr{H}_0^2)$ is the apparent susceptibility $k_{\mathscr{H}}$ in a field $\mathscr{H}$, we have

$$k_{\mathscr{H}} = k_\infty + \frac{2c'\sigma_s(\mathscr{H} - \mathscr{H}_0)}{\mathscr{H}^2 - \mathscr{H}_0^2} - \frac{2Q}{\mathscr{H}^2 - \mathscr{H}_0^2}, \quad \ldots\ldots(84)$$

and when $\mathscr{H}_0$ is small,

$$k_{\mathscr{H}} = k_\infty + \frac{2c'\sigma_s}{\mathscr{H}} - \frac{2Q}{\mathscr{H}^2}. \quad \ldots\ldots(85)$$

In general $\mathscr{H}$ is so large that the last term in (85) may be neglected, and the equation reduces to the form (81) except that $2\sigma_s$ replaces σ_s. The equation

$$k_{\mathscr{H}} = k_\infty + \frac{2c'\sigma_s}{\mathscr{H}}$$

116

appears adequate for most purposes when the Gouy method is used, and Bates and Illsley* have used it as a means of finding how σ_s varies with the concentration of iron in a series of mercury-iron amalgams.

Temperature Control. It is often necessary to determine the susceptibility of a substance over a wide range of temperature, and electric furnaces designed for this purpose must fit into the relatively narrow space between the pole pieces of a magnet. Such furnaces have been described by Föex and Forrer. They consist of a silver tube closed at one end and covered with a layer of mica on which is wound, non-inductively, two or more layers of nichrome strip or platinum wire. These are held in position by a dried paste of kaolin and water. The whole is covered with layers of mica and slipped inside a cooling jacket consisting of two concentric tubes of brass through which water is circulated; the function of the cooling jacket is, of course, to prevent heat from reaching the pole pieces, thus causing fluctuations of the magnetic field, and to prevent convection currents in the measuring apparatus. The temperature distribution in such a furnace must be known, and when used with its closed end uppermost, it is customary to fit a metal thimble inside the upper end of the furnace to assist in providing a uniform temperature distribution. The latter may also be attained by winding the furnace in sections and supplying different currents to the several sections.

For temperatures up to 400° C., a thick copper tube on which is a layer of China clay, or of asbestos coated with "Purimachos" cement, and wound non-inductively with layers of nichrome strip embedded in the same cement, may be used. The whole is wound with asbestos cord and placed inside a cooling jacket; a satisfactory form of jacket which can be detached and used with different furnaces has been described by Bates and Baqi.† When a long furnace such as is necessary for Gouy measurements is used with the open end of the tube uppermost, uniformity of temperature distribution is assisted by placing a short tube of copper foil inside the

* L. F. Bates and P. F. Illsley, *Phys. Soc. Proc.* 49, 611, 1937.
† L. F. Bates and A. Baqi, *Phys. Soc. Proc.* 48, 781, 1936.

upper portion of the furnace and allowing it to project for a short distance. Very satisfactory furnaces* for use up to 600° C. can be made by coating thick copper tubes, inside and out, with "Insalute" cement, the windings being directly embedded in this cement, layers of mica being unnecessary. For very high temperatures "alundum" tubes wound with molybdenum, embedded for convenience in magnesia, must be used. It is important to note that, in using the Kapitza and Webster distorted field method of measuring magnetic susceptibilities, particular care must be taken in selecting non-ferromagnetic furnace materials, and the furnace should be heated by alternating current. The stabilisation of a furnace current is always important, and a method used by Potter with direct current is described on p. 89. Arrangements for low-temperature measurements are described in detail in the following chapter. See Note 7.

Susceptibilities of Gases. In measuring the susceptibility of a gas two main difficulties are encountered. First, while volume susceptibilities ranging from 10^{-6} to 10^{-3} e.m. units per c.c. are usual in the cases of solids and liquids, values of the order of 10^{-10} are found for many gases at N.T.P.; it is true that a gas can be compressed and its volume susceptibility proportionately increased, but there is a limit to the degree of compression and a limit to the dimensions of the containers which can be used in magnetic measurements. Secondly, whereas oxygen is strongly paramagnetic with a volume susceptibility of 0.162×10^{-6} e.m. units per c.c. at N.T.P., the majority of gases are weakly diamagnetic. Consequently, a minute trace of oxygen present as an impurity may profoundly affect the value obtained for the susceptibility of a gas. For example, the presence of 0·01 per cent of oxygen in hydrogen would mean a possible error of about 9 per cent in the determination of the susceptibility of the latter. Hence, the purification of the gas is a matter of great importance. It follows that in dealing with gases relative measurements are the rule, and we shall now consider some of the experiments.

* L. F. Bates, R. E. Gibbs and D. V. Reddi-Pantulu, *Phys. Soc. Proc.* **48**, 665, 1936.

Soné* used the Gouy method with a particularly sensitive balance, which was well damped and provided with a Kelvin mirror system to indicate its equilibrium position. Because of its high susceptibility, air was confined under pressure in the upper portion of a tube divided like that shown in Fig. 23b on p. 98, the lower portion being evacuated, and the magnetic pull found in the usual way. The magnetic susceptibility of the air was measured in terms of the susceptibility of water, with which the upper portion of the tube was later filled. The susceptibility of the air being known, it was now confined under atmospheric or slightly higher pressure in the upper portion and the magnetic pull compared with that found when the air was replaced by a gas under pressure. The method gives valuable results, particularly in the case of a gas like oxygen, which can be used as a standard in comparison measurements.

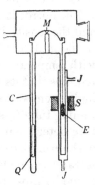

Fig. 32. Stössel's apparatus for gases.

Stössel's Experiments. A modified form of the Gouy method was also used by Stössel† in a study of the magnetic susceptibility of nitric oxide, NO, which is of considerable theoretical interest. Stössel's apparatus is shown diagrammatically in Fig. 32. From the left-hand side of the special balance is suspended an evacuated quartz tube Q, and from the right-hand side a glass tube E containing erbium oxide, which is strongly paramagnetic. The balance chamber and the tubes surrounding the suspensions can be evacuated. The lower portion of the quartz tube hangs in the field of an electromagnet. If the apparatus is evacuated, then Q is pushed upwards when the field is excited, and it can be restored to its original position by passing a suitable current through the solenoid S in order to pull the tube E upwards. The square of the current is a direct measure of the force on the quartz tube. The necessary arrangements for the initial adjustment

* T. Soné, *Phil. Mag.* **39**, 305, 1930.
† R. Stössel, *Ann. der Phys.* **10**, 393, 1931.

of the balance are not shown in the diagram. After a gas has been admitted to the apparatus a different current in the solenoid will be required to give equilibrium when the field is on, and the difference in the squares of the two currents measures the susceptibility of the gas. The apparatus is calibrated by using oxygen as standard.

The mirror M is used to magnify the motion of the balance, and the sensitivity is necessarily very high. The main interest in the study of nitric oxide is the variation of its susceptibility with temperature, for at low temperatures Van Vleck's theory requires that its susceptibility be zero rising to a high value at higher temperatures, as described on p. 47. Hence the containing tube C is immersed in liquid oxygen for low-temperature determinations. It is imperative that the susceptibility of the erbium oxide be maintained constant, and the container is water-cooled by means of the jacket JJ. Solid NO has been found by Lips* to possess a weak paramagnetism independent of the temperature.

Modern Forms of Faraday's Method. Most of the modern methods for comparing the susceptibilities of different gases are really elaborations of a method originally designed by Faraday for comparing the susceptibilities of different substances in the form of rods. Each rod was suspended in turn from the same torsion fibre to hang horizontally between the poles of an electromagnet; uniform or non-uniform fields may be used. The magnitude and direction of the rotation was observed when the field was excited.

For example, Glaser,† in his measurements on the susceptibilities of diamagnetic gases, used a small composite glass rod or test body, which he suspended from a long fibre so that in the absence of field it set at an angle of 45° with the line of centres of a pair of conical pole pieces. It was only slightly paramagnetic and made by attaching a strip of paramagnetic glass to a rod of diamagnetic glass. When the field was established the rod tended to set along the line of centres and was restored to its original position by turning the torsion head.

---

* E. Lips, *Helv. Phys. Acta*, **8**, 247, 1935.
† A. Glaser, *Ann. der Phys.* **75**, 459, 1924.

The magnitude of the torsion couple depends on the nature of the gas surrounding the test body. If θ_0 is the angle of torsion when the test body is *in vacuo* and θ_1 when in hydrogen at known pressure and temperature, then $\theta_1 - \theta_0$ is proportional to the susceptibility of hydrogen under these conditions, and the apparatus may thus be calibrated in terms of hydrogen.

Serious errors may arise in such measurements through radiometric action, unless the walls of the vessel containing the gas are kept at a uniform temperature. This requirement means that the pole gap must not be too narrow and the poles must be kept cool. Temperature control is also necessary, since the paramagnetism of the test body is a function of temperature, which must therefore be kept constant to within $0 \cdot 01°$ C. The evacuation and refilling of the container may produce electrostatic charges upon the body which the fibre, if of quartz, will not conduct away. Finally, the field must be kept constant, e.g. by means of a bismuth spiral, as described on p. 86.

In connection with Glaser's work brief mention must be made of what is known as the *Glaser anomaly*. He found that the volume susceptibility of a diamagnetic gas was not a linear function of its pressure. This result has not been verified by other workers, and various suggestions have been made to account for it. It has been shown that Glaser's results can be reproduced by the presence of moisture, but no satisfactory explanation of the mode of action of the moisture has been given. The anomaly is not found with pure and efficiently dried gases.

Special Test Bodies. The accuracy which may be attained in measurements by the method just described has been considerably improved by Bitter,* by Hammar† and others, by the use of specially designed test bodies. The special design used by Bitter is shown in Fig. 33a; it consists of a cylindrical vessel of pyrex glass (diamagnetic) divided into four equal chambers or quadrants and mounted on the stout glass fibre RR. One pair of opposite chambers AA are open, the other pair

* F. Bitter, *Phys. Rev.* **35**, 1572, 1930.
† G. W. Hammar, *Nat. Acad. Sci. Proc.* **12**, 597, 1926.

BB are evacuated and sealed. Round the apparatus a single turn of copper wire is placed to provide electromagnetic damping. The system is suspended from a quartz fibre so that the cylinder lies between the poles of an electromagnet. The small glass rod *D*, fixed with a tungsten spring, is so manipulated that, when the containing vessel, and, hence, *AA*, are evacuated, the system sets as shown in Fig. 33*b*, whether the field is off or on. When gas is admitted it enters *AA* and upsets the magnetic equilibrium, so that a deflection results when the

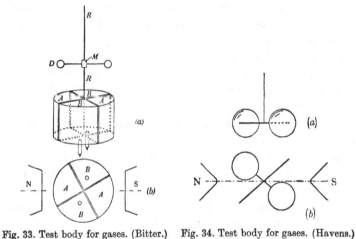

Fig. 33. Test body for gases. (Bitter.)   Fig. 34. Test body for gases. (Havens.)

magnet is excited. If the gas is paramagnetic the system tends to turn anticlockwise, as seen in Fig. 33*b*, and clockwise if the gas is diamagnetic. The angle through which the torsion head is turned to restore the cylinder to its original position in the absence of the field is therefore a direct measure of the susceptibility of the gas under the observed conditions of pressure and temperature.

Havens* has made a careful study of the behaviour of test bodies of different designs, that finally preferred being shown in Fig. 34. It consists of two evacuated spherical bulbs of quartz connected together and compensated by a small quartz rod. The whole system weighs only 0·1 gm., and is coated with gold so that it can be earthed to remove electro-

* G. Havens, *Phys. Rev.* **41**, 337, 1932; **43**, 992, 1933.

static charges from its surface. Indeed, Havens coated the whole of his apparatus with gold to avoid electrostatic disturbances. He paid much attention to the accurate centering of the fibre with respect to the torsion head. Extremely high sensitivity was attained; e.g. with oxygen at a pressure of 18·3 cm. of mercury the twist of the fibre was 454·65 complete revolutions for a field of 7500 oersteds, and it seems clear that these results are the most accurate so far obtained.

Measurements with Vapours. Very few experiments have been made with vapours mainly because of the lack of knowledge of vapour densities and because in many cases they attack the walls of the containing vessel. Measurements on paramagnetic vapours have been made by the Curie method by Lallemand* and Néel.† The liquid whose vapour is to be investigated is sealed inside an evacuated bulb, less than 1 c.c. in internal volume, with stout walls. The bulb is attached to a translation balance like that described on p. 107, and the temperature of the bulb is raised until all the liquid is converted into vapour, when magnetic measurements are commenced.

The main difficulties of experiment are at once apparent. The small quantity of the liquid, some few mg., must be weighed accurately before the bulb is sealed. The walls of the bulb are so thick that the necessary corrections are large. Moreover, in all cases the experimenter is dependent upon data for the dissociation phenomena exhibited by the vapour, which are often unreliable and inadequate. For example, Néel in working with sulphur vapour had to allow for the presence of S_6 and S_8 molecules. He found from three concordant sets of observations that sulphur vapour is paramagnetic, the susceptibility decreasing more rapidly than the Curie law predicts and giving p_{eff} at 600° C. equal to 2·3. Lallemand used iron perchloride $FeCl_3$ and found that its molar susceptibility decreased from $9·98 \times 10^{-3}$ at 125° C. to $4·59 \times 10^{-3}$ units at 327° C., where all the material was in the form of vapour; from 327° to 430° C. the susceptibility was constant at $4·59 \times 10^{-3}$, after which it again decreased with rise in temperature.

* A. Lallemand, *Compt. Rend.* **194**, 1726, 1932.
† L. Néel, *Compt. Rend.* **194**, 2035, 1932.

Apparently the above data can only be supplemented by some observations on the behaviour of potassium vapour between 300° and 500° C. obtained by a circulation method devised by Lehrer.* In principle the method is easy to follow, but its experimental working must be very difficult. Over the above range of temperature giving potassium vapour pressures of 0·5 to 30 mm. of mercury, Lehrer found the atomic susceptibility to be $33·6 \times 10^{-6}$ with $C = 0·38$, with an accuracy of about 10 per cent. It would appear that the Rankine method described on p. 110 would be most useful for the investigation of vapours.

Discussion of Experimental Data. From the theoretical considerations of Chapter I it is clear that measurements of susceptibility in the case of weakly magnetic materials are generally of restricted use unless they are made over a sufficient range of temperature and at different field strengths to eliminate the effects of ferro-magnetic impurity, if necessary, and to see that the susceptibility is not a function of the field. The experimental values of the mass susceptibilities of the elements are reproduced in Table II.

A few interesting points may be brought out by marking the distribution of the diamagnetic and paramagnetic elements on a curve of atomic volume plotted against atomic number. In general, the peaks of the curve are occupied by paramagnetic elements of the alkali metal group, while in the troughs the paramagnetics are arranged on the sides of lower atomic number and the diamagnetics on the sides of higher atomic number. Some interesting features are also brought out when the logarithms of the atomic susceptibilities of the diamagnetic elements are plotted against their atomic numbers, for many of the elements in each vertical column of the periodic table lie accurately upon one of a series of straight lines.

Much of the theoretical discussion in Chapter I was devoted to the significance of the spectroscopic states of atoms and ions, and we must now see how the effective magneton number of a paramagnetic ion of any element may be obtained from

---

* E. Lehrer, *Ann. der Phys.* **81**, 229, 1927; cf. W. Gerlach, *Rep. of the Internat. Congress of Physics*, Como, 1937 and K. E. Mann, *Zeit. für Phys.* **98**, 548, 1936.

experimental measurements. The importance of these measurements in physics and chemistry lies in the fact that they are made upon atoms and ions in the normal and not in an excited state, although, of course, the influence of neighbouring particles may produce disturbances, while spectroscopic measurements can only be made on particles which are in an excited state.

There are two methods of attack. First, we may make measurements on hydrated salts containing the metallic ion in question, in which case, cf. p. 48, it is supposed that the molecules of the water of crystallisation have little effect on the magnetic properties. The measurements must be made at a series of known temperatures to determine whether the Curie law is obeyed after correction has been made for the diamagnetism of the non-metallic ion, water of crystallisation and the core of the metallic ion; fortunately, the corrections are usually small. The corrected values of the molar susceptibility are then plotted against the reciprocals of the corresponding absolute temperatures, or, the reciprocals of the corrected molar susceptibilities are plotted against the temperatures. The former method of plotting has the advantages that any errors in correcting for the diamagnetic contributions are less important and any errors in the individual measurements are less likely to be misinterpreted. Special features of various methods of plotting have been discussed in detail by Klemm[*] and Vogt.[†]

If the Curie law is obeyed, then from either graph the Curie constant C can be deduced and inserted in the equation

$$(\chi_M)_+ = C/T = p_{eff}^2 \mu_B^2 (N^2/3RT),$$

corresponding to equation (11) of p. 16, where p_{eff} is the *effective number of Bohr magnetons* to which the magnetic moment of the ion is equivalent. Hence

$$p_{eff} = \sqrt{3RC}/N\mu_B = \sqrt{3RC}/5565 = 2\cdot839\sqrt{C}. \quad ......(86)$$

In the literature results are frequently stated in terms of *Weiss magnetons*, in which case the effective number of Weiss

[*] W. Klemm, *Magnetochemie*, p. 113, 1936.
[†] E. Vogt, *Ann. der Phys.* **29**, 358, 1937.

magnetons, p_W, to which the magnetic moment of the ion is equivalent is given by

$$p_W = \sqrt{3RC}/N\mu_W = \sqrt{3RC}/1125 \cdot 6 = 14 \cdot 05 \sqrt{C}, \quad ...(87)$$

but, as the Weiss unit has no theoretical significance, we shall not employ it.

If a Curie-Weiss law is obeyed, then

$$(\chi_M)_+ = C/T - \theta$$

and

$$p_{eff} = \frac{\sqrt{3R(\chi_M)_+ (T - \theta)}}{5565} = 2 \cdot 839 \sqrt{(\chi_M)_+ (T - \theta)}. \quad ...(88)$$

The second method of attack involves measurements with aqueous solutions of different concentrations of the ion in question at a fixed temperature, say 20° C. The molar susceptibility of the dissolved salt may then be calculated, assuming Wiedemann's law, p. 13. If it is found that the molar susceptibility of the salt is independent of its concentration, then it is further assumed that the paramagnetic ion obeys Curie's law. Let (χ_M') be the molar susceptibility of the salt in solution. Then $(\chi_M)_+ = (\chi_M') - (\chi_M)_-$ (the value of $(\chi_M)_-$ being negative since the non-metallic ion is diamagnetic), and we may write

$$(\chi_M)_+ = C/293,$$

so that

$$p_{eff} = \frac{\sqrt{3R(\chi_M)_+ 293}}{5565} = 2 \cdot 839 \sqrt{293 \, (\chi_M)_+}. \quad ......(89)$$

However, few experiments have been made with solutions over a sufficient temperature interval to justify the assumption that Curie's law holds generally, and those of Liquier-Milward* with solutions of cerium chloride over the range 7° to 70° C. give $(\chi_M)_+ (T + \theta) = 0 \cdot 762$, where θ varies from 45 to 63 over a range of concentration of 8 to 37·5 per cent of the anhydrous salt. Here, assuming Curie's law to hold, p_{eff} would vary in this range of concentration from 2·31 to 2·27, but, when the above Curie-Weiss law is assumed, the value of p_{eff} is independent of concentration and temperature and equal to 2·50.

* Jeanne Liquier-Milward, *Phys. Soc. Proc.* **47**, 559, 1935. See also G. Dupouy and G. Haenny, *Compt. Rend.* **199**, 843, 1934.

Ions of the Rare Earth Group. It is unfortunate that some of the most interesting materials from a magnetic standpoint are the rare earth salts of which many are very expensive and difficult to prepare in a pure state. The paramagnetic properties of the ions arise from the $4f$ electrons which are situated well inside the atom. The experimental data have been thoroughly

TABLE IV. MAGNETIC AND SPECTROSCOPIC DATA
FOR IONS OF THE RARE EARTHS*

No. of elec- trons	Ion	Normal state	L	S	J	p_{eff} exp.	$g\sqrt{J(J+1)}$	p_{eff} Van Vleck
54	La^{+++}	$^1S_0$	0	0	0	dia.	0	0
55	Ce^{+++}Pr^{++++}	$^2F_{\frac{5}{2}}$	3	$\frac{1}{2}$	$\frac{5}{2}$	$\left.\begin{array}{c}2\cdot50\\2\cdot62\end{array}\right\}$	$2\cdot54$	$2\cdot56$
56	Pr^{+++}	$^3H_4$	5	1	4	$3\cdot61$	$3\cdot58$	$3\cdot62$
57	Nd^{+++}	$^4J_{\frac{9}{2}}$	6	$\frac{3}{2}$	$\frac{9}{2}$	$3\cdot80$	$3\cdot62$	$3\cdot68$
58	Il^{+++}	$^5J_4$	6	2	4	—	$2\cdot68$	$2\cdot83$
59	Sm^{+++}	$^6H_{\frac{5}{2}}$	5	$\frac{5}{2}$	$\frac{5}{2}$	$1\cdot51?$	$0\cdot84$	$\left.\begin{array}{c}1\cdot55\\1\cdot65\end{array}\right\}$
60	Eu^{+++}Sm^{++}	$^7F_0$	3	3	0	$3\cdot63?$	$0\cdot00$	$\left.\begin{array}{c}3\cdot40\\3\cdot51\end{array}\right\}$
61	Gd^{+++}Eu^{++}	$^8S_{\frac{7}{2}}$	0	$\frac{7}{2}$	$\frac{7}{2}$	$\left.\begin{array}{c}7\cdot79\\7\cdot90\end{array}\right\}$	$7\cdot94$	$7\cdot94$
62	Tb^{+++}	$^7F_6$	3	3	6	$9\cdot65$	$9\cdot70$	$9\cdot70$
63	Dy^{+++}	$^6H_{\frac{15}{2}}$	5	$\frac{5}{2}$	$7\frac{1}{2}$	$10\cdot50$	$10\cdot60$	$10\cdot60$
64	Ho^{+++}	$^5J_8$	6	2	8	$10\cdot50$	$10\cdot60$	$10\cdot60$
65	Er^{+++}	$^4J_{\frac{15}{2}}$	6	$\frac{3}{2}$	$7\frac{1}{2}$	$9\cdot55$	$9\cdot60$	$9\cdot60$
66	Tu^{+++}	$^3H_6$	5	1	6	$7\cdot27$	$7\cdot60$	$7\cdot60$
67	Yb^{+++}	$^2F_{\frac{7}{2}}$	3	$\frac{1}{2}$	$\frac{7}{2}$	$4\cdot55$	$4\cdot50$	$4\cdot50$
68	Cp^{+++}	$^1S_0$	0	0	0	dia.	$0\cdot00$	$0\cdot00$

* Values of the Curie constant for Dy^{+++}, Ho^{+++} and Er^{+++} for the powdered oxides have been given recently by M. B. Cabrera (*Compt. Rend.* **205**, 400, 1937) and for Dy^{+++}, Er^{+++} and Yb^{+++} for the octahydrated sulphates by L. C. Jackson (*Phys. Soc. Proc.* **48**, 741, 1936). For work on Eu^{++} and Sm^{++} see P. W. Selwood, *Journ. Amer. Chem. Soc.* **55**, 4861, 1933, and **56**, 2392, 1934; for work on Nd^{+++} see L. C. Jackson, *Roy. Soc. Proc.* A, **170**, 270, 1939.

sifted by Gorter[†] and by Van Vleck,[‡] and the final weighted mean values of p_{eff} are given in Table IV, together with the relevant spectroscopic data and the values of p_{eff} calculated from the expression $g\sqrt{J(J+1)}$ of p. 40. In the final column

† C. J. Gorter, *Arch. du Musée Teyler*, Série III, **7**, 183, 1932.
‡ J. H. Van Vleck, *Electric and Magnetic Susceptibilities*, p. 243, 1932.

are given the values of p_{eff} calculated from the more complete Van Vleck theory. The experimental and theoretical values are plotted in Fig. 35. They show that the values of p_{eff} calculated from equation (27) of p. 40, derived for the case of wide multiplets, agree well with experiment except in the cases of Sm^{+++} and Eu^{+++}, where special conditions obtain. Here, the values of L and S are such that when they are antiparallel the value of $L - S$ is small and the energy difference between the

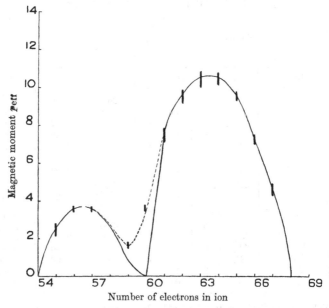

Fig. 35. Calculated and observed magnetic moments.

1 —— $p_{\text{eff}} = g\sqrt{J(J+1)}$.
2 - - - - p_{eff} calc. Van Vleck.

The thick lines represent the range of observed values. (Gorter.)

states of an ion for the two lowest values of J is small, because it is proportional to the difference in the values of $J(J+1)$, even though the overall multiplet width, i.e. the difference in energy between the states with the highest and with the lowest value of J, is large. Moreover, the temperature-independent paramagnetism of p. 43 is important in these cases. In Table IV the two separate values for p_{eff} calculated by Van

Vleck for Sm^{+++} and Eu^{+++} differ in the values assumed for the screening coefficients, representing the effect of the electrons surrounding the nucleus in reducing the action of the nuclear charge upon the $4f$ electrons, which is used in calculating the term separations.

The expression for the susceptibility which must be used in the cases of Eu^{+++} and Sm^{+++} is

$$\chi = \frac{N \sum_{J} \{[\mu_B^2 g_J^2 J(J+1)/3kT] + \alpha(J)\}(2J+1) e^{-W_J/kT}}{\sum_{J} (2J+1) e^{-W_J/kT}},$$

where the temperature-independent term $\alpha(J)$ is given by

$$\left(\frac{\mu_B^2}{6(2J+1)h}\right)\left[\frac{f(J)}{\nu_{J-1;\,J}} + \frac{f(J+1)}{\nu_{J+1;\,J}}\right],$$

where $f(J) = J^{-1}[(S+L+1)^2 - J^2][J^2 - (S-L)^2]$,

and the smallness of the frequency intervals $\nu_{J-1;\,J}$ and $\nu_{J+1;\,J}$ make $\alpha(J)$ appreciable. The experimental data have been discussed in detail by Frank,* and the agreement between experiment and theory is as good as can be expected, the theory in the case of Sm^{+++} predicting a minimum value of χ in the region of $120°$ C., which has been observed. In the case of Eu^{+++} the theory indicates that χ should become constant at low temperatures, but experimental data to test this are not yet available.

The effect of crystalline fields, p. 49, is very marked in the case of certain ions of the rare earth group. Penney and Schlapp† considered the behaviour of Pr^{+++} and Nd^{+++} in their hydrated sulphates and showed that good agreement between experiment and theory is obtained on the assumption that the crystal field has cubic symmetry. Both ions obey a Curie-Weiss law over an extensive range of temperature above $-150°$ C., but whereas the curve of $(1/\chi, T)$ for Pr^{+++} turns upwards at very low temperatures that for Nd^{+++} turns downwards apparently to run through the origin.

---

* Amelia Frank, *Phys. Rev.* **39**, 119, 1932.
† W. G. Penney and R. Schlapp, *Phys. Rev.* **41**, 194, 1932.

Ions of the Iron Group. The experimental data for ions of the iron group have also been examined by Gorter* and the main experimental results are given in Table V, together with

TABLE V. MAGNETIC AND SPECTROSCOPIC DATA
FOR IONS OF THE IRON GROUP

No. of electrons	Ion	Normal state	L	S	J	p_{eff} exp.	$g\sqrt{J(J+1)}$	$\sqrt{\dfrac{L(L+1)}{+4S(S+1)}}$	$\sqrt{4S(S+1)}$
18	K^+V^{+++++}	$^1S_0$	0	0	0	dia.	0	0	0
19	$Sc^{++}Ti^{+++}V^{++++}$	$^2D_{\frac{3}{2}}$	2	$\frac{1}{2}$	$\frac{3}{2}$	1·73	1·55	3·01	1·73
20	$Ti^{++}V^{+++}$	$^3F_2$	3	1	2	2·83	1·63	4·49	2·83
21	$V^{++}Cr^{+++}Mn^{++++}$	$^4F_{\frac{3}{2}}$	3	$\frac{3}{2}$	$\frac{3}{2}$	3·82	0·70	5·21	3·87
22	$Cr^{++}Mn^{+++}$	$^5D_0$	2	2	0	4·81	0·00	5·50	4·91
23	$Mn^{++}Fe^{+++}$	$^6S_{\frac{5}{2}}$	0	$\frac{5}{2}$	$\frac{5}{2}$	5·85	5·93	5·94	5·92
24	Fe^{++}	$^5D_4$	2	2	4	5·52 to 5·22	6·71	5·50	4·91
25	Co^{++}	$^4F_{\frac{9}{2}}$	3	$\frac{3}{2}$	$\frac{9}{2}$	5·20 to 4·43‡	6·63	5·21	3·87
26	Ni^{++}	$^3F_4$	3	1	4	3·23	5·59	4·49	2·83
27	Cu^{++}	$^2D_{\frac{5}{2}}$	2	$\frac{1}{2}$	$\frac{5}{2}$	2·02 to 1·81	3·55	3·01	1·73

‡ The value for Co^{++} has been shown by Chatillon (*Ann. de Phys.* **9**, 187, 1928) to depend on the mode of preparation and history of the material.

the relevant spectroscopic data and the values of p_{eff} calculated from the expression $g\sqrt{J(J+1)}$. It is clear that these calculated values in no wise represent the experimental results. It is likewise seen that the values of p_{eff} calculated from

$$\sqrt{L(L+1)+4S(S+1)},$$

i.e. assuming narrow multiplets, are also unsatisfactory. Now, in these ions the paramagnetism arises from the $3d$ electrons which are not deep-seated in the atom, and Stoner† suggested that interaction between neighbouring atoms caused the orbital moments to be completely quenched, leaving only the spins effective, as described on p. 47. Hence, as on p. 48, we may write $p_{eff} = \sqrt{4S(S+1)}$, and then the values given in Table V show that agreement between experiment and theory is much more satisfactory, although in many cases it is obvious that the quenching of the orbital moments is by no

* Gorter, *loc. cit.*
† E. C. Stoner, *Phil. Mag.* **8**, 250, 1929.

means complete. It should be pointed out that the experimental values for p_{eff} differ widely according to the material used and the results of different observers do not always agree. An indication of the extent of experimental variations is shown in the curve of Fig. 36, and it would seem that more measurements by absolute methods are desirable.

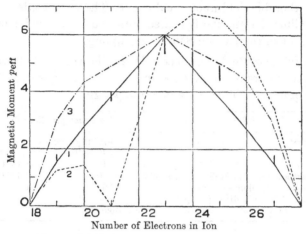

Fig. 36. Calculated and observed magnetic moments.

1 ——— $p_{\text{eff}} = \sqrt{4S(S+1)}$.

2 ······ $p_{\text{eff}} = g\sqrt{J(J+1)}$.

3 —·—· $p_{\text{eff}} = \sqrt{L(L+1)+4S(S+1)}$.

The thick lines represent the range of observed values. (Gorter.)

We have seen that in salts of the iron group the orbital momenta are quenched, and it is pertinent to enquire whether a strong crystalline field can do even more. It can be shown, if it disrupts the Russell-Saunders coupling, that the lowest energy state of the ion becomes one of smallest spin, instead of one of maximum spin under the Hund rule. Van Vleck* has shown that this accounts for the low values of the susceptibilities of iron cyanides and certain cobaltamines. The complex ions such as $Fe(CN)_6^{++++}$ are diamagnetic when the total number of electrons is even, and complex ions like $Fe(CN)_6^{+++}$ with an odd number of electrons give susceptibilities corresponding to $S = \frac{1}{2}$, $L = 0$, while the Hund rule would require

* J. H. Van Vleck, *Journ. of Chem. Phys.* **3**, 807, 1935.

susceptibilities corresponding to ^6S states. For instance, Howard* found that $K_3Fe(CN)_6$ gave a susceptibility corresponding to one electron spin only, plus a contribution from unquenched orbital momentum which practically doubled this value except at low temperatures, and showed large anisotropy which could be explained by the superposition of a small rhombic field upon a cubic one. The magnitude of the crystalline field in the hydrated sulphate is not sufficient to break down the Russell-Saunders coupling.

Useful Magnetic Data. The following data are given because of their importance in experimental determinations. The units of susceptibility are e.m. units per gm.

Substance	Mass susceptibility × 10⁶	
Water†	-0.72145 ± 0.00048	
	(0.000108) $(t-20)$ at $t°$ C.	
Mercury	-0.168	at 18° C.
Manganese sulphate, $MnSO_4 . 4H_2O$ (solid)	65.20	at 17° C.
Ferrous ammonium sulphate,	32.6	at 17·2° C.
$FeSO_4(NH_4)_2SO_4 . 6H_2O$ (solid)		
Potassium Chloride (in solution)	-0.528	

Apparent (Ionic) Susceptibilities of Ions in Aqueous Solution (Hoare and Brindley)‡

Ion	Molar susceptibility × 10⁶
H+	$+ 3.8$
Na+	$- 5.6$
K+	-14.0
Cl−	-25.1
NH_4+	-16.0
SO_4−−	-43.0

Corrections for Diamagnetism of Atomic Cores are given by Van Vleck, *Electric and Magnetic Susceptibilities*, p. 225, 1932, by Sommerfeld and Bethe, *Geiger-Scheel Handbuch der Physik*, **24**, Part II, 278, and by Pacault, *La Revue Scientifique*, III, 169, 1947. See Chapter XIII.

* J. B. Howard, *Journ. Chem. Phys.* **3**, 813, 1935.

† Cf. H. Auer, *Ann. der Phys.* **18**, 593, 1933, and S. Seely, *Phys. Rev.* **49**, 812, 1936; cf. also P. Sève, *Congr. Intern. d'Électricité*, **3**, 231, 1932, for résumé and survey of measurements to 1932.

‡ F. E. Hoare and G. W. Brindley, *Trans. Farad. Soc.* **33**, 268, 1937, and *Phys. Soc. Proc.* **49**, 619, 1937.

CHAPTER IV

THE MAGNETIC PROPERTIES OF CRYSTALS

Theoretical Considerations. Measurements of the susceptibilities of isotropic or polycrystalline solids have been described
in the preceding chapter, where it was assumed that the direction of magnetisation always coincides with the direction of
the applied field. When, however, a small single crystal of a
weakly magnetic substance is placed in a magnetic field, in
general the direction of magnetisation does not coincide with
the direction of the applied field. In other words, there are
usually directions of maximum and minimum susceptibility
in the crystal, which is then said to be *magnetically anisotropic*.
In all cases to be dealt with in this chapter it will be assumed,
unless otherwise stated, that the induced magnetisation is not
sufficient to produce marked distortion of the applied field.

The simple energy equation of a crystalline substance in a
magnetic field in an evacuated space is

$$E = -\tfrac{1}{2}\int[k_{11}\,\mathscr{H}_x^2 + k_{22}\,\mathscr{H}_y^2 + k_{33}\,\mathscr{H}_z^2 + 2k_{12}\,\mathscr{H}_x\mathscr{H}_y$$
$$+ 2k_{23}\,\mathscr{H}_y\mathscr{H}_z + 2k_{31}\,\mathscr{H}_z\mathscr{H}_x]\,dv, \qquad \text{......(90)}$$

where k_{11}, k_{22}, etc. represent the susceptibilities of the crystal
referred to the system of coordinates x, y, z, and the integral is
to be taken over the whole volume of the crystal whose shape is
not yet specified. When the system of coordinates is so chosen
that the direction of the applied field coincides with the x axis,
then $\mathscr{H}_y = \mathscr{H}_z = 0$ and equation (90) is simplified, giving

$$E = -\tfrac{1}{2}\int k_{\mathscr{H}}\mathscr{H}^2\,dv, \qquad \text{......(91)}$$

where the value of $k_{\mathscr{H}}$, the susceptibility of the crystal in the
x direction, clearly depends on the direction of $\mathscr{H}$.

Now, the magnetic potential energy cannot be changed
merely by manipulation of the axes of reference; hence, it
follows that

$$k_{11}\,\mathscr{H}_x^2 + \ldots + \ldots + 2k_{12}\,\mathscr{H}_x\mathscr{H}_y + \ldots + \ldots = k_{\mathscr{H}}\mathscr{H}^2,$$
$$\text{......(92)}$$

or, if l_1, m_1 and n_1 are the direction cosines of $\mathscr{H}$ referred to the system of coordinates x, y, z,

$$k_{11}l_1^2 + k_{22}m_1^2 + k_{33}n_1^2 + 2k_{12}l_1m_1 + 2k_{23}m_1n_1 + 2k_{31}n_1l_1 = k_{\mathscr{H}},$$
$$\dots\dots(93)$$

an equation which gives the relation between the susceptibilities k_{11}, etc. and the effective value $k_{\mathscr{H}}$ observed in the direction of the applied field, $\mathscr{H}$.

Equation (93) may be simplified by introducing the experimental fact that there exist in the crystal three mutually perpendicular axes, *the principal magnetic axes*, which are defined by the relations $k_{12} = k_{23} = k_{31} = 0$ and $k_{11} = k_1$, $k_{22} = k_2$, $k_{33} = k_3$, where k_1, k_2 and k_3 are termed *the principal susceptibilities* along these axes. If the field $\mathscr{H}$ makes angles with these axes whose direction cosines are l, m and n, then

$$k_1 l^2 + k_2 m^2 + k_3 n^2 = k_{\mathscr{H}}. \qquad \dots\dots(94)$$

Corresponding to k_1, k_2 and k_3 we have mass susceptibilities χ_1, χ_2 and χ_3, and we may note that measurements with powdered crystals give us $(k_1 + k_2 + k_3)/3$ or $(\chi_1 + \chi_2 + \chi_3)/3$.

The Significance of Measurements on Single Crystals. In earlier chapters the importance of susceptibility measurements on single crystal specimens has been indicated to some extent, but it is now desirable to consider in detail the nature of the information which they may be expected to provide. It is convenient to treat diamagnetic compounds, paramagnetic compounds and metallic single crystals separately, as the information we obtain is quite different in the three cases.

Diamagnetic anisotropy, which is in general small, must be conditioned by the structure of the individual molecules and their arrangement in the crystal lattice. Consequently, if, for example, in the case of an organic compound there exist adequate X-ray data to decide how the molecules in the crystal are arranged, then a knowledge of the susceptibilities along chosen directions in the crystal ought to be sufficient to determine the magnetic properties of the individual molecules. In this way, the principal susceptibilities of naphthalene and anthracene molecules may be found. Alternatively, a knowledge of the magnetic properties of the individual molecules derived from other measurements, such as the magnetic double refraction of the molecules in the liquid state, may be used to suggest the arrangement of the molecules in the crystal, and so to supplement the results of X-ray analysis. Indeed, the help which can be thus obtained from magnetic measurements is very considerable.

Paramagnetic anisotropy, which may be very large, must arise from entirely different causes. The hypothesis of the existence of crystalline fields of definite symmetry has been discussed on p. 47, where it was explained that they would account for the quenching of orbital momenta in the case of salts of the iron group. If these fields are not symmetrical then the crystal must possess definite directional properties. Obviously, the directional properties of representative paramagnetic crystals must be known before the nature of the crystalline fields which would account for them can be decided. The changes in these directional properties with temperature must also be determined, for it may happen that by the superposition of two fields, for example, a cubic field with one set of axes and a rhombic field with a different set of axes, the principal axes of the susceptibility change with temperature. While the presence of marked magnetic anisotropy may thus be taken to give definite information concerning the nature of the crystalline fields, the presence of weak magnetic anisotropy may also give interesting information. Thus, the magnetic anisotropy of the Mn^{++} ion in manganese ammonium sulphate is exceedingly small—practically of the order of diamagnetic anisotropy—showing that the ion is in an S state.

The magnetic anisotropy of metallic single crystals has recently become of considerable interest in connection with the electron theory of metals and with modern work on atomic order and disorder. A knowledge of the anisotropy properties of single crystals of ferromagnetic metals enables us to account for many phenomena of technical importance, such as, for instance, why certain alloys of ferromagnetic metals are markedly sensitive to strain and to special types of heat treatment. In the following sections we shall discuss the

135

experimental procedure for the determination of the magnetic anisotropy of crystals, beginning first with experiments on diamagnetic and paramagnetic compounds.

Jackson's Experiments. It follows from equation (94) that if the directions of the principal magnetic axes in a crystal are known, then we may apply a field with a gradient parallel to each in turn and measure each principal susceptibility directly, provided that the shape of the crystal does not produce spurious volume effects; the latter can be avoided if necessary by cutting the crystal to a spherical shape. This direct method was used by Jackson* in an investigation of the principal susceptibilities down to liquid-air temperatures. His apparatus is shown in Fig. 37.

The small crystal C was attached with non-magnetic cement to the quartz frame carrier with a standard face of the crystal vertical and perpendicular to the horizontal field of an electromagnet. The frame was attached to a Sucksmith ring balance R which measured the vertical forces. The weights w could be adjusted without opening the case surrounding R, and the motion was damped by a ring r dipping in oil. Sideway motion of the suspended system was prevented by a flat phosphor bronze spiral s. The apparatus was calibrated by replacing C by powdered salt of known susceptibility.

The crystal was hung inside a thick copper tube, to promote uniformity of temperature; this was sealed to an upper tube of thin nickel silver supported as shown. The temperature of the crystal was measured by the platinum resistance thermometer Th. The Dewar flask was joined to the upper portion of the apparatus by the wired-on rubber ring RR, and the ground glass joint G enabled the upper portion of the apparatus to be removed when a crystal had to be replaced. This portion could be evacuated through T and then filled with hydrogen. The lowest temperatures were obtained by boiling liquid oxygen under reduced pressure in V, the evaporated liquid leaving via F. Measurements were also made at room temperature, with a mixture of solid CO_2 and alcohol, and with liquid ethylene.

* L. C. Jackson, *Roy. Soc. Proc.* A, **140**, 695, 1933.

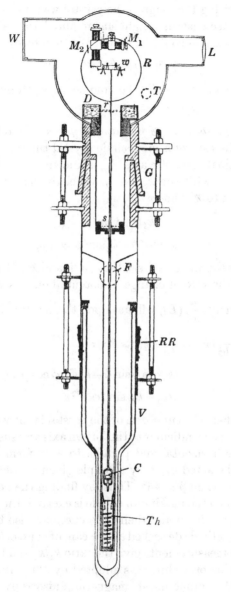

Fig. 37. Jackson's apparatus for measuring the
susceptibilities of crystals.

On exciting the magnet the crystal was pulled downwards and the decrease in weight δm required to restore it to its original position could be found. We then have

$$\delta m . g = k\mathscr{H} v d\mathscr{H}/dx,$$

where $k = k_1$ or k_2 or k_3, the volume susceptibility parallel to the field.

Experiments with Monoclinic Crystals. It often happens that in the case of a monoclinic crystal the direction of only one principal magnetic axis, k_3, is known. Now, if the crystal is mounted with the k_3 axis parallel to the z axis and perpendicular to $\mathscr{H}$, then $n = 0$ and

$$k_1 l^2 + k_2 m^2 = k_{\mathscr{H}}$$

or
$$k_1 \cos^2\phi + k_2 \sin^2\phi = k_{\mathscr{H}}, \qquad \ldots\ldots(95)$$

where ϕ is the angle between the k_1 axis and $\mathscr{H}$. Hence a couple, equal to the rate of change of potential energy with angle of displacement, $\dfrac{d}{d\phi}(E)$, will act upon the crystal. Now

$$\frac{d}{d\phi}(E) = \frac{d}{d\phi} - \tfrac{1}{2}(k_{\mathscr{H}}\mathscr{H}^2)v$$
$$= \{k_1 \cos\phi \sin\phi - k_2 \sin\phi \cos\phi\}\mathscr{H}^2 v$$
$$= \tfrac{1}{2}(k_1 - k_2)\sin 2\phi \mathscr{H}^2 v. \qquad \ldots\ldots(96)$$

Hence, when a circular disc of the crystal is cut with the plane of the disc perpendicular to the known axis and suspended with this plane horizontal and parallel to a uniform field $\mathscr{H}$, the disc will be acted upon by a couple given by (96), which is a maximum when $\phi = 45°$. Thus, by finding the position of the disc in which the maximum couple is exerted, the directions of k_1 and k_2 are fixed and their difference can also be measured. The two methods described above can, of course, be combined, when the measurements give the ratio k_1/k_2 and the difference $k_1 - k_2$. This procedure was adopted by Dupouy,* who made use of the pronounced magnetic anisotropy of siderose ($FeCO_3$) in constructing a gaussmeter.

* M. G. Dupouy, *Ann. de Phys.* 15, 495, 1931.

Sometimes, however, couples cannot readily be measured and only translational forces can be utilised. The determination of k_1 and k_2 then becomes much more complicated, since measurements of the forces in a given direction for at least three different positions of the disc with respect to $\mathcal{H}$ are required. This procedure was adopted by Jackson* in his earlier work on paramagnetic crystals down to liquid hydrogen temperatures. A cylindrical section of the crystal with its axis vertical was attached to a small horizontal aluminium disc suspended by a long quartz fibre from one side of a Pettersson† quartz microbalance. The requisite non-homogeneous magnetic field was produced by a water-cooled coil. A coil was used because the cryogenic arrangements were rather large. $\mathcal{H} \, d\mathcal{H}/dx$ was therefore small, but it was accurately known. The results are somewhat laborious to work out.

Immersion Experiments. The effects of any surrounding medium have been omitted from the preceding discussion, but, clearly, when the surrounding medium has a volume susceptibility equal to k_1, say, then no translational force can act on the crystal when the lines of force are parallel to the k_1 axis. Hence, if the crystal be immersed in a solution whose concentration can be adjusted until the force parallel to the k_1 axis vanishes, k_1 must then be equal to the known susceptibility of the solution. This method was devised by Rabi‡ and possesses the advantage that the crystal need not be very accurately shaped or as accurately mounted in the magnetic field as in other methods.

Rabi suspended the crystal vertically from a long glass fibre as shown in Fig. 38, so that it was situated in a non-homogeneous field of a strong electromagnet. The fibre was sufficiently strong to prevent rotation of the crystal under the influence of couples described above, and it could be turned about the vertical axis to orientate the crystal. Hence, the

* L. C. Jackson, *Roy. Soc. Phil. Trans.* A, **226**, 107, 1926.
† H. Pettersson, *Phys. Soc. Proc.* **32**, 209, 1920.
‡ I. I. Rabi, *Phys. Rev.* **29**, 174, 1927.

crystal was only free to move laterally under the influence of the field. It was immersed in an aqueous solution whose susceptibility could be varied over the range of the principal susceptibilities. The solution was saturated with the crystal material so that the specimen did not dissolve during the measurements. The susceptibility of the solution was increased by adding similarly saturated manganese chloride solution or decreased by adding potassium iodide solution.

Lateral motion of the fibre was observed by a microscope M. In general, it occurred both parallel and perpendicular to the axis of the pole tips. The fibre was turned until the former motion along the x axis no longer occurred. The remaining motion was nullified by adjusting the susceptibility of the solution, which was later measured separately by the Gouy method. Surface tension effects occurred where the fibre broke the surface of the solution, and differences in susceptibility much less than 10^{-8} could not be detected. A systematic error arose because the fibre had a different susceptibility from that of the crystal. Rabi was able, however, to give values of the principal susceptibilities of many crystals belonging to the monoclinic double sulphate hexahydrate isomorphous series, denoted by the formula $MSO_4R_2SO_4.6H_2O$.

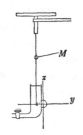

Fig. 38. Rabi's apparatus for susceptibilities of a crystal.

Oscillation Experiments. Let us suppose that a crystal is suspended by a torsion fibre parallel to the k_3 axis and oscillates with a period T_0. If, now, a uniform field $\mathscr{H}$, is applied parallel to the k_1 axis, then from (96), on rotating the crystal through an angle $\delta\phi$ about the axis of the fibre, a restoring couple $\{C + (k_1 - k_2)\mathscr{H}^2 v\}\delta\phi$ acts upon the crystal, instead of the couple $C\delta\phi$ in the absence of the field, where C is the torsion constant of the fibre, and a new period of oscillation T_1 is observed.

Hence $$CT_0^2 = \{C + (k_1 - k_2)\mathscr{H}^2 v\}T_1^2, \qquad \ldots\ldots(97)$$

so that
$$k_1 - k_2 = \frac{T_0^2 - T_1^2}{T_1^2} \frac{C}{\mathscr{H}^2} \frac{1}{v} \qquad \ldots\ldots(98)$$

or
$$\chi_1 - \chi_2 = \frac{T_0^2 - T_1^2}{T_1^2} \frac{C}{\mathscr{H}^2} \frac{1}{m}, \qquad \ldots\ldots(99)$$

where m is the mass of the crystal.

This method of finding $k_1 - k_2$ was first used by Stenger and by König and has been extensively used by Krishnan, Guha and Banerjee* in their study of diamagnetic crystals. They attached each crystal to the lower end of a very short piece of glass fibre suspended by a vertical quartz fibre from a torsion head, and by finding the periods of oscillation required in equation (98) $k_1 - k_2$ was found. It was assumed that no complications arose from the shape of the specimen, and special attention was paid to the uniformity of the field. The value of the greatest principal susceptibility was found by the Rabi immersion method, a quartz fibre being used, as motion along the x axis (Fig. 38) did not occur when the greatest susceptibility was being measured. The solution used was weak manganese chloride.

The oscillation method is not satisfactory when the magnetic anisotropy is so small that effects of asymmetry of shape cannot be neglected. Krishnan, Chakravorty and Banerjee† then immerse the crystal in a liquid whose volume susceptibility is equal to that of the powdered crystal, and measure the couples acting upon it in a uniform field. This procedure was very satisfactory, for example, in the case of $MnSO_4(NH_4)_2SO_4 . 6H_2O$, where $\chi_1 - \chi_2 = 11\cdot4 \times 10^{-6}$, $\chi_1 - \chi_3 = 6\cdot8 \times 10^{-6}$, while $1/3(\chi_1 + \chi_2 + \chi_3) = 13,830 \times 10^{-6}$ e.m. units per gm. molecule. Krishnan and Banerjee‡ have also developed two methods for dealing with very small crystals which will not be described here.

Experimental Results. Among the more important results may be noted those of Jackson for crystals containing ions

* K. S. Krishnan, B. C. Guha and S. Banerjee, *Roy. Soc. Phil. Trans.* A, **231**, 235, 1933.

† K. S. Krishnan, N. C. Chakravorty and S. Banerjee, *Roy. Soc. Phil. Trans.* A, **232**, 99, 1933.

‡ K. S. Krishnan and S. Banerjee, *Roy. Soc. Phil. Trans.* A, **234**, 267, 1935.

in which only electron spins effectively contribute to the magnetic moment, of which salts of the iron group are excellent examples. For instance, the following results were obtained with $NiSO_4 . 7H_2O$ at $15°$ C., $\chi_1 = 16·66 \times 10^{-6}$, $\chi_2 = 16·53 \times 10^{-6}$ and $\chi_3 = 16·63 \times 10^{-6}$ e.m. units per gm., illustrating the general rule that these crystals are magnetically isotropic to within about 1 per cent. The principal susceptibilities of $NiSO_4 . 7H_2O$ followed a Curie-Weiss law down to temperatures not lower than $160°$ K., the Curie constant being the same for all three. Schlapp and Penney have shown that a crystalline field practically cubic with a small superimposed rhombic field with the same axes accounts satisfactorily for these results. With cobalt ammonium sulphate, $CoSO_4(NH_4)_2SO_4 . 6H_2O$, this statement was true for temperatures down to that of liquid nitrogen. Cobalt sulphate, $CoSO_4 . 7H_2O$ is peculiar in that the curves of $1/\chi_1$, etc. against temperature are linear but not parallel, and potassium ferricyanide, $K_3[Fe(CN)_6]$, shows a complicated behaviour. Dupouy found that the principal susceptibilities of siderose, $FeCO_3$, and dialogite, $MnCO_3$, obey Curie-Weiss laws, but oligist, Fe_2O_3, is peculiar in that χ_1 and χ_2 both increase with temperature to a maximum in the neighbourhood of $60°$ C. and then decrease with further rise in temperature.

Van Vleck* has examined in detail the paramagnetic anisotropy of different salts of the iron group. Taking the ammonosulphates of Mn, Co, Ni and Cu he points out that their anisotropies are respectively 1, 30, 1·5 and 20 per cent approximately. The nearly perfect isotropy of the manganous salt is readily explained if the normal Mn^{++} ion is in a ^6S state, for here, if the Russell-Saunders coupling is intact there is no orbital moment to play a part. The great difference in the behaviour of the Co^{++} and Ni^{++} ions, both in F states, can, surprisingly enough, be explained if the crystalline field has only rhombic symmetry but with relatively small departure from cubic symmetry, because in the Stark spectrum of the d^8 ^3F state of Ni^{++} the single level is lowest, while in the d^7 ^4F state of Co^{++} the triple level is lowest. In the case of the

* J. H. Van Vleck, *Phys. Rev.* **41**, 208, 1932.

Cu salts the directions of the principal magnetic axes change with temperature.* Jordahl† has shown that this can be explained to some extent by a monoclinic field, made up mainly of a cubic field with a superimposed rhombic field having only one of its axes coincident with one of the former.

Crystallographic and magnetic data for rare earth salts are at present very meagre. The absorption spectra in these salts consist of sharp lines and it should be possible to correlate them with the magnetic data. However, while Spedding‡ and his collaborators conclude from optical data that the crystalline fields of the sulphate have cubic symmetry, Krishnan and Mookherji§ find marked magnetic anisotropy.

For a discussion of the relation between the magnetic properties of crystals and X-ray data reference should be made to the works quoted above and to the papers by Krishnan and Lonsdale;|| only a few examples can be briefly mentioned here.¶ For example, it is found that a number of crystalline nitrates, carbonates, etc. show marked diamagnetic anisotropy, which, it must be concluded, arises from the non-metallic ions in the crystal lattice, since the metallic ion is bound to be more or less magnetically isotropic. Thus, sodium and potassium nitrates, in which the three oxygen atoms are distributed symmetrically about the N atom as centre with all of them in the same plane, exhibit a greater susceptibility in a direction perpendicular to this plane than in the plane itself. The difference in susceptibility in the two directions is the same for the two crystals. The magnitude of the observed anisotropy, if attributed entirely to the NO_3^- ion, can be correlated with the observed magnetic double refraction of NO_3^- in nitric acid solutions.

Again, on the basis of published X-ray data, Krishnan calculated the principal susceptibilities of naphthalene and

* B. W. Bartlett, *Phys. Rev.* **41**, 818, 1932; **44**, 687, 1933.
† O. M. Jordahl, *Phys. Rev.* **45**, 87, 1934.
‡ F. H. Spedding, *Journ. of Chem. Phys.* **5**, 191, 316 and 416, 1937.
§ K. S. Krishnan and A. Mookherji, *Nature*, **140**, 549, 1937.
|| K. S. Krishnan and Kathleen Lonsdale, *Roy. Soc. Proc.* A, **156**, 597, 1936; A, **159**, 149, 1937.
¶ See also R. Schlapp and W. G. Penney, *Reports on Prog. in Phys.* **3**, 64, 1937, and Kathleen Lonsdale, *ibid.* **4**, 332, 1938.

anthracene molecules. Two of the magnetic axes of the molecules lie in the plane of the benzene rings and the other lies perpendicular thereto. It is found that as one proceeds from benzene to naphthalene and from the latter to anthracene the increase in diamagnetic susceptibility takes place almost entirely perpendicular to the plane of the rings. Proceeding further, from the fact that in 1, 3, 5 triphenylbenzene the increase in diamagnetic susceptibility of the molecule along the normal to the (001) plane of the crystal is not as great as it would be if the benzene rings were all in this plane, it may be concluded that the rings are inclined to the plane at an angle whose magnitude can be calculated from the magnetic data. Following a theory due to Pauling,* since elaborated by London,† Lonsdale‡ has shown that the magnetic behaviour of these aromatic compounds can be explained on the hypothesis that three of the valency (s) electrons of each aromatic carbon atom are perfectly normal and precess to give the usual diamagnetic contributions expected on Larmor's theory, p. 11, while the fourth (π) electron is free to precess only in its own plane. The π electrons may be pictured as describing or, rather, contributing to a large plane, not necessarily circular, orbit in the plane of the atomic nuclei. The electron attached to the hydrogen atom in each CH group behaves normally. The radius of the π-electron orbit in the case of benzene appears to be somewhat greater than that of the ring of carbon atoms in the benzene molecule. See Chapter XIII.

Metallic Single Crystals. The rapid development of the technique for the production of metallic single crystals§ during recent years has made possible the preparation of long rods of material each formed of a single crystal. In general a crystalline axis does not coincide with the long axis of the rod, but it will be assumed, for the moment, that such coincidence is obtained. The crystal can then be mounted with the long axis

---

* L. Pauling, *Journ. of Chem. Phys.* 4, 673, 1936.

† F. London, *Journ. de Phys. et le Radium*, 8, 397, 1937.

‡ Kathleen Lonsdale, *Roy. Soc. Proc.* A, 159, 149, 1937.

§ For an account of methods of preparation see P. W. Bridgman, *Proc. Amer. Acad. Arts and Sci.* 60, 305, 1925; E. N. da C. Andrade and R. Roscoe, *Proc. Phys. Soc.* 49, 152, 1937; and E. N. da C. Andrade, *Proc. Roy. Soc.* A, 163, 16, 1937.

perpendicular to the lines of force of a magnetic field and measurements of the susceptibility parallel to the lines of force can be made by the Gouy method. It is therefore necessary to know how the remaining axes of the crystal are orientated with respect to the field. Let these two axes be those of k_1 and k_2. Then, referring to Fig. 39, it follows that the force acting upon any element of the rod of length dx is given by

$$\alpha k_{\mathscr{H}} \mathscr{H} \frac{d\mathscr{H}}{dx} dx,$$

where α is the area of cross-section and $k_{\mathscr{H}}$ is the susceptibility of the rod in the direction parallel to the line of centres of the pole pieces. But, by equation (95) of p. 138, it follows that

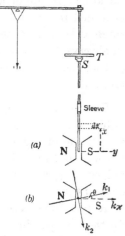

Fig. 39. Focke's apparatus for susceptibilities of metal single crystals.

$$k_{\mathscr{H}} = k_1 \cos^2 \theta + k_2 \sin^2 \theta$$

$$= k_2 + (k_1 - k_2) \cos^2 \theta,$$

so that the total magnetic pull is given by

$$F_x = \{k_2 + (k_1 - k_2) \cos^2 \theta\} \frac{\alpha}{2} \mathscr{H}^2$$

$$= \{\chi_2 + (\chi_1 - \chi_2) \cos^2 \theta\} \frac{\alpha}{2} \rho \mathscr{H}^2, \quad \ldots\ldots(100)$$

where ρ is the density of the metal, assuming the field at the upper end of the rod to be extremely small. Equation (100) shows that the magnetic pull on the crystal will vary as the crystal is rotated about its long axis, and a series of measurements can be made for different values of θ when k_1 differs from k_2. For this purpose the flexible suspension normally employed with the Gouy method must be replaced by a rigid connection in which a small aluminium turn-table or divided torsion head is inserted, as shown in Fig. 39. The latter represents the arrangement used by Focke* and later by Goetz and Focke† in an important series of experiments on the effects of added impurities on the magnetic anisotropy of metallic single crystals. Hoge‡ used a similar method in the case of single crystals of tin, but as the volume susceptibility of white (tetragonal) tin is so small a sensitive microbalance was employed.

Focke's Experiments. In his experiments on single crystals of bismuth Focke used rods in which the principal axis of the crystal was either parallel to or perpendicular to the long axis of the rod. In the first case there is no difference between the values of χ_1 and χ_2 to be inserted in equation (100), and a constant magnetic force is obtained whatever the value of θ, corresponding to a constant value of $\chi_{\mathscr{H}} = \chi_\perp = -1 \cdot 482 \times 10^{-6}$ for the susceptibility in a plane perpendicular to the principal axis. In the second case, the magnetic force varies considerably with the angle θ, so that, when the lines of force are parallel to the principal axis the force is a minimum, corresponding to a value of $\chi_{\mathscr{H}} = \chi_\| = -1 \cdot 053 \times 10^{-6}$, and when the lines of force are perpendicular to the principal axis the magnetic force is a maximum, with $\chi_{\mathscr{H}} = \chi_\perp = -1 \cdot 482 \times 10^{-6}$. The ratio of the last two values $\chi_\perp/\chi_\|$, known as the *magnetic anisotropy*, is in this case $1 \cdot 425$; this quantity changes considerably when certain impurities are present in the crystal. The mean value of the susceptibility of pure bismuth obtained from these measurements is

$$-\tfrac{1}{3}(2 \times 1 \cdot 482 + 1 \cdot 053)\, 10^{-6} = -1 \cdot 339 \pm 0 \cdot 013 \times 10^{-6},$$

a value in very good agreement with the mean value obtained from experiments on polycrystalline material.

In making absolute measurements the main difficulty lies in the determination of the area of cross-section, for, owing to its method of preparation, the shape of the rod is peculiar, and as Focke computed α by determining the weight, length and density of his specimens, an error of about 1 per cent was

* A. B. Focke, *Phys. Rev.* **36**, 319, 1930.
† A. Goetz and A. B. Focke, *Phys. Rev.* **45**, 170, 1934.
‡ H. J. Hoge, *Phys. Rev.* **48**, 615, 1935.

possible. Since it is now clear that we have only two principal susceptibilities to deal with, we may carry out the experiments as follows. In Fig. 40 let the single crystal rod have its principal crystalline axis along the direction OP making an angle ϕ with the long axis. Then the susceptibility along OQ is $\chi_\| \sin^2\phi + \chi_\perp \cos^2\phi$, and that along OR is $\chi_\perp$, so that the observed susceptibility $\chi_{\mathscr{H}}$ is given by

$$(\chi_\| \sin^2\phi + \chi_\perp \cos^2\phi)\cos^2\theta + \chi_\perp \sin^2\theta. \quad \ldots\ldots(101)$$

Consequently, if the magnetic force is now measured for different values of θ we again get a $\cos^2$ curve, i.e. a sine curve with a period of $180°$. The peak values of $\chi_{\mathscr{H}}$ will be obtained at $\theta = 0°$ and $\theta = 90°$; that at $\theta = 90°$ gives $\chi_\perp$ alone, while that at $\theta = 0°$ gives $\chi_\| \sin^2\phi + \chi_\perp \cos^2\phi$, so that ϕ must be obtained from separate determinations of the orientation of the principal axis.

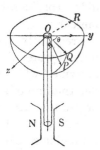

Fig. 40. Directional properties of a single crystal rod.

When the positions of the axes are not known, then the angle θ is known only in terms of arbitrary settings of the torsion head, and, consequently, as the experimental curve does not tell us which values of $\chi_{\mathscr{H}}$ correspond to $\theta = 0°$ and $\theta = 90°$ respectively, we cannot determine the values of $\chi_\perp$ and $\chi_\|$ unless experiments are made with a number of crystals each with a different value of ϕ, the angle between the principal axis and the long axis. Focke's experiments with trigonal bismuth crystals represent the ideal method of procedure. It is pertinent to remark that we should expect k_1, k_2 and k_3 to have the same value in the case of a cubic single crystal; Montgomery[*] found this to be the case for diamagnetic crystals of copper to within 1 per cent. The same statement does not, of course, hold for cubic crystals of ferromagnetic metals.

Vogt's Experiments. In his study of the properties of single crystals of mercury at liquid air temperatures Vogt[†] used the

[*] C. Montgomery, *Phys. Rev.* **36**, 498 and 1661, 1930.
[†] E. Vogt, *Ann. der Phys.* **21**, 791, 1934.

above method. In addition to the complications due to lack of knowledge of the orientations, Vogt had to make allowance for the slight magnetic effects of the mercury container; cf. Fig. 23, p. 98. These effects cannot be ignored, for in the case of mercury $\chi_\perp$ is -0.121×10^{-6} and $\chi_\parallel = -0.112 \times 10^{-6}$. It would appear possible to use a third equation, viz.

$$\chi_{mean} = \tfrac{1}{3}(2\chi_\perp + \chi_\parallel), \qquad \ldots\ldots(102)$$

in addition to those given above, in cases where $\chi_\parallel$ and $\chi_\perp$ differ considerably; this is, naturally, not the case with mercury.

Frequently the position of the principal axis may readily be found by breaking the end of the rod, as in the case of bismuth and zinc,* which break easily along the basal plane. To prevent deformation of soft crystals they may be dipped in liquid air before fracture. In other cases the Bridgman† procedure can be followed. The crystal is threaded through a sphere and rotated in a beam of light until the latter is reflected. A mirror is then held in such a position that it also reflects the light. If the back of the mirror is wetted with printer's ink and touched on the sphere the position of a crystal face is thus determined.

The de Haas-van Alphen Effect. Shoenberg and Uddin‡ made very small spherical single crystals of bismuth and its alloys merely by allowing a tiny bead, containing about 0.3 gm. of the molten metal, to cool very slowly under electrical control through its melting point, in the absence of mechanical disturbances. The position of the trigonal axis was readily found by etching and the bead was attached to a Sucksmith ring balance as in Jackson's arrangement shown on p. 137, with its axis making a known angle with a horizontal field of suitable vertical gradient. As the crystal was very small the field was considered constant throughout the volume occupied by the crystal. This is an important point, for, below 50° K., it was found that $\chi_\perp$ depended upon the field strength. More-

* J. C. McLennan, R. Ruedy and E. Cohen, *Roy. Soc. Proc.* A, **121**, 9, 1928.
† P. W. Bridgman, *Proc. Amer. Acad. Arts and Sci.* **60**, 305, 1925; see also B. Chalmers, *Proc. Phys. Soc.* 47, 732, 1935 and *Phil. Mag.* **14**, 612, 1932.
‡ D. Shoenberg and M. Z. Uddin, *Roy. Soc. Proc.* A, **156**, 687, 1936; *Camb. Phil. Soc. Proc.* **32**, 500, 1936.

over, it was found that $\chi_\perp$ and $\chi_\parallel$ varied linearly with temperature over a wide range, the linear relation in the case of $\chi_\parallel$ ceasing somewhat sharply below 75° K., and rather more gradually below 200° K. in the case of $\chi_\perp$, the linear temperature coefficients being

$$d\chi_\perp/dT = 19 \cdot 8 \times 10^{-10} \text{ and } d\chi_\parallel/dT = 7 \cdot 1 \times 10^{-10};$$

i.e. the diamagnetism decreases with rise in temperature. A similar linear variation with temperature of $\chi_\parallel$, but not of $\chi_\perp$, is found in the case of antimony. It is difficult to explain on theoretical grounds.

The fact that at low temperatures the susceptibility of bismuth is not independent of field strength is known as the *de Haas and van Alphen effect* in honour of its discoverers.* It has been studied in great detail by Shoenberg and Uddin,† who have also shown that it does not exist in antimony, at any rate above 4° K. In the case of single crystals of bismuth it is fortunate that $\chi_\parallel$ is also independent of $\mathscr{H}$ down to 4° K., but the value of $\chi_\perp$ changes as the crystal is rotated about its trigonal axis in a field applied perpendicular thereto, being a maximum when the field is parallel to the binary axis. The dependence on the field is of a very complicated nature and follows a power series which converges very slowly, and measurements have therefore been confined to settings of the binary axis parallel and perpendicular to the field. On plotting the susceptibilities for these two settings at a chosen temperature as a function of $\mathscr{H}$, a pair of curves of irregular form (something like distorted sine curves) is obtained. They intersect, forming a series of loops whose areas increase with $\mathscr{H}$. The singularities or points of intersection are more pronounced, and those for lower fields more prominent, at lower temperatures, although the values of $\mathscr{H}$ at which they occur are practically the same.

Effects of Alloying on Crystal Diamagnetism. On account of its great magnetic anisotropy an extensive series of measurements have been made on the effects of impurity

* H. de Haas and van Alphen, *Comm. Phys. Lab. Leiden*, No. 212a, 1930; 220 d, 1932.
† D. Shoenberg and M. Z. Uddin, *Roy. Soc. Proc.* A, **156**, 701, 1937.

added to bismuth. Such measurements are important because from them we hope to deduce the reasons for the temperature variation of the diamagnetism of bismuth. Goetz and Focke showed that when the impurity dissolved in the bismuth very pronounced changes were caused, the magnetic anisotropy being increased when the impurity is electropositive with respect to bismuth and decreased when electronegative. Shoenberg and Uddin have made similar measurements but over a more extended temperature range, and, contrary to the finding of Goetz and Focke, they report that the reciprocal of the anisotropy, $\chi_{\parallel}/\chi_{\perp}$, is not a linear function of the temperature. Shoenberg and Uddin emphasise the importance of low-temperature measurements, for here temperature complications are less and the effects of alloying more pronounced.

The de Haas-van Alphen effect is very much affected by alloying, the singularities moving to higher fields when tellurium, with one electron more than bismuth, and to lower fields when lead, with one electron less, is added. The singularities disappear when the amounts of lead and tellurium respectively exceed 0·08 and 0·009 per cent. Peierls* has given a theory, based on the assumption that there are only a few electrons in the highest energy band in bismuth, which explains these results qualitatively but not quantitatively, and Jones† has discussed the results in terms of the formation of Brillouin zones.

Collected Results. Table VI (p. 151) gives the principal atomic susceptibilities so far obtained with single metal crystals.

Ferromagnetic Crystals. Up to 1918 nearly all the published work on the properties of ferromagnetic substances was restricted to polycrystalline specimens. Experiments with single crystals had indeed been made, but they were confined to pyrrhotite and magnetite, of which relatively large crystals can be obtained in the natural state. In the case of a polycrystalline material, the induction or *macroscopic field*

---

* R. Peierls, *Zeit. für Phys.* **81**, 186, 1933.
† H. Jones, *Roy. Soc. Proc.* A, **147**, 396, 1934.

TABLE VI. SINGLE CRYSTAL ATOMIC
SUSCEPTIBILITIES

Metal	Temperature °C.	$-(\chi_\parallel)_A \times 10^6$	$-(\chi_\perp)_A \times 10^6$	Difference $(\chi_\perp)_A - (\chi_\parallel)_A \times 10^6$	Anisotropy $\chi_\perp/\chi_\parallel$
Antimony	17	170·6	66·1	+104·5	0·387
Bismuth	17	219·0	308·0	− 89·0	1·405
Cadmium	17	29·4	18·0	+ 11·4	0·612
Mercury	−183	22·5	24·3	− 1·8	1·080
Thallium (α)	30	84·2	33·7	+ 50·5	0·406
Tin	17	20·9	23·4	− 2·5	1·120
Zinc	17	12·4	9·5	+ 2·9	0·766

intensity, B, which may be defined as the average value of the
field intensity taken throughout a small volume of the substance, is considered to be made up of two parts according to
the statement

$$B = \mathcal{H} + 4\pi I. \qquad \ldots\ldots(103)$$

In general, however, $\mathcal{H}$ is itself made up of two parts, or,

$$\mathcal{H} = \mathcal{H}_f - \mathcal{H}_d, \qquad \ldots\ldots(104)$$

where $\mathcal{H}_f$ represents the intensity of the external field produced outside the specimen, and $\mathcal{H}_d$ represents the demagnetising field. As mentioned on p. 61 $\mathcal{H}_d$ is mainly
dependent on the shape and slightly on the imperfections in
the specimen. We may therefore write

$$B = \mathcal{H}_f - \mathcal{H}_d + 4\pi I, \qquad \ldots\ldots(105)$$

an equation in which all four quantities are vectors. When
conditions of homogeneity obtain, $\mathcal{H}_d$ is a linear vector function of I, and there are three mutually perpendicular axes
along which the relations between I and $\mathcal{H}_d$ are

$$\mathcal{H}_{d_1} = n_1 I_1, \quad \mathcal{H}_{d_2} = n_2 I_2 \text{ and } \mathcal{H}_{d_3} = n_3 I_3, \quad \ldots\ldots(106)$$

where n_1, n_2 and n_3 are demagnetisation factors which remain
constant as long as the body is homogeneously magnetised.
Now, in a ferromagnetic crystal all five vectors B, $\mathcal{H}$, $\mathcal{H}_f$, $\mathcal{H}_d$
and I can, and generally do, lie along different directions,
although we naturally try to arrange the experimental conditions so that they coincide in direction. Fortunately, in

every cubic crystal there are always three mutually perpendicular axes along which I is parallel to $\mathscr{H}$, when we may write

$$I_1 = k_1\mathscr{H}_1, \quad I_2 = k_2\mathscr{H}_2 \text{ and } I_3 = k_3\mathscr{H}_3, \quad ......(107)$$

where k_1, k_2 and k_3 are termed the principal ferromagnetic susceptibilities, and we thus get a model in which I is always a radius vector of an ellipsoid—which may degenerate into a spheroid or a sphere—while $\mathscr{H}$ is a radius vector of a sphere. In what follows, then, we have to discuss the experimental investigation of $\mathscr{H}$ and I, bearing in mind the three-dimensional nature of the problems and the existence of $\mathscr{H}_f$ and $\mathscr{H}_d$.

Experimental Procedure. From an experimental point of view a discussion of the investigation of the magnetic properties of pyrrhotite is a very convenient mode of approach to the whole subject, for this crystal behaves in a relatively simple manner. It consists of iron and sulphur in quantities which vary from specimen to specimen, and it may contain traces of nickel; its chemical constitution may best be denoted by the formula $n(\text{FeS}) + \text{FeS}_2$. It belongs to the rhombohedral class of crystals, and if it is brought into a magnetic field in such a manner that the plane of its hexagonal base is perpendicular to the lines of force, it is scarcely affected by the field, whereas it becomes strongly magnetised when introduced in any other position. This shows that along one direction in the crystal magnetisation is almost impossible, while in a plane perpendicular to this direction, termed the *plane of easy magnetisation*, the substance can be strongly magnetised. Moreover, in this plane there are two directions at right angles; parallel to one magnetisation is obtained with ease, but along the other with some difficulty; these directions are therefore respectively known as the *directions of easy and of difficult magnetisation*.

Weiss' Experiments. In his pioneer investigations of these properties Weiss[*] used two methods which have frequently been used by later workers, namely, a ballistic method and a

---

[*] P. Weiss, *Journ. de Phys.* **3**, 194, 1904; **4**, 469 and 829, 1905 and *Arch. des Sci. Phys. et Nat.* **19**, 537, 1905; **20**, 213, 1905.

torsion balance method. The latter method, being the more sensitive, may be used for the investigation of small specimens, but its use entails considerable uncertainty and possible error in the cases of materials other than pyrrhotite. As depicted in Fig. 41 a stirrup of non-magnetic material is suspended by a torsion fibre from a graduated torsion head T between the poles of an electromagnet. A disc of pyrrhotite about 1 to 2 cm. in diameter and about 0·5 mm. thick cut parallel to the base is

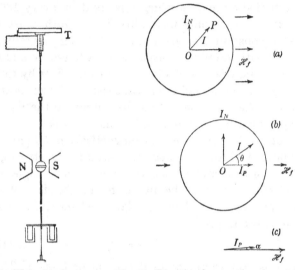

Fig. 41. Weiss torsion apparatus for I_N, $\mathscr{H}$ measurements.

Fig. 42. Directional properties of crystal discs.

placed in the stirrup; the latter is connected by a long wire to a damping-vane system immersed in oil. In fact, Fig. 41 also represents the apparatus used by Beck* in his pioneer experiments on iron crystals.

Let us first suppose that the plane of the disc is horizontal and that the torsion head is set so that the crystal occupies a known position in the absence of a magnetic field. On exciting a magnetic field $\mathscr{H}_f$, suppose that the crystal acquires an intensity of magnetisation I along the direction OP, Fig. 42a,

* K. Beck, V. Jahr. Sch. der Nat. Ges. in Zürich, 63, 116, 1918.

which makes an angle θ with the direction of the lines of force. Then, a couple equal to $\mathcal{H}_f I \sin \theta = \mathcal{H}_f I_N$, per unit volume, will act upon the crystal, where I_N thus represents the intensity of magnetisation normal to the external field. The couple may be balanced by turning the torsion head through an angle ϕ, such that

$$\mathcal{H}_f I_N = C\phi, \qquad \ldots\ldots(108)$$

where C is the torsion constant of the fibre. This method readily permits the study of I_N, and for pyrrhotite I_N is found to be a periodic function which repeats itself every 180°. In such experiments some instability in azimuth must occur unless the constant C is large, so that ϕ, being generally small compared with θ, may be measured by a telescope and scale. The angle θ may be varied either by turning T or by rotating the whole magnet about a vertical axis; the latter method is preferable, the wire connecting the stirrup to the damping-vanes passing through a hole in the magnet yoke.

To obtain the component of magnetisation I_P parallel to the applied field, the disc may be placed in the stirrup, with its plane vertical and turned through a small angle α, say 3° or 4°, with respect to the lines of force. The state of affairs may then be represented by Figs. 42b, c, whence it follows that the couple is now given by

$$\mathcal{H}_f I_P \sin \alpha = C\phi', \qquad \ldots\ldots(109)$$

where ϕ' is now the angle of deflection or torsion. The behaviour of I_P with θ can thus be studied by rotating the disc in the vertical plane, and again a curve which repeats itself every 180° is obtained. In this treatment it is assumed that the angle α is so small that the deflections of the suspended system measured by a telescope and scale or by an angle of torsion are a measure of I_P, although the lines of force are not parallel to the plane of the disc. While this is quite accurate in the case of pyrrhotite, it is not reliable for other substances, and its use with a material which does not possess a direction in which it is almost impossible to magnetise it has been severely condemned by some workers.*

* W. Sucksmith, H. H. Potter and L. Broadway, *Roy. Soc. Proc.* A, **117**, 471, 1928.

In the ballistic method used by Weiss the field $\mathcal{H}_f$ was produced by a solenoid inside which was a coaxial coil joined in series with a ballistic galvanometer. The pyrrhotite disc was suddenly placed inside the coil with its plane parallel to the lines of force of the field $\mathcal{H}_f$. The resulting deflection of the ballistic galvanometer measured the change of flux associated only with the component of magnetisation I_P parallel to the lines of force. This method is much more satisfactory than the torsion method, in spite of its apparent lesser sensitivity, and can clearly be adapted to measure I_N as well as I_P.

Now, $\mathcal{H}_f$ enters into equations (108) and (109), and we really wish to know how I_N and I_P vary with $\mathcal{H}$, but in order to find $\mathcal{H}$ we must know $\mathcal{H}_d$. The latter is equal to nI_P, and for a disc of radius a and thickness $2c$ we may take $n = \dfrac{\pi^2 c}{a}$ with sufficient accuracy.* We then have $\mathcal{H} = \mathcal{H}_f - \dfrac{\pi^2 c}{a} I_P$. It is well to remember, however, that the demagnetisation factor is rarely known with certainty, and that in the interpretation of many experimental results there is a tendency to overlook this lack of certainty. Curves showing how I_P and I_N vary with θ in the case of nickel crystals are shown on p. 159.

Experiments with Iron Crystals. Experiments with ferromagnetic crystals other than pyrrhotite differ from those on the latter because the specimens available are generally smaller with correspondingly larger demagnetisation factors, and, as there is no direction in which it is practically impossible to magnetise the metal, special precautions have to be taken to keep the vectors I and $\mathcal{H}$ in the same plane. The latter is often done by using thin discs of the metal, but, while I_N can then be safely found by the torsion method, it is essential that I_P be found by an induction method. Moreover, experiments cannot be confined to a single plane, but must be made in at least three planes.

These points are well brought out in Beck's† pioneer work

* J. Clerk Maxwell, *A Treatise on Electricity and Magnetism*, **2**, arts. 437–438, 1892.

† K. Beck, *loc. cit.*

on single crystals of iron. At room temperatures, iron has a body-centered cubic structure, and he cut discs 0·06 to 0·1 mm. thick parallel to the (100), (110) and (111) planes. Beck's torsion apparatus for the study of I_N as a function of θ in a constant $\mathscr{H}_f$ has already been described, and a similar apparatus was used by Webster.*

Honda and Kaya† prepared large iron crystals by the strain-anneal process, and, by careful sawing and filing, prepared from them three oblate spheroids with their diametrical planes respectively parallel to the (100), (110) and (111) planes; the diameter $2a$ was about 2 cm., and $2c$ about 0·045 cm. Their properties were studied by the special apparatus represented in Fig. 43.

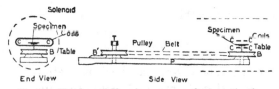

Fig. 43. Honda and Kaya's apparatus for investigating single crystals of ferromagnetics.

A circular brass disc forms a table which is mounted on a spindle coaxial with and attached to the pulley B, which can rotate as indicated at the one end of a stout brass plate P. At the other end of P is a pulley B' identical with B, and the two are connected by a belt under tension. P is mounted on a rigid stand and can be moved so that the table lies in the middle of a magnetising solenoid. The specimen is cemented to the table with canada balsam and on turning the head T by a known amount, B and the specimen rotate through a known angle. Two similar parallel coils CC, each of 1000 turns, are placed only 1·5 mm. apart on opposite sides of the centre of the specimen. They are attached to a small table, not shown, and can be turned about the axis of B, so that they can be placed parallel or perpendicular to the field of the solenoid.

When the I_P component is to be measured the coils are placed perpendicular to the field, the specimen is initially

* W. L. Webster, *Roy. Soc. Proc.* A, **109**, 570, 1925.
† K. Honda and S. Kaya, *Sci. Rep. Tôhoku*, **17**, 1157, 1928.

placed with a known direction parallel to the field. B' is then turned suddenly so that the principal axis of the crystal coincides with the direction of the field. The deflection of a ballistic galvanometer to which CC are joined in series measures the change in the parallel component of magnetisation with respect to that along the principal axis. In measuring the I_N component the coils are placed parallel to the field, and the crystal placed with a known direction parallel to the field.

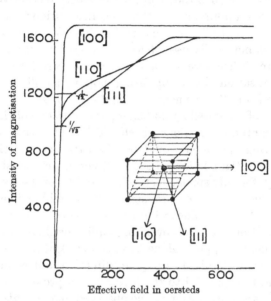

Fig. 44. Magnetisation curves for iron.

B' is then suddenly rotated until the principal axis coincides with the field, when the galvanometer deflection is a direct measure of I_N, since $I_N = 0$ when the principal axis is parallel to the field. The results obtained by these methods differ somewhat from those described earlier, presumably because the hysteresis changes are different.

Fig. 44 shows the $I, \mathcal{H}$ curves obtained when $\mathcal{H}$ is applied parallel to the (110) plane along the three axes; the relevant crystal structure is also shown in the figure. It is clear that the [100] direction is one of easy magnetisation. The initial rapid increase of I with $\mathcal{H}$ along the direction of easy magnetisation is a characteristic feature of single crystals, the more perfect the crystal, the more rapid the initial rise, and many workers hold that a perfect crystal would have an infinitely large initial permeability. The kinks in the curves for $\mathcal{H}$ parallel to the [110] and [111] directions arise because the magnetic axes of the elementary particles change from one cubic axis to another, and the changes should theoretically become apparent at $I = 1/\sqrt{2}$ and $1/\sqrt{3}$ of the saturation intensity respectively and this we see is approximately the case. In addition to the sharp kinks shown in the figure, some workers have found other kinks, which will not be discussed here, in "suspicious abundance".

Williams* has recently avoided the uncertainties of the demagnetisation factor by using hollow parallelogram or "picture-frame" specimens of single crystals of silicon iron, each side of a parallelogram being parallel to one of the three crystallographic directions—it should be noted that a single crystal ring specimen would not be satisfactory. He thus found the initial portions of the $I, \mathcal{H}$ curve with considerable accuracy, for only small effects of the corners of the specimens could influence the measurements. The initial portions of the $I, \mathcal{H}$ curve were not linear, but the observed anisotropy at low magnetisations can, however, be readily explained, because when a field is applied along, say, the [110] direction it has a component parallel to the [100] direction and results in a magnetisation parallel to the latter whose projection on the [110] gives a contribution to the observed magnetisation. The curves of Fig. 44 are similar to those obtained by Williams after allowing for this effect. The maximum permeability along the [100] direction found in these experiments was 1,380,000; the latest value is 1,450,000.

It is generally held that the more perfect the crystal, the lower its coercivity, but Coiffi, Williams and Bozorth† found a high coercivity with a very pure crystal of iron. Kaya‡ has made a special study of the retentivity of single crystals of iron.

* H. J. Williams, *Phys. Rev.* **52**, 747, 1937.
† P. P. Coiffi, H. J. Williams and R. M. Bozorth, *Phys. Rev.* **51**, 1009, 1937.
‡ S. Kaya, *Ann. der Phys.* **84**, 705, 1933.

He defines the *true retentivity* I_r as that value of I at which the sudden, almost rectangular kink, cf. p. 324, is found with decreasing field in the neighbourhood of $\mathscr{H}_f = 0$. Such kinks only occur when the specimen has not been tempered or aged for some time at about 130° C. With the above definition, Kaya finds that the true retentivity in any direction is given by the relation

$$I_r = I_\infty / l + m + n, \qquad \ldots\ldots(110)$$

where I_∞ is the saturation intensity of magnetisation, and l, m and n the direction cosines of $\mathscr{H}_f$ with respect to the axes of the crystal cube; this relation has been examined by Becker. See Chapter XII.

Experiments with Nickel Crystals. Sucksmith, Potter and Broadway* investigated single crystals of nickel which have a face-centred cubic structure. Sample curves for a disc cut parallel to the (110) plane are shown in Fig. 45. Here, in finding I_N it is necessary to take the mean of two sets of readings with the crystal, or field, rotated in opposite directions, to eliminate the marked effects of rotational hysteresis observed with these crystals. In the case of nickel the [111] direction is that of easy magnetisation. Theoretically, kinks should occur in the I, $\mathscr{H}$ curves for the [110] and [100] directions at $\sqrt{\tfrac{2}{3}}$ and $\sqrt{\tfrac{1}{3}}$ of the saturation value of I, and this condition is approximately satisfied.

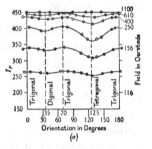

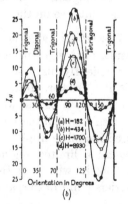

Fig. 45. I, $\mathscr{H}$ curves for single crystals of nickel. (Sucksmith, Potter and Broadway.)

Experiments with Cobalt Crystals. Single crystals of cobalt, which have a hexagonal close-packed structure below 470° C. and a face-centred cubic structure

* W. Sucksmith, H. H. Potter and L. Broadway, *Roy. Soc. Proc.* A, **117**, 471, 1928.

above, have been investigated by Honda and his collaborators. Kaya* cut discs about 0·45 cm. in diameter and 0·028 cm. thick parallel to the (1000) and (0001) planes; these are not equivalent as the six-fold axis is unique. He found the base plane of the hexagonal prism to be a plane of difficult magnetisation, and that the hexagonal axis along the [0001] direction was the only direction of easy magnetisation. Along the [1010] and [1020] directions the approach to saturation is very gradual; and this readily explains the difficulty of producing saturation in polycrystalline cobalt. $I, \mathscr{H}$ curves at higher temperatures have been obtained by Honda and Masumoto† with the ballistic method.

Single Crystals of Alloys. Shih‡ has studied the properties of single crystals of iron-cobalt alloys, using a pendulum magnetometer§ designed by McKeehan, who has also written a valuable account‖ of the work on single crystals. In these body-centred cubic lattices the direction of easiest magnetisation was found to change from the [100] to the [111] direction as the percentage of cobalt increased. This indicates that the direction of easiest magnetisation does not depend on the crystal structure alone, but also on the type of atom of which it is formed.

Theoretical Aspects of Ferromagnetic Anisotropy. The results obtained with single crystals of ferromagnetics can be conveniently discussed in terms of the theory developed by Akulov, Becker, Heisenberg, Bozorth¶ and others. This gives a conception of the way in which the magnetic vectors are rotated from a direction of easy magnetisation to lie along the direction of an applied field. Let us picture within the cubic crystal a limited region, or a Weiss domain, described on p. 238, in which all the magnetic vectors of the magnetic carriers point in the same direction. Then with an intensity of

* S. Kaya, *Sci. Rep. Tôhoku*, **17**, 1157, 1928.
† K. Honda and H. Masumoto, *Sci. Rep. Tôhoku*, **20**, 323, 1931.
‡ J. W. Shih, *Phys. Rev.* **46**, 139, 1935.
§ L. W. McKeehan, *Rev. Sci. Instr.* **5**, 265, 1934.
‖ L. W. McKeehan, "Ferromagnetism in Metallic Crystals", *Tech. Publ. Amer. Inst. Mining Eng.* No. 554, 1934.
¶ R. M. Bozorth, *Phys. Rev.* **50**, 1076, 1936; *Journ. App. Phys.* **8**, 575, 1937. References to work of others will be found in these papers.

magnetisation I_0 of this domain there is assumed to exist a magnetic potential energy E per unit volume whose magnitude depends upon the direction of the magnetisation. If the latter makes angles whose direction cosines are S_1, S_2, S_3 with respect to the axes of the crystal, we write

$$E = K_0 + K_1(S_1^2 S_2^2 + S_2^2 S_3^3 + S_3^2 S_1^2) + K_2(S_1^2 S_2^2 S_3^3), \quad \ldots (111)$$

neglecting all terms of lower power on account of the symmetry of the cubic crystal. Suppose now that a field $\mathscr{H}$ is applied at an angle θ with respect to I_0, then the additional energy stored in unit volume of the domain will be $-\mathscr{H} I_0 \cos \theta$, and the direction of magnetisation will then take up the position of minimum energy. Let α and α_0 be the angles between the axis of easy magnetisation in the plane of I_0 and $\mathscr{H}$ and the directions of I_0 and $\mathscr{H}$ respectively, then α is fixed by the condition

$$\frac{d}{d\alpha}(E - \mathscr{H} I_0 \cos \theta) = 0. \qquad \ldots\ldots(112)$$

Writing $\alpha_0 - \alpha = \theta$, then $d\alpha = -d\theta$, so that

$$\mathscr{H} = \frac{dE/d\alpha}{I_0 \sin \theta} = \frac{dE/d\alpha}{I_0 \sin (\alpha_0 - \alpha)}, \qquad \ldots\ldots(113)$$

and the magnetisation parallel to $\mathscr{H}$ is given by

$$I_{\|} = I_0 \cos (\alpha_0 - \alpha), \quad \text{and} \quad I_{\perp} = I_0 \sin (\alpha_0 - \alpha).$$

Consequently the I, $\mathscr{H}$ curve can be constructed if the *anisotropy constants*, K_1, K_2, ..., are known. Alternatively, if we know the experimental magnetisation curves for the [100], [110] and [111] directions of the crystal, we may calculate K_1 and K_2 by adjusting the theoretical magnetisation curves to fit them.

Now if we take $\mathscr{H}$ in the (100) plane along the [110] direction in a crystal of iron, in which the [100] direction is that of easy magnetisation, then

$$\alpha_0 = 45°, \quad S_1 = \cos \alpha, \quad S_2 = \sin \alpha, \quad S_3 = 0,$$

$$E = K_0 + \frac{K_1}{8}(1 - \cos 4\alpha), \quad \frac{dE}{d\alpha} = \frac{K_1}{2}\sin 4\alpha,$$

$$\mathscr{H} = K_1 \sin 4\alpha / 2 I_0 \sin (45 - \alpha) \quad \text{and} \quad I_{\|} = I_0 \cos (45 - \alpha).$$

It will be noted that for the [110] direction $\mathscr{H}$ and I are independent of K_2.

The theoretical curves agree with the experimental curves as closely as could be reasonably expected,* except for a peculiar discontinuity which Akulov first predicted for magnetisation along the [111] direction. In the neighbourhood of $\mathscr{H} = 350$ to 400 oersteds the magnetisation curve bends upwards and $dI_{111}/d\mathscr{H}$ becomes infinite. The energy relations around this value of the field must therefore be very ill-defined over a considerable range. Direct evidence for such a discontinuity has been provided by the experiments of Bates and Healey which are described later.

From equation (111) it follows that the potential energies E for the principal directions are

$$\left.\begin{array}{l} E_{100} = K_0 \\ E_{110} = K_0 + K_1/4 \\ E_{111} = K_0 + K_1/3 + K_2/27 \end{array}\right\}, \qquad \ldots\ldots(114)$$

from which it is seen that the directions of easiest, intermediate and most difficult magnetisation are determined by the magnitudes and signs of K_1 and K_2. This accounts for the difference between iron and nickel in which the orders of ease of magnetisation are [100], [110], [111] and [111], [110] and [100] respectively, and for changes in the order which occur in alloy series in which K_1 changes sign, and for changes in nickel or ferromagnetic alloys at high temperatures when K_1 approaches zero.†

On p. 59 an expression for the energy of magnetisation was derived and from this it is seen that the energies given in equations (114) are equal to the values of $\int_0^{I_0} \mathscr{H} dI$, where $\mathscr{H}$ and I are measured along the appropriate directions. Hence

$$E_{100} = \int_0^{I_0} \mathscr{H}_{100} dI_{100}, \ldots, \qquad \ldots\ldots(115)$$

---

* H. B. G. Casimir, W. J. de Haas and D. de Klerk, *Physica*, **6**, 365, 1939.
† Cf. K. Honda, H. Masumoto and S. Kaya, *Sci. Rep. Tôhoku*, **17**, 111, 1928 and N. S. Akulov, *Zeit. für Phys.* **100**, 202, 1936.

so that the anisotropy constants can be found from the areas on the three I, $\mathscr{H}$ graphs, for

$$
\left.\begin{aligned}
K_0 &= E_{100} \\
K_1 &= 4(E_{110} - E_{100}) \\
K_2 &= 27(E_{11\bar{1}} - E_{100}) - 36(E_{110} - E_{100})
\end{aligned}\right\} \quad \ldots\ldots(116)
$$

K_1 can also be measured by a special torque magnetometer.* From a survey of the chemical composition and crystal structure of iron-cobalt-nickel alloys McKeehan† concluded that there is no striking change in K_1 when the crystal structure changes from face-centred to body-centred cubic or when the specimens are worked or annealed; in other words, K_1 is not very structure-sensitive. Both K_1 and K_2 change considerably with rise in temperature and this provides an explanation for the striking variation in permeability with a weak field observed by Hopkinson. Brukhatov and Kirensky‡ state that, for nickel and certain copper-nickel alloys, $K_1 = K'_0 e^{-\alpha T^2}$, where K'_0 is a constant which may be obtained by extrapolation to absolute zero, and which decreases rapidly with increase in copper content.

In the case of cobalt crystals, which are hexagonal at room temperature, equation (111) becomes

$$
E = K_0 + K_1 S_1^2 + \ldots, \qquad \ldots\ldots(117)
$$

where S_1 is the direction cosine of I with respect to the hexagonal axis; higher terms appear to be unnecessary in (117). It follows that there is little or no anisotropy in the plane perpendicular to this axis, and that

$$
\begin{aligned}
K_0 &= E_{1000}, \\
K_1 &= E_{0001} - E_{1000}.
\end{aligned}
$$

Further reference to the above theory will be made in Chapter X. For iron it is found that K_1 and K_2 are respectively 420×10^3 and 150×10^3 erg per c.c., while for nickel the corresponding values are -50×10^3 and 0 erg. per c.c.

Rotational hysteresis problems are discussed in Chapter XII.

* H. J. Williams, *Rev. Sci. Instr.* **8**, 56, 1937.
† L. W. McKeehan, *Phys. Rev.* **51**, 137, 1937.
‡ N. L. Brukhatov and L. V. Kirensky, *Phys. Zeit. Sowjet.* **12**, 602, 1937.

CHAPTER V

EXPERIMENTS WITH SINGLE PARTICLES

Theoretical Considerations. We have seen on p. 21 that according to the theory of spatial quantisation, the magnetic axis of a free atom or molecule may occupy only one of a certain number of discrete positions with respect to the lines of force of a magnetic field. In this chapter an account will be given of the experiments which have been performed in investigating this phenomenon and which give us direct information concerning the magnetic behaviour of isolated particles.

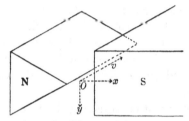

Fig. 46. Principle of Stern and Gerlach experiment.

The experimental procedure is based on the theory given by Stern* and first adopted by Stern and Gerlach. A strong non-uniform magnetic field is established between the poles of an electromagnet; we may suppose one pole tip to be plane and the other wedge-shaped as shown in Fig. 46. Let a single uncharged magnetic particle start to move from O with velocity v parallel to the edge of the wedge; then it must move with its magnetic axis orientated in one of a limited number of positions with respect to the lines of force. The most simple case is that of an alkali metal atom with a magnetic moment of one Bohr magneton, when only two positions are possible, viz. with the magnetic axis parallel or anti-parallel to the field.

At any point in its path the particle will accordingly be acted upon by a force $\pm \mu_B d\mathcal{H}/dx$ parallel to the x direction

* O. Stern, *Zeit. für Phys.* **7**, 249, 1921.

shown in the figure, $d\mathscr{H}/dx$ being the gradient of the field along this direction, which is parallel to the line of centres of the pole tips. If the mass of the atom is M it will experience an acceleration $f = \pm \dfrac{\mu_B}{M} \dfrac{d\mathscr{H}}{dx}$; and if the atom moves through a distance l parallel to the edge of the wedge in a time t, it will finally be displaced through a distance s parallel to x given by

$$s = \tfrac{1}{2}ft^2 = \pm \frac{1}{2M} \mu_B \frac{d\mathscr{H}}{dx} \left(\frac{l}{v}\right)^2. \qquad \ldots\ldots(118)$$

Now, in practice, the atom is separated from a cloud of vapour at a temperature T, and, therefore, we may assume as a basis for discussion that the most probable value of v is v_α given by the kinetic theory equation,

$$v_\alpha^2 = 2\frac{kT}{M} = 2\frac{RT}{NM},$$

where N is Avogadro's constant and R is the gas constant for one gram atom. Hence, the most probable deflection

$$s_\alpha = \pm \frac{N\mu_B}{4RT} \frac{d\mathscr{H}}{dx} l^2. \qquad \ldots\ldots(119)$$

But $N\mu_B = 5565$ c.g.s. units, and, consequently, a narrow beam of atoms entering the field at O should be split into two narrow beams separated at the other side of the field by a distance

$$2s_\alpha = 2\frac{5565}{4RT} \frac{d\mathscr{H}}{dx} l^2. \qquad \ldots\ldots(120)$$

Thus, taking $T = 600°$ K., $l = 10$ cm. and $d\mathscr{H}/dx = 25{,}000$ units per cm., $2s_\alpha$ should be approximately equal to 0.14 cm. If the path in the magnetic field is l but the receiver is placed at an additional distance l_1 outside the field, then the above expression for $2s_\alpha$ must be multiplied by $(1 + 2l_1/l)$.

The early experiments of Stern and Gerlach showed conclusively that such separated beams are formed, and thus proved that the individual atoms of metals, such as silver, possess magnetic moments associated with angular momenta which are spatially quantised in a magnetic field.

165

EXPERIMENTS WITH SINGLE PARTICLES

Experimental Aspects. A simple diagram of the experimental arrangements used by Stern and Gerlach is shown in Fig. 47, where O represents an oven in which a metal is vaporised and the vapour atoms raised to a temperature T, so that a fine beam passes through the narrow slit S_1 into an evacuated space. A slit S_2 limits the beam, which is further limited by a collimator slit, S_3, placed immediately before the gap between the special pole tips, which are of the type shown. Finally, the separated beams of atoms fall upon a receiver R, a cold glass plate, to form a deposit of the shape D in the simple case considered above. It is to be emphasised that the atoms must not suffer collisions with their neighbours

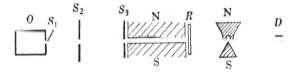

Fig. 47. Diagram of Stern-Gerlach apparatus.

in the beam, or with other atoms in their passage through the field; consequently, the apparatus must be mounted in a chamber, or series of chambers, which can be thoroughly evacuated.

Now, in the light of the foregoing theory certain necessary experimental conditions for the accurate measurement of the final separation of the beams can readily be laid down.* Obviously, the parent beam has to be very narrow, i.e. it must issue from an oven slit whose jaws are very close together. From this it might appear that the beam would be very weak and a deposit therefore obtained after an unduly prolonged exposure, during which the magnetic field would have to be maintained constant. However, by keeping constant the product of the vapour pressure of the atoms in O and the width of the slit S_1, it is possible to obtain an intense but very fine beam, whose fineness is limited by the mechanical difficulties of slit construction.

* See O. Stern, *Zeit. für Phys.* **39**, 751, 1926.

Since s is proportional to l^2, it is necessary that l be made as long as possible, and, finally, $d\mathscr{H}/dx$ must be as large as possible, so that pole tips of a wedge and trough design are used. This raises the question concerning the region over which $d\mathscr{H}/dx$ may be treated as constant, a question which can only be answered by direct experiment, by measuring $\mathscr{H}$ and $\mathscr{H}\,d\mathscr{H}/dx$ at various points.

Exact alignment of the slits S_1, S_2 and S_3 is of great importance, and for this purpose S_2 and S_3 must be mounted on the same support as S_1, the support being sufficiently robust to withstand the large stresses set up by the attraction of the pole pieces. Knauer and Stern* were the first to use the atomic beam itself in aligning the slits, by arranging a prism and microscope system to view deposits made by mercury atoms on a silvered plate receiver.

The receiver R of Fig. 47, in the early experiments, was a glass plate cooled on the far side by liquid air, and the impinging atoms stuck to the glass. The exposures then lasted many hours and the deposits had to be intensified by washing the plate with a special solution consisting of 1 per cent hydrokinone developer, a small amount of silver nitrate and a fair amount of gum arabic. It is now possible with the better alignment in later technique directly to obtain visible deposits with exposures of some few minutes. The visible deposits have more or less the same appearance in the cases of different metals. At first they are bright brown, becoming darker and then black with long times of exposure, finally taking on a bluish and metallic appearance. Other, and more effective, ways of detecting the separated beams have latterly been developed, and these are described as occasion warrants in the following sections.

The Rabi Method of Separation. The magnetic separation of the beam may also be produced by a method due to Rabi.† The deflection of a beam of spatially quantised atoms in the Stern and Gerlach experiment may be likened to the gradual deviation of a ray of monochromatic light as it passes through

---

* F. Knauer and O. Stern, *Zeit. für Phys.* **39**, 764, 1926.
† I. I. Rabi, *Zeit. für Phys.* **54**, 190, 1928.

a medium whose refractive index varies continuously in a direction perpendicular to the path of the rays. The rate of change of refractive index is analogous to the gradient of the magnetic field. Now, when a ray of light passes suddenly from one medium into a thin layer of another medium, then the total bending is practically independent of the way in which the refractive index varies within the second medium, and it is this picture which is to be borne in mind in discussing the Rabi method of separation.

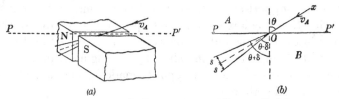

Fig. 48. Rabi's method of separating spatially quantised beams.

Suppose, then, that a beam of atoms, each of magnetic moment $1\mu_B$, of velocity v_A and kinetic energy E, is travelling in the equatorial plane of the gap between two flat, rectangular, pole pieces as depicted in Fig. 48 a, b. The path of the beam is always perpendicular to the lines of force, and, in the simple case under discussion, the atomic moments set parallel or antiparallel to the field. Now, the force on each atom in a direction parallel to PP' is zero, consequently the components of the atomic velocity parallel to PP' before entering and after entering the gap must be equal. Yet, the kinetic energy of an atom before entering and after entering must differ by an amount $\pm \mu_B \mathscr{H}$, because of the change in magnetic potential energy of the atom. There will therefore result two sets of atoms with velocities v_B equal either to $\sqrt{\left(\dfrac{2}{M}(E + \mu_B \mathscr{H})\right)}$ or to $\sqrt{\left(\dfrac{2}{M}(E - \mu_B \mathscr{H})\right)}$ according as the atoms set with their magnetic axes respectively parallel or antiparallel to the field.

The conservation of momentum parallel to PP' therefore requires, as seen from Fig. 48b, that

$$v_A \sin \theta = v_B \sin (\theta \mp \delta),$$

whence

$$\frac{\sin\theta\cos\delta \mp \cos\theta\sin\delta}{\sin\theta} = \frac{v_A}{v_B} = \frac{\sqrt{\dfrac{2E}{M}}}{\sqrt{\left(\dfrac{2}{M}(E \pm \mu_B \mathscr{H})\right)}}.$$

Since δ is small, we may now write

$$\delta = \pm\tan\theta\left\{1 - 1\Big/\sqrt{\left(1 \pm \frac{\mu_B \mathscr{H}}{E}\right)}\right\},$$

and when $\mu_B \mathscr{H}/E \ll 1$, then

$$\delta = \pm\frac{\mu_B \mathscr{H}}{2E}\tan\theta.$$

Consequently, when the atoms have travelled a distance l within the field, the separation of the two component beams is $2s = 2l\delta = \dfrac{\mu_B \mathscr{H}}{E} l\tan\theta$, thus being directly proportional to the strength of the field $\mathscr{H}$.

In practice, however, errors must arise because the transition from zero field to field $\mathscr{H}$ does not take place in a very restricted region close to the edges of pole pieces as is postulated in the theory. Moreover, the finite length of the slits through which the atoms enter the field do not permit all atoms to travel in the equatorial plane. Hence, as Meissner and Scheffers* found, this method of separation, while quite suitable for comparison measurements, is not suitable for absolute or precision measurements.

Modern Technique. We now turn to a consideration of the more important refinements in experimental technique which finally culminated in precision measurements of the magnetic moments of individual atoms. Most of them are described in a paper by Leu,† whose apparatus is shown diagrammatically in Fig. 49. The oven O consisted of a copper block mounted in a water-cooled constantan tube C, heated by a platinum filament. The metal to be vaporised was placed in the small metal chamber M communicating with O, so that the evaporated metal was at a lower temperature than that of the oven,

* W. Meissner and H. Scheffers, *Phys. Zeit.* **34**, 48 and 245, 1933.

† A. Leu, *Zeit. für Phys.* **41**, 551, 1927.

and, consequently, no condensation occurred in the oven and upon the slit S_1. The latter was 4 mm. across and less than 0·02 mm. wide. A thermo-junction placed inside O served to measure the temperature of the vapour.

Having passed through S_1 the atoms traversed a distance of some 6 cm. and then passed through a slightly wider collimating slit S_2 parallel to S_1 and mounted rigidly in the end of C. The whole of the arrangements so far described were contained in a glass bulb which communicated by a narrow glass tube T''

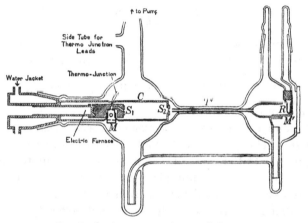

Fig. 49. Leu's apparatus for atomic beams.

with a second glass bulb in which was mounted the cooled glass plate receiver R. The surface of R was usually chemically silvered and polished, and could be viewed by means of a prism arrangement, not shown. Both bulbs were highly evacuated. A copper tube ran through T' and was cooled by conduction of heat along a metal strip M' which made contact with a metal attachment on the receiver R. T' lay between the wedge and trough pole tips.

With this arrangement, excellent visible deposits were obtained with times of exposure varying from 5 minutes to 2 hours. Fig. 50 shows a typical result for sodium. If the separation as shown in Fig. 50 is measured, the magnetic moment of the sodium atom can be calculated from equation (120). The accuracy of the result depends mainly on the evaluation of $d\mathcal{H}/dx$, and is said to be of the order of 2 per cent, but there remains for consideration the manner in which the true separation $2s_\alpha$ should be computed from the somewhat diffuse traces of Fig. 50.

This point has been discussed in detail by Stern,* who has shown how to take account of the density distribution in the deposits. In deriving the expression for $2s_\alpha$, it was virtually assumed that v_α was the same for every atom in the initial beam. Actually, the atoms must possess velocities more or less in accord with a Maxwell distribution, and hence the deposits must be wide. The evaluation of s_α is of great importance in precision measurements and is fully discussed in connection with those of Meissner and Scheffers, below.

(a)   (b)

Fig. 50. Leu's result for sodium.
(a) undeflected beam;
(b) deflected beams.

Experimental Results. The experimental results so far obtained are given in Table VII, where the relevant spectroscopic data are also given.

TABLE VII.

Atoms	Normal state	$2J+1$	μ/μ_B exp.
H, Li, Na, K	$^2S_{\frac{1}{2}}$	2	1
Cu, Ag, Au	$^2S_{\frac{1}{2}}$	2	1
Zn, Cd, Hg	$^1S_0$	1	0
Sn, Pb	$^3P_0$	1	0
Tl	$^2P_{\frac{1}{2}}$	2	$\frac{1}{3}$
Bi	$^4S_{\frac{3}{2}}$	4	0·72
Fe	$^5D_4$	9	Traces?
Co	$^4F_{\frac{9}{2}}$	10	Traces
Ni	$^3F_4$	9	Traces

The results in the first five rows of this table prove that atoms like silver possess magnetic moments and give results in

* O. Stern, *Zeit. für Phys.* **41**, 563, 1927.

complete agreement with the theory of spatial quantisation. Now we saw on p. 21 that when an atom of known L, S and J is placed in a magnetic field, the components of the magnetic moment parallel to the lines of force were not given by $g\sqrt{M_J(M_J+1)}$ but by gM_J, where $M_J = J$, $J-1$, ..., $-(J-1)$, $-J$. This means that the separation of the two outermost of the $2J+1$ traces observed in a Stern-Gerlach experiment should not be proportional to $g\sqrt{J(J+1)}\mu_B$ but to $gJ\mu_B$. This is definitely proved to be the case, for in the case of the silver atom, whose normal state is $^2S_{\frac{1}{2}}$, we have $g\sqrt{J(J+1)}\mu_B = 2\sqrt{\frac{1}{2}\cdot\frac{3}{2}}\mu_B = 1\cdot73\,\mu_B$ and $gJ\mu_B = 2\cdot\frac{1}{2}\mu_B$, and experiment gives $1\mu_B$ in agreement with the latter. The case of thallium is still more interesting, for the normal state is $^2P_{\frac{1}{2}}$. We find, as expected, two traces in the Stern-Gerlach experiment, but they correspond to an atomic moment of $\frac{1}{3}\mu_B$. Now, for thallium

$$g\sqrt{J(J+1)}\mu_B = \tfrac{2}{3}\sqrt{\tfrac{1}{2}\cdot\tfrac{3}{2}}\,\mu_B = 0\cdot58\,\mu_B,$$

while $gJ\mu_B = \frac{2}{3}\cdot\frac{1}{2}\mu_B = \frac{1}{3}\mu_B$; here again we have very striking agreement between experiment and theory.

In the case of bismuth, Leu and Fraser had to use a very hot oven to ensure the absence of Bi_2 molecules from the initial beam and they then obtained two traces, corresponding to $\mu = \pm\,0\cdot72\,\mu_B$ or to $g = 1\cdot45$. Now, it happens that the normal state of the Bi atom is $^4S_{\frac{3}{2}}$, but the coupling in the atom is not the normal Russell-Saunders type, for which $g = 2$, but rather of the complex jj type, for which $g = \frac{4}{3}$. As we get by experiment a value for g which lies between the above two theoretical values we may consider that agreement between theory and experiment is satisfactory. The apparent lack of agreement between experiment and theory in the case of the iron group metals is attributed to experimental defects, and particularly to overlapping of the several traces.

Precision Measurement of the Bohr Magneton. The numerical value of the Bohr magneton was measured with considerable accuracy by Meissner and Scheffers,* and will be

* W. Meissner and H. Scheffers, *Phys. Zeit.* **34**, 48 and 245, 1933.

considered in some detail as an excellent example of modern technique. The success of the work is due to the efficient manner in which the conditions prescribed by Leu for the production of very narrow but intense beams of atomic rays were applied, and to the method, first devised by Taylor,[*] employed in their detection. The apparatus is shown diagrammatically in Fig. 51. It consisted of two brass compartments C and R. The first contained the oven O in which potassium or lithium was evaporated, and it was joined to R by the brass tube TT. The whole apparatus was very efficiently evacuated, and water-cooled blocks WW condensed stray

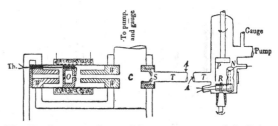

Fig. 51. Apparatus for precision measurement of the Bohr magneton. (Meissner and Scheffers.)

alkali atoms. The evaporated atoms passed through the narrow oven slit, through the fine, 0·03 mm., collimating slit S and on to the receiver N, being exposed to the action of a non-uniform field of the Stern-Gerlach type during their passage through TT. Inside N was a stretched tungsten filament 0·02 mm. in diameter, so mounted that it could be moved across the incident beams of atomic rays, parallel to the slits. It was heated electrically, and alkali metal atoms falling upon it lost electrons and could leave the surface as positive ions, the resulting ionisation current between the filament and a collecting electrode or plate being a measure of the number of atoms received in unit time.

Special precautions were taken to avoid fluctuations of the oven temperature, and the gradient of the non-uniform field was carefully mapped by means of a tiny search coil mounted on a brass tube, fitted to a cathetometer with horizontal and

[*] J. B. Taylor, *Zeit. für Phys.* **57**, 242, 1929.

vertical movements. The initial beam of alkali metal atoms was split into two by the action of the field, the distributions of the atoms in the deflected and undeflected beams being evaluated from the ionisation current measurements, while the separation of two beams into which an infinitely thin initial beam of single velocity atoms would be split was computed on the basis of the theoretical treatment given by Stern.*

Stern supposes that in the absence of the field $I_0\,ds_0$ atoms pass through each element of width ds_0 of the collimating slit placed immediately before the pole tips. The original intensity

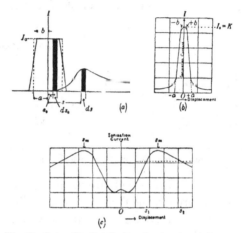

Fig. 52. Intensity distribution in atomic ray beams.

distribution in the beam entering the gap between the pole tips may therefore be represented by the dotted rectangle in Fig. 52a, and the intensity distribution in the beam falling on the receiver by the trapezium in that figure. The change in the intensity distribution from that represented by the rectangle to that represented by the trapezium is, of course, occasioned by the slight spreading of the undeflected beam, and can be calculated from the geometry of the apparatus.

Now, when the field is excited, owing to the Maxwell velocity distribution of the atoms the very narrow undeflected beam is replaced in the case of alkali metal atoms by two wide

* O. Stern, *Zeit. für Phys.* **41**, 563, 1927.

and rather diffuse beams, in which the intensity distribution of the one will now be represented by a continuous curve like that of Fig. 52a. Let us consider what happens to the group of atoms which pass through an element ds_0 at a distance s_0 from the centre of the undeflected beam when the field is excited. Those atoms in this group which possess velocities between v_1 and $v_1 + dv_1$ will be displaced to a distance s to pass through a strip of the deflected beam ds wide. The separation of the elements ds and ds_0 is therefore $s - s_0$. Hence, as from equation (118), p. 165, the displacement is inversely proportional to the square of the velocity, $s - s_0 \propto 1/v_1^2$. But the most probable deflection $s_\alpha \propto 1/v^2$, so that

$$s - s_0 = s_\alpha \frac{v^2}{v_1^2} \qquad \ldots\ldots(121)$$

and
$$ds = d(s - s_0) = -2s_\alpha \frac{v^2}{v_1^3} dv_1. \qquad \ldots\ldots(122)$$

Now the number of atoms in the group $I_0 ds_0$ whose velocities lie between v_1 and $v_1 + dv_1$ is, from kinetic theory, given by

$$I_0 ds_0 \frac{2}{v^4} e^{-v_1^2/v^2} v_1^3 dv_1; \qquad \ldots\ldots(123)$$

consequently, the number of such atoms per unit width in the deflected beam at s is equal to

$$\tfrac{1}{2} I_0 ds_0 \frac{v_1^6}{v^6} \frac{e^{-v_1^2/v^2}}{s_\alpha}, \text{ from (122)},$$

$$= \tfrac{1}{2} I_0 ds_0 \frac{s_\alpha^2}{(s - s_0)^3} e^{-s_\alpha/s - s_0}.$$

Hence, the total intensity I_s at a point in the displaced beam distant s from the centre is given by

$$I_s = \tfrac{1}{2} I_0 \int_{-a}^{+a} e^{-s_\alpha/s - s_0} \frac{s_\alpha^2}{(s - s_0)^3} ds_0, \qquad \ldots\ldots(124)$$

and $\quad I_s = \dfrac{K}{2(a - b)} [e^{-s_\alpha/s + a} (s + a) + e^{-s_\alpha/s - a} (s - a)$

$$- e^{-s_\alpha/s + b} (s + b) - e^{-s_\alpha/s - b} (s - b)], \qquad \ldots\ldots(125)$$

for an initial distribution represented by the trapezium in Fig. 52b, where I_0 is a function of s_0 between $s_0 = b$ and $s_0 = a$.

The type of ionisation current-displacement curve for the undeflected ray obtained by Meissner and Scheffers when the magnetic field is off is shown in Fig. 52b, where the full line represents the experimental results and the trapezium an assumed intensity in order to make use of equation (125). The "feet" on the experimental curve, which lie outside the trapezium, are due to scattered atoms produced mainly in the neighbourhood of the hot filament and, hence, are of little consequence in a discussion of the intensity distribution when the field is applied, as they would be outside its range. The peak of the curve also departs from the trapezium configuration, and this departure also is relatively unimportant, for with large values of s_α the precise shape of the undeflected beam is not of great importance.

The type of ionisation current-displacement curve for the deflected beams is shown in Fig. 52c where the small "hump" in the undisplaced position is attributed to undeflected particles. There are, clearly, two methods by which s_α may be evaluated from Fig. 52b, c. First, it may be calculated from the positions s_m of maximum intensity in the deflected beams, for, by differentiating (125) and putting $s = s_m$, we obtain the equation

$$e^{-s_\alpha/s_m+a}\left(1+\frac{s_\alpha}{s_m+a}\right)+e^{-s_\alpha/s_m-a}\left(1+\frac{s_\alpha}{s_m-a}\right)$$

$$= e^{-s_\alpha/s_m+b}\left(1+\frac{s_\alpha}{s_m+b}\right)+e^{-s_\alpha/s_m-b}\left(1+\frac{s_\alpha}{s_m-b}\right),$$

from which s_α may be found. Secondly, two positions s_1 and s_2 of equal intensity in each deflected beam may be found and inserted directly in equation (125). In this way an approximate formula, used by Leu, can be derived for s_α. The first method is the more satisfactory, provided that sufficient points lie in the regions of the maxima of the ionisation current-displacement curves, since no correction need be applied for the presence of undeflected particles.

The final values obtained for the Bohr magneton from four sets of determinations with potassium and five sets with lithium were $0\cdot916 \pm 0\cdot005$ and $0\cdot916 \pm 0\cdot003 \times 10^{-20}$ respectively, the mean being 0.916×10^{-20} c.g.s. units, correct to 5 parts in 1000. These values were calculated on the basis of Millikan's value of e, 4.774×10^{-10} e.s.u.; using the revised value 4.802×10^{-10}, given on p. 54, they become 0.921×10^{-20}.

Measurements with Atomic Hydrogen. Attention has so far been confined to experiments in which metallic atoms have been used, and it might seem a priori that gases would be particularly easy to employ in Stern-Gerlach experiments. However, the only monatomic gas which possesses a magnetic moment is atomic hydrogen, and great difficulties are encountered when the few paramagnetic diatomic gases are used because of the thermal rotation of the molecules.

Experiments with atomic hydrogen were made by Wrede* and by Phipps and Taylor,† and both sets of experiments showed the isolated hydrogen atom to possess a magnetic moment of $1\mu_B$. Phipps and Taylor prepared atomic hydrogen by Bonhoeffer's‡ method in a Wood§ discharge tube and by Langmuir's‖ method of evaporation from a hot wire. The stream of atoms passed through a special pyrex slit system, traversed a non-uniform field and fell upon a glass receiver coated with a fine yellow-white deposit of molybdenum oxide, MoO_3, formed from the white smoke obtained by burning a piece of molybdenum in an oxygen-gas flame. The impinging atoms reacted with the MoO_3 to give a blue deposit, which usually reached its maximum density after about four hours' exposure.

Experiments with Diatomic Gases. The special difficulties encountered when experiments are made with diatomic gases may be appreciated by considering the behaviour of oxygen. The oxygen molecule, as we saw on p. 46, normally exists in a $^3\Sigma$ state with $\Lambda = 0$, $S = 1$ and $J = 1$. The molecule rotates rapidly about N, Fig. 11, and there is a strong coupling between the spin and the rotation. A reasonable value to take for the

* E. Wrede, *Zeit. für Phys.* **41**, 569, 1927.
† T. E. Phipps and J. B. Taylor, *Phys. Rev.* **29**, 309, 1927.
‡ K. F. Bonhoeffer, *Zeit. für Phys. Chem.* **113**, 199, 1934.
§ R. W. Wood, *Phil. Mag.* **42**, 729, 1921.
‖ I. Langmuir, *Journ. Amer. Chem. Soc.* **38**, 2021, 1916.

most probable value of N is 9, and this would give rise to $\sum_N (2S+1)(2N+1) = 136$ beams in a Stern-Gerlach experiment, assuming that odd values of N alone are effective. In other words, so many discrete positions are possible that classical conditions of an infinite number of orientations in the field would be simulated; consequently, the pattern would be a broad band and very difficult to interpret. Such a band was obtained by Shaw and Phipps* and was considered to provide proof that the non-rotating sulphur molecule possessed a magnetic moment of two Bohr magnetons corresponding to a $^3\Sigma$ state.

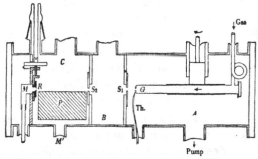

Fig. 53. Apparatus for molecular ray beams. (Schnurmann.)

A comprehensive series of experiments on O_2 and NO molecular beams have been made by Schnurmann† using an arrangement very similar to that designed by Knauer and Stern‡ for measurements with light gases, and shown in Fig. 53. The gas entered an evacuated chamber A through the fine slit G, its temperature being measured by a thermocouple, and some molecules passed through the fine slits S_1 and S_2 to form the molecular beam which passed through a non-uniform field of the Rabi type between pole tips P. The broadened beam then fell upon a receiving slit R and part of it entered the tube M. R was parallel to the other slits, and could be moved across the beam. M communicated with a chamber containing a Pirani hot wire manometer and the molecules passing through

* E. J. Shaw and T. E. Phipps, *Phys. Rev.* **38**, 174, 1931.
† R. Schnurmann, *Zeit. für Phys.* **85**, 213, 1933 and *Journ. de Phys.* **6**, 99, 1935.
‡ F. Knauer and O. Stern, *Zeit. für Phys.* **53**, 766, 1929.

R caused the pressure to rise therein until the number of molecules entering was equal to the number leaving; thus the Pirani gauge measured the relative numbers of molecules in the various portions of the beam, while a compensating gauge attached to M' eliminated fluctuations due to changes in the pressure of gas in the chamber C. The pattern or distribution of molecules in the beam varied greatly with the Rabi field and the temperature of the beam. Curve I in Fig. 54 shows the

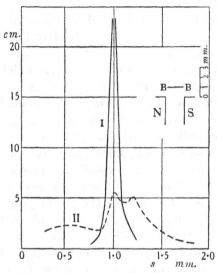

Fig. 54. Schnurmann's results for oxygen.

pattern obtained with zero field and Curve II the pattern obtained with an initial beam of oxygen molecules of section BB passing as shown above the pole tips NS, the temperature being 300° K. The strong asymmetry of curve II is due to the non-uniformity of the field, for while the undeflected molecules moved in this case in a field of 6800 oersteds, those corresponding to the left-hand portion moved in a stronger field than those corresponding to the right-hand portion, so that the degrees of coupling of electronic and molecular spins were different. Following a procedure due to Einaudi,* the pattern to be expected on the assumption of fixed values of S and N

* R. Einaudi, *R.C. Acad. Lincei*, **16**, 132, 1932.

with the type of Rabi field actually used was calculated and compared with the experimental pattern. The agreement between experiment and theory with the above values of S and N is good.

The rotational effects are even more complicated in the case of the NO molecule. The normal state of the molecule is $^2\Pi_{\frac{1}{2}}$, where the spin and orbital momenta are antiparallel and the molecule is non-magnetic. At high temperatures many molecules pass into the $^2\Pi_{\frac{3}{2}}$ state, which exhibits a magnetic moment of two Bohr magnetons. Molecules in this state should give rise to five beams in a Stern-Gerlach or Rabi experiment if rotational effects are absent. Actually, on account of these effects it is advisable to use strong fields in which decoupling occurs and $\Sigma 2(2J+1)$ magnetically deflected and one un-deflected beam of non-magnetic molecules should be obtained, the most probable value of J being $\frac{15}{2}$. So many components could not possibly be unravelled, but a relatively simple procedure is possible both for O_2 and NO. As before, the molecule is assumed to have a specified magnetic moment and known partition of the rotational states. The shape of the beam pattern curve is obtained from the Pirani gauge readings in the absence of a magnetic field, and the intensity of the un-displaced component is calculated for known conditions of the magnetic field and compared with that obtained by experiment. For example, in the case of oxygen the experimental value for I/I_0 was 0·287, whilst calculation based on $N = 1$, to 15 gave 0·299. On the whole, agreement between experiment and theory is as good as can be expected.

CHAPTER VI

NUCLEAR SPINS AND MAGNETIC MOMENTS

Introduction. In the experiments described in Chapter v we have confined attention to the magnetic moment of the electron system of the atom or molecule, and we must now consider what information the method of molecular rays can give concerning the magnetic properties of the nucleus. It is known that certain *hyperfine structures* of spectral lines or components of the individual multiplets, which are less than one wave number apart, arise from a spin of the nucleus of the atom by which they are emitted. With this nuclear spin there must, in general, be associated a magnetic moment.

Now, the mass of an atomic nucleus is great in comparison with that of the electron, so that its magnetic moment must be quite small even when the angular momentum associated with the spin is large. It is therefore found convenient to express the values of nuclear magnetic moments in terms of a quantity equal to $\frac{1}{1840}\mu_B$, called the *unit of nuclear magnetic moment* or *nuclear magneton*, and denoted by μ_n, while the spin is still measured in terms of the unit $h/2\pi$. Hence, in the case of an alkali atom the nuclear magnetic moment can only be a very small fraction of the magnetic moment of the atom, and is thought to play no appreciable part in the precision determination of μ_B by the method of Meissner and Scheffers. It also follows that it is impossible to determine by direct deflection methods the magnetic moment of the nucleus of an atom whose electron system possesses a moment of the order of a Bohr magneton. Direct measurements are only possible with atoms or molecules where the magnetic moment arising from external electron configurations is practically zero. Such a case is provided by the orthohydrogen molecule, but here, of course, the determination of the nuclear moment is complicated by the presence of molecular rotations.

Fortunately, there generally exists a weak coupling between the nuclear spin and the spin of the external electron system of an atom which may be profoundly modified by the application of relatively small external magnetic fields. In the strong fields employed in all the experiments described in the previous chapter, this coupling is broken down and the nuclear and electron spins are orientated independently of each other—in fact, the behaviour of the spins is analogous to that of spin and orbital momenta in the Paschen-Back effect—and the particle behaves practically as if it possessed no nuclear spin. In weak fields, however, the coupling plays an important role, and was shown by Breit and Rabi* and by Cohen and Rabi† to provide a method for the measurement of nuclear spins.

In the theory underlying the Stern-Gerlach experiments it was assumed that an atom whose electron system possessed a total angular momentum of $Jh/2\pi$ units could set in $2J + 1$ positions with respect to the lines of force of an applied field. If, in addition, the atomic nucleus possesses I units of spin momentum, the two being normally coupled to give $F = I + J$ and the field not sufficiently strong to decouple the nuclear and electron spins, then $(2J + 1)(2I + 1)$ positions become possible. We get $2(J + I) + 1$ states with total angular momentum quantum number $J + I$, $2(J + I - 1) + 1$ states with total angular momentum quantum number $J + I - 1$, and so on to $2(J - I) + 1$ states with total angular momentum quantum number $J - I$, and it is easily seen that the number of states adds up to $(2J + 1)(2I + 1)$.

Now, in the state with total angular momentum quantum number $J + I$ we do not know a priori whether the axes of magnetic moment of the nucleus and the electron system are set parallel or antiparallel, or, we may say, *whether the magnetic moment of the nucleus is positive or negative*. We must therefore realise at the outset that the energy of each of these states when a magnetic field is applied depends on whether the nuclear magnetic moment is positive or negative. Thus, in a magnetic

* G. Breit and I. I. Rabi, *Phys. Rev.* **38**, 2082, 1931. See also H. Kopfermann, *Ergeb. der exakt. Wissen.* **15**, 242, 1936.

† V. W. Cohen and I. I. Rabi, *Phys. Rev.* **43**, 582, 1933.

field $\mathscr{H}$ the component of F parallel to $\mathscr{H}$, is $M = I + J$, where I takes the values $I, I-1, ..., -I$, and J the values $J, J-1, ... -J$, so that for a $^2S_{\frac{1}{2}}$ state where $J = \frac{1}{2}$, the energy is changed by an amount

$$E_{+\mathscr{H}} = -\frac{\Delta E}{2(2I+1)} \pm \frac{\Delta E}{2}\left(1 + \frac{2M}{I+\frac{1}{2}}y + y^2\right)^{\frac{1}{2}} \quad ......(126)$$

or by

$$E_{-\mathscr{H}} = +\frac{\Delta E}{2(2I+1)} \mp \frac{\Delta E}{2}\left(1 - \frac{2M}{I+\frac{1}{2}}y + y^2\right)^{\frac{1}{2}}, \quad ......(127)$$

according as the nuclear magnetic moment is positive or negative. The upper sign applies to $M = I + J$ and the lower to $M = I - J$. In equations (126) and (127)

$$y = g\mu_B\mathscr{H}/\Delta E,$$

where ΔE is the energy difference, $hc\Delta\nu$, corresponding to an overall separation $\Delta\nu$ between the hyperfine structure components of the multiplet term, or the difference in the energy levels when the nuclear and electron angular momenta are set parallel and antiparallel respectively, and g is the Landé splitting factor for the electron system.

The force which acts upon an atom along the x axis due to the presence of a non-homogeneous field is, therefore, either

$$F_{+x} = -\frac{dE}{dy}\frac{dy}{d\mathscr{H}}\frac{d\mathscr{H}}{dx}$$

$$= \mp \frac{1}{2}\frac{\dfrac{2M}{2I+1}+y}{\left(1 + \dfrac{4M}{2I+1}y + y^2\right)^{\frac{1}{2}}}g\mu_B\frac{d\mathscr{H}}{dx}$$

$$= \mu_{+\text{eff}}\frac{d\mathscr{H}}{dx}, \quad ......(128)$$

or

$$F_{-x} = \pm \frac{1}{2}\frac{y - \dfrac{2M}{2I+1}}{\left(1 - \dfrac{4M}{2I+1}y + y^2\right)^{\frac{1}{2}}}g\mu_B\frac{d\mathscr{H}}{dx}$$

$$= \mu_{-\text{eff}}\frac{d\mathscr{H}}{dx}, \quad ......(129)$$

with the same sign convention as before.

In other words, the atom appears to possess a series of effective magnetic moments $\mu_{+\text{eff}}$ or $\mu_{-\text{eff}}$, according as the nuclear magnetic moment is positive or negative, whose values depend on the magnitude of $\mathscr{H}$. The experiments now

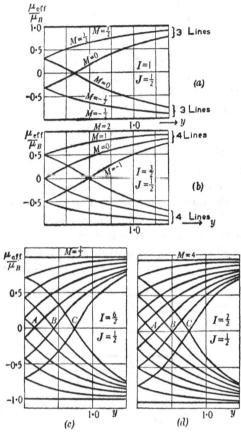

Fig. 55. Hyperfine structure of atomic rays.

to be described deal with the measurement of this effective magnetic moment and lead to a knowledge of M and ΔE, whence, assuming that J is known, the value of I can be found. Moreover, the sign of the nuclear magnetic moment can be directly deduced from these experiments, and its magnitude can be calculated from the experimental value of ΔE on the basis of wave-mechanics formulae derived by Goudsmit and others, with which we shall not deal.

Let us now take the particular case of an hydrogen atom with $I = \frac{1}{2}$, $J = \frac{1}{2}$ and $g = 2$. Then whether we assume the nuclear magnetic moment positive or negative we get the same four values of μ_{eff}, corresponding to $M = 1, 0, 0$ and -1, namely,

$$1\mu_B, \quad \frac{y}{(1+y^2)^{\frac{1}{2}}}\mu_B, \quad \frac{-y}{(1+y^2)^{\frac{1}{2}}}\mu_B, \quad \text{and} \quad -1\mu_B.$$

Hence by deflections of the atoms in a strong Stern-Gerlach field alone we cannot distinguish between positive and negative nuclear magnetic moments.

Two of the above values contain y and are, consequently, functions of $\mathscr{H}$, as shown in Fig. 58. In a weak field, which may be said to correspond to $y < 0\cdot1$, a beam of atoms after passage through the field should therefore give four lines or traces on a receiving surface, two deflected and two so little deflected that the pattern should practically consist of three traces As the energy differences between the various positions in the field are so small, the four positions are almost equally probable, and, consequently, the central trace should have twice the intensity of either of the displaced ones. In a strong field, say, where $y > 3$, the coupling is destroyed, and only two traces should be observed. In fields of intermediate strength, say, where $y = 1$, four separate traces should be found.

Turning to equations (128) and (129) we see that for certain values of y, i.e. of $\mathscr{H}$, $\mu_{+\text{eff}} = 0$, or an atom with total magnetic number M will appear to possess zero magnetic moment. Taking the appropriate values of M and plotting the values of the effective magnetic moment against the corresponding values of y, we obtain a set of curves which illustrates this behaviour in a striking manner, and shows how the several traces arise. Thus, if we take $I = \frac{2}{2}$ and $J = \frac{1}{2}$, the appropriate values of M are $\frac{3}{2}, \frac{1}{2}, 0, 0, -\frac{1}{2}, -\frac{3}{2}$, and these give the set of curves of Fig. 55a, showing that a pattern of six lines corresponding to $(2 \times \frac{2}{2}+1)(2 \times \frac{1}{2}+1) = 6$ is possible in moderate fields, and these close together to form a two component pattern in strong fields. Similarly, for $I = \frac{3}{2}$ the curves of Fig. 55 b are obtained, giving a pattern of eight lines in moderate fields. Sets of curves for $I = \frac{6}{2}$ and $I = \frac{7}{2}$ are likewise shown in Fig. 55 c and d respectively.

Experiments of Rabi and Cohen. The above sets of curves show that the determinations of nuclear spins may be made by two distinct methods. In one method, of which the experimental arrangements are shown in Fig. 56, a beam of sodium atoms passed through the vertical slit S_1 of a nickel furnace. The beam then passed through the slit S_2 to pass some 6 cm.

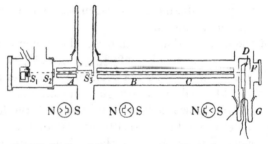

Fig. 56. Rabi and Cohen's apparatus for nuclear spins.

within a strong Stern-Gerlach field which spread it into two deflected beams. The slit S_3 allowed atoms of approximately the same velocity from one of these beams to pass through two more non-uniform fields. The first was a weak analysing field B, between pole tips 18 cm. long in which the nuclear and electron spins were no longer decoupled and the atoms were sorted into $(2I+1)$ components. The second was a stronger field C between pole tips 18 cm. long; it acted to about the same extent on all the components separated by B, displacing them more or less equally towards the position occupied by the original undisplaced beam. C thus produced a focusing effect and removed any scattered atoms.

A tungsten filament was moved across the final beam and was allowed to collect atoms, on account of their small number, for a known interval of time, whereupon the filament was then suddenly heated by the passage of a current and the positive ion emission to the plate P was measured by a valve amplifier and ballistic galvanometer. In this way four well-defined peaks were obtained in the curve showing the number of atoms received per second by the filament plotted against the position of the filament as it was moved across the final beam by turning the ground joint G. Since the beam consisted of four branches, then $2I + 1 = 4$, or $I = \frac{3}{2}$. The experiment, of course, gives no direct information concerning the magnetic moment of the nucleus.

The Method of Zero Moments. In the second method of measuring nuclear spins by the use of weak fields, a beam of atoms of approximately the same velocity is separated by a strong field and allowed to pass through a weak non-uniform field, whose strength may be continuously varied as required, to fall upon a fixed receiver placed to collect the maximum number of atoms when the field is not excited. Then, starting with a very weak field and increasing it gradually, the number of atoms collected by the receiver should decrease at first but later rise to a maximum, and with continued increase in the field successive maxima corresponding to the points A, B and C should be obtained in the cases of atoms represented by the curves of Fig. 55c or 55d. This is because those atoms for which $\mu_{\text{eff}} = 0$ are not deflected by the field and hence at every value of the field for which a particular set of atoms possesses zero effective magnetic moment a maximum number must in general be collected by the receiver. In the case represented by Fig. 55c we see that, if a maximum number is recorded for a value of the field H_1, successive maxima should also occur at values $3H_1$ and $5H_1$ only, i.e. the three successive maxima should be unequally spaced. Similarly, in the case represented by Fig. 55d, the maxima should be equally spaced at H_1', $2H_1'$ and $3H_1'$ only. The number of successive maxima and their spacing is therefore sufficient to fix the value of I uniquely, provided that I is large enough to give two or more maxima. This was the method used by Cohen* in the determination of the nuclear spin of caesium, by Millman† for the nuclear spin

* V. W. Cohen, *Phys. Rev.* **46**, 713, 1934.
† S. Millman, *Phys. Rev.* **47**, 739, 1935.

of potassium and by Fox and Rabi* for the nuclear spins of lithium, potassium and sodium.

Cohen used the same apparatus as that used by Rabi and Cohen for sodium, except that in order to deal with caesium S_1 was made rather wider, S_2 narrower and S_3 was made a fine collimator slit. The deflecting field was provided by B alone. The intensity of the beam was measured for different values of the current in the field coils of B, with the receiver placed in the position in which it received the maximum number of atoms when this field was not excited. The experimental curve, after allowance for an estimated background of atoms with finite moment, showed the existence of three equally spaced maxima, i.e. the experiment showed directly that I was equal to $\frac{7}{2}$ and not $\frac{5}{2}$. It was unnecessary to know the value of the field B in absolute measure—only the proportional change of the field with magnetising current was required and this was found by simple search-coil measurements. The main source of uncertainty would appear to be in the estimation of the background of atoms with finite moment, but this should have little effect on the final conclusion concerning the value of I.

The resolution obtained in measurements by the method of zero moments depends directly upon the length of the path of the atoms in the weak field B. This length is of course severely restricted when wedge and trough pole pieces are used. Rabi and his collaborators have shown, however, that a weak non-uniform magnetic field may readily be produced by using a system of two (or four) long parallel wires carrying the same current in opposite directions. The cross-sections of two water-cooled conducting tubes used for the passage of heavy currents are shown in Fig. 57a, where the position of the beam with respect to the conductors is indicated by the line BB. The conductors were rigidly mounted in a duralumin block and insulated by mica. With this arrangement the field and its gradient along the z axis can be calculated from the geometry of the system, and the need for field measurements disappears.

Let $2a$ be the distance between the centres of the two tubes

* M. Fox and I. I. Rabi, *Phys. Rev.* **48**, 746, 1935.

and r_1 and r_2 the distances from the two centres of a point in BB distant $1\cdot2a$ from the line of centres as shown in the figure; then the field at this point is

$$\mathscr{H} = 2C\frac{2a}{r_1 r_2},$$

and

$$\frac{d\mathscr{H}}{dz} = -2C\frac{2a}{r_1^3 r_2^3}(r_1^2 + r_2^2)z,$$

where C is the current in absolute units and z is the distance of the point from the plane containing the axes of the two tubes.

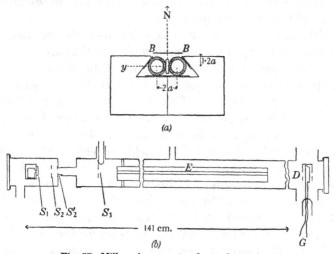

Fig. 57. Millman's apparatus for nuclear spins.

For a value of $z = 1\cdot2a$ both the field and its gradient are reasonably constant over the region from $y = -0\cdot7a$ to $+0\cdot7a$, the region occupied by BB in the figure.

Such a system was incorporated in the apparatus used by Millman in the determination of the nuclear spin of potassium, shown diagrammatically in Fig. 57b. The oven slit S_1 was about $0\cdot013$ mm. wide, S_2 about $0\cdot2$ mm. wide, S_2' about $0\cdot15$ mm. wide, and the collimating slit S_3 about $0\cdot02$ mm. wide. All these slits were of course parallel to the plane containing the axes of the wires or tubes. Currents up to 200 amperes were passed in a system of four water-cooled tubes E used in producing the field, the diameter of the conductors being $0\cdot248$ cm.

189

This dimension is of some importance, for Fox and Rabi using the same apparatus in a later examination of lithium, potassium and sodium obtained greater precision by the use of conductors of twice this diameter. If Δz is the error in measuring the distance between the atomic beam and the conductors, then

$$\frac{d\mathscr{H}/dz}{\mathscr{H}}\,\Delta z = \frac{\Delta\mathscr{H}}{\mathscr{H}},$$

and hence precision is increased by reducing the ratio of the gradient to the field. To counteract the decrease in resolution Fox and Rabi used narrower beams.

Although Millman's curve showing how the number of atoms collected by the receiver depended on the current in the tubes bore remarkable similarity to that deduced theoretically on the assumption that for K^{39}, $I = \frac{3}{2}$, it did not permit a definite statement to this effect, since curves for $I = \frac{3}{2}$ and $I = \frac{2}{2}$ both give single maxima for fields greater than zero. It was, however, possible to decide definitely in favour of $I = \frac{3}{2}$ by fixing the value of the field current at that value at which the maximum number of atoms was collected, and then obtaining the intensity distribution of the atoms in the beam under these conditions by displacing the detector D by rotating the ground-glass joint G. The experimental curve was then compared with curves calculated for the two values of I. An irregularity in the curve of the number of atoms against current in the wires was attributed to the presence of K^{41} atoms, and under conditions of high resolution this irregularity was found to give a clear maximum from which Manley* deduced that the nuclear spin of K^{41} was $\frac{3}{2}$.

The Determination of the Hyperfine-Structure Separation. The hyperfine-structure separation ΔE is contained in the parameter y of (126) and (127) and can be determined from the experiments of Millman and of Fox and Rabi provided that I and the values of $\mathscr{H}_0$ and $d\mathscr{H}_0/dz$, where $\mathscr{H}_0$ is a value of the deflecting field for which $\mu_{\text{eff}} = 0$, are accurately known, since $\Delta E = 2\mu_B\,\mathscr{H}_0/y = 2\mu_B\,\mathscr{H}_0/(2M/2I + 1)$. This requires an

* J. H. Manley, *Phys. Rev.* **49**, 921, 1936.

accurate knowledge of the position of the beam with respect to the field wires, and is best obtained by a special method adopted by Fox and Rabi. In their work the intensity of the atomic beam in the absence of the deflecting field was first found as a function of the position of the receiver, to provide a comparison curve. The end of the block supporting the field wires was then gradually raised until a fine wire, or stopping filament, mounted on the receiver end of the block parallel to, and at a known small distance above the surface of the wires, blocked out the middle portion of the atomic beam, or rather, produced a minimum intensity in the shadow cast by the filament with a maximum on either side. Now, the collimating slit S_3 was previously set by means of a telescope parallel to the wires and at a measured distance from their surface; since the middle of the atomic beam passed through both that of S_3 and of the stopping filament, the position of the beam with respect to the wires was accurately known.

The stopping filament could now be removed either by burning it out by the passage of a heavy current, or, the collimating slit could be moved a known distance either towards or away from the surface of the wires until the stopping filament no longer blocked the beam. The second method was obviously the more desirable. The apparatus was re-set many times in order to avoid systematic errors in the measurement of the position of the beam. The velocity of the selected set of atoms had to be maintained constant and the oven was accordingly kept at a steady temperature for some hours before measurements were commenced. The effect of stray magnetic fields on the value of $\mathscr{H}_0$ was eliminated by taking the mean value for the field current in opposite directions. Allowance was made for end effects of the field wires. The main experimental error, of course, lay in the measurement of z, and Fox and Rabi estimated it to be of the order of 1 per cent.

The values obtained for the hyperfine separations $\Delta\nu$ of alkali atoms are given in the table on p. 199, the values being expressed in cm.$^{-1}$, using the relation $\Delta\nu = \Delta E/hc$; the values of the nuclear magnetic moment calculated therefrom are also given.

The Magnetic Moment of the Proton. Experiments with Molecular Hydrogen. It has already been remarked that a direct determination of a nuclear magnetic moment is possible only when the permanent magnetic moment of the electron system is zero. Now, molecular hydrogen is diamagnetic, and, therefore, the molecule possesses no permanent moment, so that direct measurements of the magnetic moment of the nuclear system of the hydrogen molecule should be possible. Direct determinations by the molecular ray method have been made by Frisch and Stern* and by Estermann and Stern,† and the magnetic moment of the proton deduced from them. An apparatus very similar to that shown in Fig. 53 was used, with a Stern-Gerlach instead of a Rabi type of field for producing the magnetic deflections.

Unfortunately, ordinary hydrogen is a mixture of para-hydrogen, in which the two protons are set with their spins antiparallel, and orthohydrogen, in which they are set with their spins parallel with an associated magnetic moment of $2\mu_n$, or 2 nuclear magnetons (1 nuclear magneton = $1/1840$ Bohr magneton). At very low temperatures, about 99 per cent of the parahydrogen molecules possess no rotational energy, but the orthohydrogen molecule possesses at least one $h/2\pi$ unit of momentum even at liquid air temperatures, although the coupling between the rotation and nuclear spins is very small and is readily destroyed in a Stern-Gerlach field. The pattern for a beam of orthohydrogen molecules should therefore, theoretically, consist of two deflected and one undeflected beam, each beam consisting of three components corresponding to the magnetic moment due to the molecular rotation being aligned parallel or perpendicular or antiparallel to that due to nuclear spin. The determination of the magnetic moment of the proton thus requires the determination of the magnetic moments of the orthohydrogen molecule and the magnetic moment $l\mu_R$ due to the molecular rotation, the difference between them being twice the value of the magnetic moment of the proton. The quantity μ_R is determined from experiments

* R. Frisch and O. Stern, *Zeit. für Phys.* **85**, 4, 1933.
† I. Estermann and O. Stern, *Zeit. für Phys.* **85**, 17, 1933.

on pure parahydrogen, for at $90°$ K. this consists of $98\cdot3$ per cent of molecules in the state $l = 0$ and $1\cdot7$ per cent in the state $l = 2$. The relative numbers of molecules in the various quantum states can be calculated at higher temperatures from Boltzmann's formula. The maximum intensity of the undisplaced pattern when the field is applied can be compared with the value calculated from an assumed value of μ_R and the distribution in the several quantum states at the temperature of the experiment and a satisfactory estimate thus made. Fortunately, the value of μ_R is relatively unimportant. A similar experiment is then performed with ordinary hydrogen and the maximum intensity of the undisplaced beam when the field is applied is compared with that calculated on the assumptions that the relative percentage of ortho- and para-hydrogen molecules are known, that the orthohydrogen molecules possess a magnetic moment $n'\mu_n$, where n' is an integer, and that the distribution of rotational states in the beam is known. In this manner μ_R was found equal to $0\cdot9\mu_n$, a value agreeing with that calculated by Wick,* and the nuclear magnetic moment of the orthohydrogen molecule equal to $5\mu_n$, whence that of the proton was $2\cdot5\mu_n$, correct to about 10 per cent.

These experiments have been continued with practically the same apparatus by Estermann, Simpson and Stern† using beams of normal H_2 and HD. The latter molecule gives no ortho and para states and at $90°$ K., 55 per cent of the molecules are in the state $l = 0$, 40 in $l = 1$ and 4 in $l = 2$, so that neglecting particles in the latter state, a fairly simple pattern for the deflected beams in a Stern-Gerlach field is obtained. In this pattern there is a conspicuous absence of an undeflected component, so that there is a pronounced minimum in the middle of the pattern, in strong contrast to the H_2 pattern, where there is a maximum. Fortunately the theoretical pattern is not much affected by the value chosen for the magnetic moment of the deuteron, μ_D. At room temperatures, however, HD behaves less satisfactorily.

* G. C. Wick, *Zeit. für Phys.* **85**, 25, 1933.
† I. Estermann, O. C. Simpson and O. Stern, *Phys. Rev.* **52**, 535, 1937.

From the ratio of the intensities I and I_0 at the centre of the undeflected beam observed when the field was on and off respectively, the deflection of a H_2 or HD molecule of the most probable velocity at $90°$ K. with a magnetic moment of μ_P was calculated, and by substitution in the formula (120) of p. 165, μ_P could be found. In the calculations it was assumed that the separations S_R and S_D of the triplet components due respectively to the rotation of H_2 and to the magnetic moment of the deuteron were each equal to $\frac{1}{3} S_P$, where S_P is the separation due to the magnetic moment of the proton.

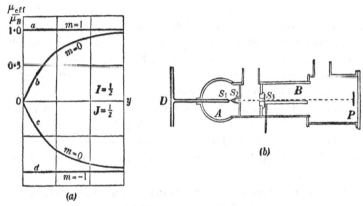

Fig. 58. Magnetic moment of hydrogen atom.

Special methods were used to measure the inhomogeneity of the field, an ingenious double search coil being used to find $d\mathscr{H}/dx$ directly. Corrections were applied to allow for departures from the Maxwell distribution of velocities caused by traces of gas in the apparatus, for the finite width of the receiver slit and tube or canal and for the residual field of the magnet. The final result was $\mu_P = 2 \cdot 46 \mu_n$, a value now known to be too small, probably due to more uncertainty in the value of $\mathscr{H} d\mathscr{H}/dx$ than was thought possible.

Experiments with Atomic Hydrogen. In the case of the neutral hydrogen atom it is known that $I = \frac{1}{2}$ and $J = \frac{1}{2}$, and, consequently, in weak fields a four-component pattern must be produced. This is shown in Fig. 58a, the four components corresponding to

$$\mu_{\text{eff}} = \pm\mu_B \quad \text{or} \quad \pm\frac{y}{(1+y^2)^{\frac{1}{2}}}\mu_B,$$

where $\quad y = 2\mu_B \mathcal{H}/\Delta E = 2\mu_B \mathcal{H}/hc\,\Delta\nu,$

$\Delta\nu$ being the separation in wave numbers of the hyperfine-structure levels corresponding to the electron spin and nuclear spin set parallel and antiparallel respectively. Now, while it is thought that the magnetic moment of the proton μ_P is equal to $\Delta\nu/0\cdot0169$, there is no optical data from which $\Delta\nu$ can be calculated; but it may be obtained if we determine the value of y for a known field $\mathcal{H}$ from atomic ray experiments now to be described.

In the experiments of Rabi, Kellogg and Zacharias[*] a beam of neutral hydrogen atoms was produced in a Wood discharge tube D, Fig. 58b, and allowed to pass through a glass slit S_1 into the antechamber A of the atomic ray apparatus, whence they passed through a steel limiting slit S_2 to traverse a highly evacuated chamber to a collimating slit S_3. On leaving S_3 the atoms passed over a wire system, producing the weak non-uniform magnetic field, mounted inside the still more highly evacuated chamber B, to fall upon the receiver P. The latter was a glass plate coated with a layer of molybdenum oxide soot, carried on a slide which could be moved across the beam by a magnetic control. The correct alignment of the limiting slit S_2 was very important and was made by setting it with the aid of a trace made on a detecting plate placed between S_1 and S_2; alignment by the usual optical method was not adequate. A deposit suitable for measurement was obtained on P with exposures of some few hours' duration. The pattern consisted of two traces or bands.

Now, no velocity selector was used in the apparatus. Consequently, each set of atoms with a particular value of μ_{eff} had a Maxwellian distribution of velocities. The faster atoms were deflected less than the slower atoms and hence each set of atoms was spread out into a band. The two experimental

* I. I. Rabi, J. M. B. Kellogg and J. R. Zacharias, *Phys. Rev.* **46**, 157, 1934.

traces were therefore formed of the superimposed deposits of atoms with $\mu_{\text{eff}} = +\mu_B$ and $+\dfrac{y}{(1+y^2)^{\frac{1}{2}}}\mu_B$, and with $\mu_{\text{eff}} = -\mu_B$ and $-\dfrac{y}{(1+y^2)^{\frac{1}{2}}}\mu_B$, respectively, and to the measurements with these deposits the theory given on p. 175 was applied.

For the deposits corresponding to $\mu_{\text{eff}} = \pm\mu_B$ equation (124) was (after integration) rewritten in the form

$$(I_s)_{\mu_B} = I_0 f(s, s_{\alpha_0}), \qquad \ldots\ldots(130)$$

where I_s is the intensity of the trace at a point with displacement s and s_{α_0} is the most probable displacement of an atom with $\mu_{\text{eff}} = \pm\mu_B$, and could at once be calculated from a knowledge of the field gradient and the geometry of the apparatus. For the deposits corresponding to

$$\mu_{\text{eff}} = \pm\frac{y}{(1+y^2)^{\frac{1}{2}}}\mu_B = \mu'$$

in like manner could be written

$$(I_s)_{\mu'} = I_0 f(s, s_{\alpha'}), \qquad \ldots\ldots(131)$$

where $s_{\alpha'}$, the most probable deflection of an atom with $\mu_{\text{eff}} = \mu'$, had to be found from the experimental measurements. I_0 occurs in the two equations (130) and (131) as all four values of μ_{eff} are equally probable.

Now, when the separation $2s_1$ of the inside edges of the traces and the separation $2s_2$ of their outer edges were measured by a microscope with micrometer eyepiece, this was equivalent to finding four regions of equal intensity on the traces. Hence these values could be substituted in equations (130) and (131), thus giving, for the total intensity at either of the four points,

$$(I_{s_1})_{\mu_B} + (I_{s_1})_{\mu'} = (I_{s_2})_{\mu_B} + (I_{s_2})_{\mu'},$$

whence $\quad f(s_1, s_{\alpha_0}) + f(s_1, s_{\alpha'}) = f(s_2, s_{\alpha_0}) + f(s_2, s_{\alpha'}). \quad \ldots\ldots(132)$

In (132) all quantities except $s_{\alpha'}$ were known, so that the latter could be calculated. By substituting $s_{\alpha'}$ in equation (124) of p. 175 the value of μ' and, hence, y and $\varDelta\nu$ were found. Correction was, of course, made for the fact that the whole path of the atoms did not lie in the magnetic field. The final value so obtained for μ_P was $3 \cdot 25 \pm 0 \cdot 3 \mu_n$, instead of the value $2 \cdot 80$ now generally accepted, presumably because of incorrect knowledge of the average velocity of the atoms in the beam.

The Magnetic Moment of the Deuteron. Rabi, Kellogg and Zacharias[*] also used the apparatus just described to determine the magnetic moment of the deuteron, by sending through it a beam of neutral deuterium atoms from a Wood discharge tube. The deuteron has a spin of $\frac{2}{2}$ unit and therefore gives a pattern with six components. The value of $\Delta\nu$ for the deuteron was found in the same way as the corresponding value for the proton, and the magnetic moment of the deuteron μ_D calculated from the theoretical result $\mu_D = \Delta\nu/0 \cdot 0127$. The experiments were difficult, because of the three deposits in each trace, the smaller value of $\Delta\nu$ and the difficulty of freeing deuterium from hydrogen. The value of μ_D was found to be $0 \cdot 75 \pm 0 \cdot 2 \mu_n$. More accurate determinations will be discussed in Chapter XIV.

The Process of Spatial Quantisation. A very interesting problem was first studied theoretically by Güttinger and later by Majorana and put to experimental test by Phipps and Stern[†] and by Frisch and Segrè.[‡] Let us suppose that a beam of neutral potassium atoms traverses a strong Stern-Gerlach field, and that by means of a selector slit one of the two beams into which it resolves is isolated. The selected beam is now allowed to enter a field which varies rapidly in magnitude and direction as the beam proceeds on its way. From the standpoint of an atom this would be equivalent to exposing a stationary atom to a field which varied in the same manner with time. The beam then enters a strong Stern-Gerlach analysing field parallel to the first, and the problem is to determine how many atoms have remained with their magnetic axes set as when they passed the selector slit.

In Frisch and Segrè's apparatus the two strong Stern-Gerlach fields were produced by the same electromagnet and between them lay a region, which was heavily shielded from magnetic fields by permalloy and iron shields, in which a

---

[*] I. I. Rabi, J. M. B. Kellogg and J. R. Zacharias, *Phys. Rev.* **46**, 163, 1934.

[†] T. E. Phipps and O. Stern, *Zeit. für Phys.* **73**, 183, 1931.

[‡] R. Frisch and E. Segrè, *Zeit. für Phys.* **80**, 610, 1933.

weak field was produced by a current flowing in a wire. Now, the theoretical discussions lead to the view that, when the angular velocity of rotation of the field was small compared to the frequency of the Larmor precession produced by the field, the atom should always adjust itself non-adiabatically to the field, e.g. maintain its magnetic axis either parallel or antiparallel to the field at any instant. If the field changes were more rapid, abrupt changes or reversals, adiabatic processes, might occur, and on entering the analysing field any changes in space quantisation should be detected.

The above experiments were inconclusive, and Rabi* showed that effects of nuclear spin were considerable and also that properly conducted experiments would give information both of the magnitude of the spin and the sign of the nuclear magnetic moment. Let us examine the particular case of hydrogen. Referring to equations (128) and (129) and to Fig. 58 a we see that a weak Stern-Gerlach field will resolve a beam of neutral hydrogen atoms into four groups a, b, c, d. Groups a, c and d all correspond to values of the total angular momentum quantum number $F = \frac{1}{2} + \frac{1}{2}$, while group b corresponds to $F = \frac{1}{2} - \frac{1}{2} = 0$, since for the H atom $I = \frac{1}{2}$ and $J = \frac{1}{2}$, *assuming the nuclear magnetic moment to be positive.*

If by means of a slit we let through the beam b, then no change can be registered in the analysing field, as transitions to another state are impossible; but if we let through beam a into a properly chosen rotating field, then transitions from the angular momentum state $\frac{1}{2} + \frac{1}{2}$ to $\frac{1}{2} - \frac{1}{2}$ are possible; some of these states will have moments of opposite sign in the analysing field and give two beams therein. If we let the beams c and d through the selector slit we should again get two beams in the analyser. Now the position of the beam b with respect to the field before the selector slit is important; if it is in the "stronger field half" or the field nearer the wedge pole piece, then the nuclear magnetic moment is positive, if in the "weaker field half" it is negative.

When the experiment was performed, Kellogg, Rabi and Zacharias† found a single trace for beam b in the analyser,

* I. I. Rabi, *Phys. Rev.* **49**, 324, 1935.
† J. M. B. Kellogg, I. I. Rabi and J. R. Zacharias, *Phys. Rev.* **49**, 421, 1936.

hence the proton moment is positive; it was equal to $2 \cdot 85 \pm 0 \cdot 15 \mu_n$. In like manner they found the deuteron moment positive and equal to $0 \cdot 85 \pm 0 \cdot 03 \mu_n$, while Torrey§ found the nuclear magnetic moments of Na^{23} and K^{39} to be positive.

TABLE VIII

Nucleus	Observed g-factor	Angular momentum I	Moment μ	Diamagnetic correction (%)*
$_1H^1$	$5 \cdot 5791 \pm 0 \cdot 0016$	$\frac{1}{2}$	$2 \cdot 7896$	0
$_1H^2$	$0 \cdot 8565 \pm 0 \cdot 0004$	1	$0 \cdot 8565$	0
$_3Li^6$	$0 \cdot 8213 \pm 0 \cdot 0005$	1	$0 \cdot 8213$	$0 \cdot 01$
$_3Li^7$	$2 \cdot 1688 \pm 0 \cdot 0010$	$\frac{3}{2}$	$3 \cdot 2532$	$0 \cdot 01$
$_4Be^9$	$0 \cdot 784 \pm 0 \cdot 003$	$\frac{3}{2}$	$-1 \cdot 176$	$0 \cdot 02$
$_5B^{10}$	$0 \cdot 598 \pm 0 \cdot 003$	1	$0 \cdot 598$	$0 \cdot 03$
$_5B^{11}$	$1 \cdot 791 \pm 0 \cdot 005$	$\frac{3}{2}$	$2 \cdot 686$	$0 \cdot 03$
$_6C^{13}$	$1 \cdot 402 \pm 0 \cdot 004$	$\frac{1}{2}$	$0 \cdot 701$	$0 \cdot 03$
$_7N^{14}$	$0 \cdot 403 \pm 0 \cdot 002$	1	$0 \cdot 403$	$0 \cdot 04$
$_7N^{15}$	$0 \cdot 560 \pm 0 \cdot 006$	$\frac{1}{2}$	$0 \cdot 280$	$0 \cdot 04$
$_9F^{19}$	$5 \cdot 250 \pm 0 \cdot 005$	$\frac{1}{2}$	$2 \cdot 625$	$0 \cdot 06$
$_{11}Na^{23}$	$1 \cdot 4765 \pm 0 \cdot 0015$	$\frac{3}{2}$	$2 \cdot 215$	$0 \cdot 08$
$_{13}Al^{27}$	$1 \cdot 452 \pm 0 \cdot 004$	$\frac{5}{2}$	$3 \cdot 630$	$0 \cdot 10$
$_{17}Cl^{35}$	$0 \cdot 547 \pm 0 \cdot 002$	$\frac{5}{2}$	$1 \cdot 368$	$0 \cdot 14$
$_{17}Cl^{37}$	$0 \cdot 454 \pm 0 \cdot 002$	$\frac{5}{2}$	$1 \cdot 136$	$0 \cdot 14$
$_{19}K^{39}$	$0 \cdot 260 \pm 0 \cdot 001$	$\frac{3}{2}$	$0 \cdot 391$	$0 \cdot 16$
$_{19}K^{41}$	$0 \cdot 143 \pm 0 \cdot 001$	$\frac{3}{2}$	$0 \cdot 215$	$0 \cdot 16$
$_{37}Rb^{85}$	$0 \cdot 536 \pm 0 \cdot 003$	$\frac{5}{2}$	$1 \cdot 340$	$0 \cdot 39$
$_{37}Rb^{87}$	$1 \cdot 822 \pm 0 \cdot 006$	$\frac{3}{2}$	$2 \cdot 733$	$0 \cdot 39$
$_{49}In^{113}$	—	$\frac{9}{2}$	$5 \cdot 480$	$0 \cdot 04$
$_{49}In^{115}$	—	$\frac{9}{2}$	$5 \cdot 490$	$0 \cdot 04$
$_{55}Cs^{133}$	$0 \cdot 731 \pm 0 \cdot 002$	$\frac{7}{2}$	$2 \cdot 558$	$0 \cdot 67$
$_{56}Ba^{135}$	$0 \cdot 558 \pm 0 \cdot 002$	$\frac{3}{2}$	$0 \cdot 837$	$0 \cdot 68$
$_{56}Ba^{137}$	$0 \cdot 624 \pm 0 \cdot 002$	$\frac{3}{2}$	$0 \cdot 936$	$0 \cdot 68$

* To be added to the moment, μ.

Values of the nuclear spins, magnetic moments and g-factors obtained by the atomic-ray method are given above; many of them were obtained by the new and powerful method described in Chapter XIV.

In addition to the values for the magnetic moment of the proton given in the above table, a valuable check is provided by the measurements of Lasarew and Schubnikow‡ on the susceptibility of solid hydrogen which give $\mu_P = 2 \cdot 3 \mu_n \pm 10$ per cent.

As a sound working rule§ it appears that the magnetic and mechanical moments of a nucleus are entirely to be attributed

† H. C. Torrey, *Phys. Rev.* **51**, 501, 1937.

‡ B. G. Lasarew and L. W. Schubnikow, *Phys. Zeit. Sowjet.* **10**, 117, 1936 and **11**, 445, 1937.

§ Th. Schmidt, *Zeit. für Phys.* **106**, 358, 1937.

to the last odd particle, neutron or proton, to be added in the formation of the nucleus; e.g. the properties of $_3Li^7$ are decided by the last proton, while those of $_{30}Zn^{67}$ are decided by the last neutron.

The Magnetic Moment of the Neutron. We have just seen that the magnetic moments of the proton and deuteron are respectively 2·79 and $0·86\mu_n$, with $I = \frac{1}{2}$ and $I = \frac{2}{2}$. The value

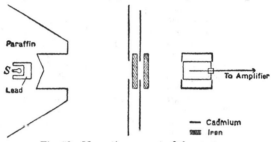

Fig. 59. Magnetic moment of the neutron.

of I for the neutron is undoubtedly $\frac{1}{2}$. Hence, if a neutron and a proton combine to form a deuteron, it would be expected that the magnetic moment of the neutron would be about $-1·9\mu_n$, since the nuclear spins would have to be directed parallel. Now the neutron has a mass 1·009 units ($O = 16$) and possesses no electric charge, and, consequently, it is hardly affected by the electric fields of the atoms of substances through which it passes.

It occurred to Bloch* that the magnetic field of the neutron might interact with the magnetic fields of the extranuclear electrons in ferromagnetic substances through which it passed. Halpern and Johnson† suggested a similar interaction between the neutron and atoms of paramagnetic substances. Several experiments based on Bloch's ideas were carried out, and, although exceedingly difficult to interpret, they proved that the neutron possesses a magnetic moment. The magnetic interactions are very small, but they act over regions large compared with the fields of force of atomic nuclei, and are, therefore, relatively more effective than those responsible for the well-established nuclear scattering.

* F. Bloch, *Phys. Rev.* **50**, 259, 1936.
† O. Halpern and M. H. Johnson, *Phys. Rev.* **51**, 992, 1937.

Thus Hoffman, Livingston and Bethe* produced a beam of slow neutrons, i.e. neutrons absorbed by cadmium, from a beryllium-radon source S of approximately 500 millicuries, collimated it by apertures in cadmium shields and allowed it to pass through two magnetised iron bars, as shown in Fig. 59, at right angles to the magnetisation, whereupon the unscattered neutrons were registered by an automatic counter filled with BF_3. According to Bloch's theory the number of neutrons scattered should depend on whether the spin of the atom is parallel or antiparallel to that of the neutron. The first iron bar acts as a polariser and the second as an analyser, and the number of neutrons transmitted should be larger when the magnetisations of polariser and analyser are parallel than when antiparallel. This was found to be the case experimentally, but the difference in numbers of the particles for parallel and antiparallel magnetisations was only of the order of $1\cdot8 \pm 0\cdot54$ per cent. This was, however, three times as great as the estimated experimental error, and in agreement with Bloch's theory, which predicted a difference of some $2\cdot3$ per cent.

Using the Rabi method described on p. 198, Powers‡ showed that the magnetic moment of the neutron is negative. Let us suppose that two sets of iron strips some 10 cm. apart are placed with their planes perpendicular to the paper and magnetised in the direction towards the reader. Let two long parallel wires, carrying equal currents in opposite directions, pass perpendicular to the paper into the region between the plates. By proper adjustment of the currents and the position of the plates, the resultant of the magnetic field of the wires and the stray field of the plates can be made to vary rapidly and so produce adiabatic transitions or orientations of the neutrons in their passage between the plates. When the currents are in such a direction that transitions for a negative magnetic moment would be facilitated, the transmission of neutrons by the system is reduced; hence the neutron must have a negative magnetic moment. The best method of measuring the magnetic moment of the neutron is described in Chapter XIV.

* J. G. Hoffman, M. S. Livingston and H. A. Bethe, *Phys. Rev.* **51**, 214, 1937.
† H. N. Powers, *Phys. Rev.* **54**, 827, 1938.

CHAPTER VII

GYROMAGNETIC EFFECTS

Introduction. In describing the more elementary properties of magnetic substances the individual magnetic particle is usually pictured as a magnetic doublet or dipole. We have seen in earlier chapters that this picture is inadequate and that the particle possesses the properties of a gyrostat. As early as 1861 Clerk Maxwell* devised an experiment in which he sought to detect the inertia of the Ampèrian currents which the molecules of a permanent magnet were then thought to contain. The experiment failed, and we know now that the effect sought is too small to be detected, even with modern equipment. In 1890, Perry† suggested that the elementary magnetic particle might be a tiny gyrostat and unsuccessfully tried to magnetise an iron bar by rotating it about its axis, but the first published recognition of the gyrostatic properties of the electron is due to O. W. Richardson,‡ whose paper in 1908 started a train of experiments which will now be described.

The Properties of the Gyroscope. In outlining these experiments it is helpful to consider the behaviour of a simple gyroscope, as illustrated in Fig. 60a. The rod AB is pivoted at O' so that it can turn about the axis XY. The flywheel I is pivoted at a and b, so that it may rotate about the axis AB, and is counterpoised by the weight W. The whole system is also free to turn in a collar about the vertical axis of support. Now, two important and fundamental experiments may be carried out by means of this apparatus, assuming frictional effects to be very small.

First, the flywheel is set into rapid rotation about its axis in a clockwise direction as viewed from A, when W is sufficient to keep the axis AB horizontal. The system is then slowly turned

---

* J. Clerk Maxwell, *Electricity and Magnetism*, § 575.
† J. Perry, *Spinning Tops*, p. 112, 1890.
‡ O. W. Richardson, *Phys. Rev.* **26**, 248, 1908.

about the vertical axis in a clockwise direction when viewed from above. Thus, if in Fig. 60*b*, *PQ* represents the angular momentum of the wheel in magnitude and direction at the start, then after a short time the axis *AB* will have turned through an angle $\delta\theta$ about the vertical axis and the line *PQ'* will then represent the angular momentum in magnitude and direction. Hence there has occurred a change in angular

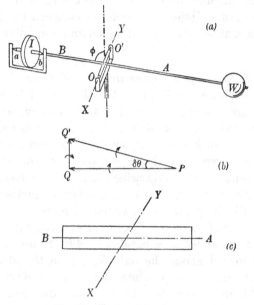

Fig. 60. Theory of the gyroscope.

momentum represented in magnitude and direction by *QQ'*, i.e. about the axis *XY*, so that an increase in the angle ϕ must have taken place. In other words, as the system is forced to rotate clockwise the flywheel must tend to dip downwards. When the direction of rotation about the vertical axis is reversed, other conditions remaining unchanged, then the wheel must tend to rise.

Secondly, while the wheel is rotating about *ab* the angle ϕ may be increased or diminished by respectively decreasing or increasing the weight *W* by an amount ΔW. It is then found that the system as a whole rotates about the vertical axis, for, referring to Fig. 60*b*, it follows that QQ' now represents in magnitude and direction the product of the applied couple and the time for which it acts, and it represents the change in angular momentum which is equal to $I\omega\, d\theta$. The couple is thus equal to $I\omega\dfrac{d\theta}{dt}$, equal to the moment of $\varDelta W$ about XY, where I is the moment of inertia of the flywheel about ab, ω is its constant angular velocity and dt is the time in which the change $d\theta$ occurs. Hence the whole system must tend to turn, or to *precess*, about the vertical axis with an angular velocity $d\theta/dt$.

Types of Gyromagnetic Experiments. With these two fundamental experiments in mind the theoretical basis of the magnetic experiments described in this chapter can readily be appreciated. In general it is sufficient to describe four different types of experiment which have been carried out, although, as Auwers* has shown, it is possible to tabulate some eight different types. To describe the first type, consider the rod AB, Fig. 60*c*, which is permanently magnetised in the direction of its long axis. This we may suppose to correspond to the rotation of the majority of the elementary gyrostats in the same sense about axes parallel to AB. Hence, if the rod AB were turned about the axis XY, then the elementary gyrostats should turn in an endeavour to counteract this change, and should thus cause the whole bar to turn about an axis perpendicular to AB and to XY. This was the unsuccessful experiment devised by Maxwell.

In the second type of experiment, imagine the rod AB to be initially unmagnetised, i.e. the elementary gyrostats have their axes orientated in all possible directions. Hence on rotating the rod about the axis AB the gyrostats will tend to set with their axes in the same sense and parallel to AB, even in the absence of an external magnetic field. This may be directly deduced from the second fundamental experiment described above. It therefore follows that the rod will be magnetised parallel to the axis AB, and the phenomenon is

* O. v. Auwers, *Naturwissen.* **13**, 202, 1935.

referred to as *magnetisation by rotation* and sometimes as the *Barnett effect*.

To explain the third type of experiment suppose that the rod is unmagnetised and suspended with the axis AB vertical. Then by means of an external applied field the rod may be magnetised parallel to this axis, i.e. the elementary gyrostats made to set with their axes of rotation parallel to AB. There will thus be a change in the angular momentum about AB to which there must be an equal and opposite reaction. If it is assumed that this reaction is experienced only by the rod— without making any conjectures as to the mechanism by which the reaction is communicated to it—then the whole rod must rotate about the axis AB in the opposite direction to that in which the majority of the gyrostats finally spin. Referring to Fig. 60a it is seen that if the angle ϕ were suddenly decreased then the system would receive an equal and opposite impulse which would cause it to precess round the vertical axis in an anticlockwise direction when viewed from above. Thus rotation should accompany magnetisation, a phenomenon known on the continent as the *Einstein-de Haas effect* and in England as the *Richardson effect*. It permits the most satis-factory investigation of the gyromagnetic aspects of magneti-sation; for, if we suddenly magnetise the rod so that A be-comes north-seeking, then the rod must rotate clockwise or anticlockwise about its axis as viewed from the right ac-cording as the magnetic particles are positively or negatively charged.

Experiments of the fourth type involve the use of a rotating magnetic field whose lines of force are always perpendicular to the axis AB. The effect of such a field in making the magnetic particle turn should correspond to that produced by a couple in turning the system of Fig. 60a about the vertical axis, resulting in an increase or decrease of the angle ϕ, thus causing a change of angular momentum about that vertical axis. In other words, the transverse rotating magnetic field should produce a longitudinal magnetisation.

In order to estimate the magnitude of the mechanical effects accompanying magnetisation we must know the relation between the angular momentum U and the magnetic moment μ of the elementary magnetic gyrostat, the ratio U/μ being termed the *gyromagnetic ratio* and denoted by ρ. We saw in Chapter I that, for an electron describing a circular orbit $\rho = 2m/e$, for a spinning electron $\rho = m/e$, and for an electron system for which g is the Landé splitting factor $\rho = \dfrac{1}{g}\dfrac{2m}{e}$. By means of gyromagnetic experiments, then, we measure $\dfrac{1}{g}\dfrac{2m}{e}$, and if we find that $g = 2$ we know that the elementary magnetic particle owes its properties to electron spin alone, whereas if we find that g lies between 0 and 2 we know that orbital as well as electron spin motion is involved. Since $2m/e = 1\cdot137 \times 10^{-7}$, it is clear that the angular momentum accompanying change of magnetisation is always quite small; to facilitate comparison with other works the value in this chapter has been taken to be $1\cdot13 \times 10^{-7}$.

Barnett's Experiments on Magnetisation by Rotation. Let us suppose that the magnetic properties of the bar in Fig. 60c are due to electron systems each of angular momentum U, and that the bar is set rotating with angular velocity σ about the axis AB. Each electron system whose axis makes an angle θ with AB will be acted upon by a couple $U\sigma\sin\theta$ tending to turn its axis of angular momentum parallel to AB, causing an increase in magnetisation parallel to AB. Now, whether the gyrostats are turned by the application of an appropriate magnetic field $\mathcal{H}$ parallel to AB or by rotation of the bar, the result is the same, so that $\mathcal{H}\mu\sin\theta = -U\sigma\sin\theta$ or $\mathcal{H} = -\sigma U/\mu = -\rho\sigma = -2\pi n\rho$, where n is the frequency of rotation of the bar; we may call $\mathcal{H}$ the equivalent magnetic field, the negative sign merely indicating reaction.

The experiment gives no information at all concerning the process of magnetisation. Indeed, we might suppose that the impressed angular velocity σ corresponds to the Larmor precession produced by a field $\mathcal{H}$, in which case, cf. p. 11, $\sigma = -\dfrac{\mathcal{H}e}{2m}$ for orbital motion, giving $\mathcal{H} = -2\pi n\rho$ for the general case.

It thus follows that if $\mathscr{H}$ and n are determined then ρ can be found, but it is impossible to measure the magnetic effects of small fields like $\mathscr{H}$ except in the case of ferromagnetic substances. The experimental procedure consists in rotating a bar of ferromagnetic material in a field-free space and comparing the intensity of magnetisation it thus acquires with that produced by the application of a known external field. Two different experimental arrangements have been used by Barnett[*]; in the first the magnetisation was detected by an

induction method, in the second by a magnetometer method. With the first arrangement large iron rods approximately 1 m. long and 7 cm. in diameter were used. Two identical rods were mounted parallel to each other and with their long axes in the magnetic meridian in a region over which the total intensity of the earth's magnetic field was neutral-ised by means of coils carrying currents. The arrangement is shown

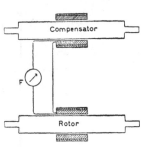

Fig. 61. Measurement of magnetisation produced by rotation. (Barnett.)

diagrammatically in Fig. 61. One of the rods, the rotor, could be rapidly rotated while the other, the stator or compensator, was at rest.

Identical coils were wound on formers placed as shown over the two rods and the change of flux produced by stopping the rotor was compared with the change of flux produced by the application of a known external field. Let θ_n be the fluxmeter deflection corresponding to the reversal of the direction of rotation of the rotor at a frequency n, and $\theta_{\mathscr{H}}$ that corresponding to the reversal of a known field $\mathscr{H}$. Then assuming magnetisation to be proportional to field, $\theta_n \propto 2\pi n \rho$ and $\theta_{\mathscr{H}} \propto \mathscr{H}$, whence

$$\rho = \frac{\mathscr{H}}{2\pi n} \frac{\theta_n}{\theta_{\mathscr{H}}}.$$

The complete neutralisation of the earth's magnetic field was necessary to avoid direct magnetisation of the rods and

* S. J. Barnett, Reviews of Modern Physics, **7**, 136, 1935; Phys. Rev. **6**, 239, 1915; Proc. Amer. Acad. Arts and Sci. **75**, 109, 1944.

also to avoid the production of eddy currents. The direction of rotation of the rotor could be reversed in order to eliminate disturbing effects from change of dimensions due to stresses set up by rotation and by temperature variations. The rotor was driven from one end, and, to counteract torsional disturbance of the magnetisation it could be reversed in its bearings. Spurious changes of induction due to variations of the earth's field were automatically eliminated by connecting the induction coils in series opposition.

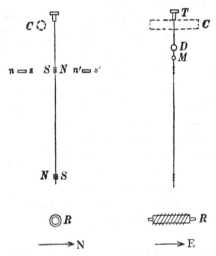

Fig. 62. Measurement of magnetisation produced by rotation. (Barnett.)

The second experimental arrangement used by Barnett[*] was much more sensitive, and possessed the advantage that large rods were not needed, a point of some importance when cobalt was used, but it was subject to much greater disturbances from fluctuations of the earth's field.

The necessary details are shown in Fig. 62. A light vertical rod of aluminium carried two groups of very small horizontal magnets NS, forming an accurately adjusted astatic system. A damping vane D and mirror M were attached and the whole suspended from a torsion head, T, by a delicate quartz fibre

---

* S. J. Barnett, *Phys. Rev.* **10**, 7, 1917; **17**, 404, 1921; **20**, 90, 1922 and more particularly *Proc. Amer. Acad. Arts and Sci.* **60**, 125, 1925.

inside a draught-proof case, containing a little uranium salt to get rid of electrostatic disturbances. Two small control magnets ns and $n's'$ respectively produced fields parallel and perpendicular to the upper magnet of the astatic system and permitted the zero to be controlled.

The rotor R was mounted with its axis perpendicular to the magnetic meridian and usually below the lower set of magnets, while the compensator C was mounted parallel to the rotor and about the same distance above the upper set, but, for convenience, displaced a little north or south. The earth's field was neutralised as before. The magnetisation acquired by rotation was compared with that acquired when the rotor was magnetised by a standard solenoid placed over it.

The experimental accuracy was limited by the minute intensity of magnetisation which is acquired at frequencies of revolution which are mechanically possible; e.g. for a frequency of 100 revolutions per second the equivalent field $\mathscr{H}$ is 10^{-5} oersted. The latest mean values of ρ obtained by Barnett[*], employing the first method, are given in the following table:

TABLE IX. MAGNETISATION BY ROTATION

Material	Armco iron	Steel	Nickel	Cobalt
$\rho \times e/m$	1·038	1·043	1·042	1·071

Experiments on Rotation by Magnetisation. The experimental measurements of the rotation which accompanies magnetisation of ferromagnetics fall into three groups. First, there is a direct method in which the mechanical impulse accompanying a sudden change of the magnetisation of a vertical cylinder causes it to turn against the torque of a delicate fibre by which it is suspended. Secondly, there is a resonance method in which the angle of twist is greatly increased by repeating the impulse at intervals of time equal to the period of the suspended cylinder. Finally, there is a null method in which the amplitude of the resonance oscillations produced by the timed impulses is reduced to zero by

[*] S. J. Barnett, *Proc. Amer. Acad. Arts and Sci.* **75**, 109, 1944.

applying an opposing set of timed impulses of known magnitude.

The Direct Method. Measurements of the gyromagnetic ratio by the direct method will now be described; they are important in that they show directly that electrons are

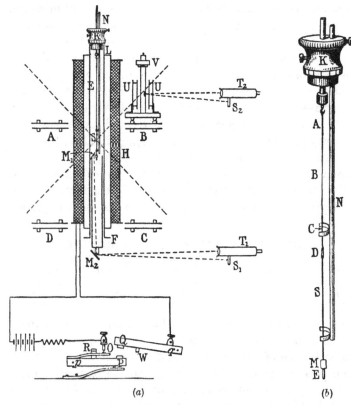

Fig. 63. Apparatus for direct determination of the gyromagnetic ratio. (Chattock and Bates.)

responsible for ferromagnetism and they avoid certain sources of error which are sometimes thought to influence the determinations by other methods. The arrangements used by Chattock and Bates,* which are shown in Fig. 63*a*, incorporated

* A. P. Chattock and L. F. Bates, *Roy. Soc. Phil. Trans.* A, **223**, 257, 1923.

many features first used by Stewart.* The specimen of ferromagnetic material S was a thin cylinder of wire about 10 cm.
long and about 0·35 mm. in diameter suspended by a fine
quartz fibre from the torsion head K, the details of the suspension being shown in Fig. 63b. K turned in a brass tube
which was mounted coaxially inside the vertical solenoid H
wound on a brass tube whose axis could be made vertical by
the lateral adjustment of sets of horizontal brass screws, not
shown, pressing against the projecting ends of the solenoid
tube. The brass rod N, mounted excentrically in K, was provided merely for the easy transport of the suspended system
when being removed for adjustment.

The magnetisation of the specimen was suddenly reversed
by passing a current through H for a time short in comparison
with the period of torsional oscillation of the system. This was
done by releasing a stop W, thus allowing the heavy bar qQ,
pivoted at q, to fall and make electrical contact between
·platinum surfaces at P and Q. A current then passed through
the solenoid via the spring strip at O until the bar pP, pivoted
at p and supported by a spring, was depressed beyond a certain
limit. The time of contact was thus determined by the height
from which qQ was allowed to fall and by the stiffness of the
several springs. In this process the magnetic moment M of
the specimen was reversed or changed by an amount $2M$, and
an amount of angular momentum $I\omega$ was delivered to the
specimen. The specimen therefore turned through an angle θ
until its kinetic energy was converted into potential energy of
the twisted fibre, i.e.

$$\tfrac{1}{2}I\omega^2 = \tfrac{1}{2}C\theta^2,$$

whence
$$U = I\omega = \theta\sqrt{IC} = \theta\frac{TC}{2\pi}$$

and
$$\rho = \frac{\theta TC}{4\pi M}, \qquad \quad ......(133)$$

where C is the torsion constant of the fibre and T is the period
of torsional oscillation of the system. It was therefore necessary
to measure the quantities θ, T, C and M.

* J. Q. Stewart, *Phys. Rev.* **11**, 100, 1918.

To obtain the deflection θ the values θ_1 and θ_2 for the initial deflections to either side of zero following the passage of the solenoid current were measured by means of the telescope T_1 and scale S_1, the light reflected from the mirror attached to the specimen being brought to and from the latter by the two 45 degree mirrors $M_1 M_2$; actually, for convenience two small mirrors were cemented back to back on the specimen. Now, the motion of the specimen was heavily damped, and the value of θ to be inserted in equation (133) was obtained from the expression

$$\theta = \theta_1 \left(\frac{\theta_1}{\theta_2}\right)^{\frac{1}{\pi} \text{arc tan} \frac{\pi}{\log_e(\theta_1/\theta_2)}}$$

In like manner the value of T was obtained from the observed time of oscillation T_1 by means of the equation

$$T = T_1 \left/ \left(1 + \frac{(\log_e \theta_1/\theta_2)^2}{4\pi^2}\right)^{\frac{1}{2}}\right. . \qquad \ldots\ldots(134)$$

The torsion constant C of the fibre was found by removing the system from the torsion head and suspending it upside down from the hook E. A brass cylinder of calculated moment of inertia was then attached to the hook A of Fig. 63b and the constant C calculated from the period of oscillation of the disc. This avoids distorting the specimen.

The magnetic moment M was measured by means of a small magnetometer which was really a small Helmholtz galvanometer with coils UU and a small control magnet at V. The magnetometer was placed in turn on special supports in the four positions A, B, C and D, so that its needle system lay upon one of the two dotted lines shown passing through the centre of the specimen at 45 degrees to the vertical; in these positions the horizontal couple on the needle due to the specimen is a maximum. The mean deflection δ thus produced was substituted in the following equation:

$$\frac{Md}{(d^2 - \frac{1}{2}l^2)^2} + \frac{M}{2(d^2 + \frac{1}{2}l^2)^{\frac{3}{2}}} = k \tan \delta, \qquad \ldots\ldots(135)$$

where d is the distance between the centre of the specimen and that of the needle and $2l$ is the magnetic length of the specimen.

212

On passing a known current i through the coils UU, whose constant Γ was accurately found in a subsidiary experiment, a deflection ϕ was obtained, given by

$$\Gamma i = k \tan \phi. \qquad \ldots \ldots (136)$$

The value of M was then found from equations (135) and (136). The necessary experiments were however much affected by fluctuations of the earth's magnetic field and could only be made at night when the local trams were stopped.

The neutralisation of the horizontal component of the earth's field over the region occupied by the specimen was important, because the magnetic axis of the specimen was, in general, not strictly vertical, and the horizontal component of the earth's field acting upon the horizontal component of the specimen's magnetism could produce much larger rotations of the specimen than those due to the gyromagnetic effect. A large rectangular coil was therefore mounted in a vertical plane perpendicular to the meridian and with its centre coincident with that of the specimen. It carried the compensation current and fine adjustments were made by slight distortion of the sides of the coil.

The effects of any uncompensated portion of the horizontal component of the earth's field were greatly reduced by mounting the specimen so that it possessed no appreciable horizontal magnetic moment. For this purpose a little paraffin wax on the joint D of Fig. 63b was melted and the joint manipulated until the horizontal magnetic moment was very small, a special apparatus termed a "gallows" being used to facilitate the process. The presence of horizontal magnetic moment and the lack of compensation of the horizontal component of the earth's field was manifested by a change of zero after reversal of magnetisation. By artificial adjustments of the compensating field a series of values of θ_1 could be obtained for different changes of zero, and, hence, the value corresponding to no change of zero could be found by extrapolation.

In general, the field of the magnetising solenoid was not strictly vertical, so that it possessed a horizontal component which acted upon the horizontal magnetic moment of the

213

specimen. Any impulses so produced were, however, always in the same direction, as long as the magnetisation was reversed with reversal of the solenoid current, and could therefore be eliminated by taking the mean value of θ_1 for equal and opposite changes in magnetisation. The effects of permanent magnetisation of the specimen always assisted or opposed the true gyromagnetic rotation, and they were eliminated by turning the torsion head K through 180 degrees, repeating the determination of θ_1 and taking the mean of all the values of θ_1. Elimination of these effects was also carried out by noting the rotations which occurred when the magnetising current was switched on and off without reversal in direction. These were added to or subtracted from the observed values of θ_1. Permanent magnetism might have been present to some small extent in these experiments as it was not considered desirable or necessary to neutralise the vertical component of the earth's field. In all cases a linear relation between angular momentum and change in magnetic moment was found.

The final values for ρ are shown in the following table:

TABLE X

Material	Soft iron	Steel	Nickel
$p \times e/m$	1·014	1·002	1·010

Simple Resonance Methods. We now turn to resonance methods for measuring the gyromagnetic ratio. A simple resonance method was first used by Einstein and de Haas* but they failed to obtain the correct experimental value of ρ. Later, the method was successfully used by Beck† who suspended a ferromagnetic rod by a single or a double German silver wire suspension inside a vertical solenoid. The latter was supplied with alternating current whose frequency ν could be varied over narrow limits about ν_0, the natural frequency of torsional oscillation of the suspended system.

* A. Einstein and W. de Haas, *Ver. der d. Phys. Ges.* **17**, 152, 1915.
† G. Beck, *Ann. der Phys.* **60**, 109, 1919.

Let the field in the solenoid at any instant be proportional to $\cos 2\pi\nu t$, when the magnetic moment of the specimen is given by $M \cos 2\pi\nu t$, where M is the maximum value of the magnetic moment of the specimen produced by the maximum current in the solenoid. Then the rate at which angular momentum is delivered to the system owing to the gyromagnetic effect is $-\rho(2\pi M\nu \sin 2\pi\nu t)$. Hence the equation of motion of the system is

$$I\frac{d^2\theta}{dt^2} + k\frac{d\theta}{dt} + C\theta = -\rho(2\pi M\nu) \sin 2\pi\nu t, \quad \ldots\ldots(137)$$

where I is the moment of inertia of the system, k the damping factor and C the torsion constant of the suspension.

This equation is of standard form and the solution for a steady state is

$$\theta = \frac{\rho M}{\sqrt{k^2 + (2\pi\nu I - C/2\pi\nu)^2}} \sin(2\pi\nu t + \phi), \quad \ldots\ldots(138)$$

where
$$\cot\phi = \frac{2\pi\nu I - C/2\pi\nu}{k}.$$

If k is small and resonance is established, then $\nu = \nu_0$, $2\pi\nu I - C/2\pi\nu = 0$, $\phi = \pi/2$, θ is in phase with the field, and the amplitude is

$$\theta_M = \frac{\rho M}{k}. \quad \ldots\ldots(139)$$

Now, k is the damping factor and is equal to $4I\lambda\nu_0$, where λ is the logarithmic decrement. Hence

$$\rho = 4I\lambda\nu_0\theta_M/M. \quad \ldots\ldots(140)$$

Beck made many measurements in which I, λ, ν_0, M and θ_M were separately determined and found ρ by substituting in equation (140).

It is possible to avoid direct measurement of ν_0 by measuring the amplitudes corresponding to a series of values of ν in the region of ν_0, provided that k is strictly constant. For, let $b\theta_M$ be the amplitude at frequency ν, then

$$(b\theta_M)^2 = \theta_M^2 \frac{k^2}{k^2 + (2\pi\nu I - C/2\pi\nu)^2}, \quad \ldots\ldots(141)$$

$$2\pi\nu_0 I = C/2\pi\nu_0,$$

and, remembering that $\nu \simeq \frac{1}{2}(\nu + \nu_0)$, we find that

$$k = 4\pi(\nu - \nu_0) I \sqrt{\frac{b^2}{1 - b^2}} \qquad \ldots\ldots(142)$$

and

$$\rho = \frac{4\pi(\nu - \nu_0) I}{M} \sqrt{\frac{b^2}{1 - b^2}} \theta_M.$$

The frequency difference $(\nu - \nu_0)$ must, however, be accurately found, and Beck used a special frequency meter for the purpose.

The motion of the specimen was recorded photographically on a moving drum, and the damping factor obtained from the measured record. As the light passed to and from a mirror on the specimen through a gap in the solenoid, disturbances due to non-uniformity of field must have been introduced. Beck showed conclusively that the resonance oscillations were in phase with the solenoid current, except for a slight phase difference due to hysteresis, i.e. the magnetic effects were due to the rotation of negative electricity. The values of $\rho \cdot e/m$ thus obtained were 1·06 and 1·14 ± 5 per cent for iron and nickel respectively. The errors to which such measurements are liable will be discussed in the following section.

A resonance method in which the substances used were in powder form and were packed inside thin glass tubes was also used by Ray-Chaudhuri.

Modified Resonance Methods. The principle of the modified resonance methods may be followed by reference to Fig. 64, which shows the arrangement used by Sucksmith and Bates.† A thin ferromagnetic rod L was suspended vertically by a German silver wire S from the torsion head T, along the axis of the vertical magnetising solenoid H. The latter was wound on an insulating tube and supplied with current from an A.C. generator at frequencies between 24 and 80 cycles per second maintained with the aid of a stroboscope. A rigid aluminium wire D was attached to the lower end of the rod and carried the small mirrors M and a number of short magnetised needles N mounted parallel to one another on a mica vane. The needles

---

* D. P. Ray-Chaudhuri, *Ind. Journ. Phys.* **9**, 383, 1935.

† W. Sucksmith and L. F. Bates, *Roy. Soc. Proc.* A, **104**, 499, 1923.

were situated midway between a pair of small Helmholtz coils *BB*. A helix *J*, coaxial with *L*, was fixed inside *H*; it was wound upon a glass tube, and, occasionally, directly upon *L*.

A similar helix, not shown, was placed close to and parallel to *J* and was connected to it in series opposition so that the mutual inductance of *H* and the two helices in the absence of *L* was zero. The coils *BB* were connected in series with a non-inductive resistance *S*, a commutator *C* and the induction helices. By means of the switch *K* a known current measured by the milliammeter *A* could be passed through *BB*. The whole system was kept taut by a weight *W*, immersed in oil, attached to the aluminium wire by a fine silk thread which was practically without effect on the torsion control of the system. The whole was mounted inside a coil system for the neutralisation of the horizontal component of the earth's field.

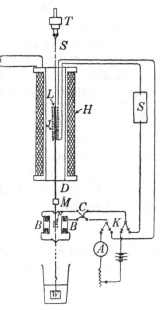

Fig. 64. Null method of measurement of the gyromagnetic ratio. (Sucksmith and Bates.)

Let us suppose that alternating current of appropriate frequency sets the specimen into resonance gyromagnetic oscillation. Since both the gyromagnetic effect and the current induced in *J* by the changes in magnetisation are directly proportional to the rate of change of the magnetisation, they will be in the same phase, provided the inductance of the circuit containing *J* is negligible. The induced current in *J* flows through the coils *BB* and by adjusting the resistance it may be arranged that the momentum delivered to the system by the action of the current on the magnet needles *N* is equal and opposite to that arising from the gyromagnetic effect; in other words, the resonance oscillation can be reduced to zero.

217

Let N_M be the number of lines of force due to the specimen's magnetism linked with the solenoid J when the magnetic moment changes by M, then

$$N_M = 4\pi(M/l)\,nl = 4\pi Mn, \qquad \ldots\ldots(143)$$

where n is the number of turns per cm. on J, which is of the same length l as the specimen. Let Γ be the couple acting on the magnets N when unit current flows in the coil BB, then the angular momentum delivered to the magnets corresponding to the change M is

$$\int \Gamma i\,dt = \Gamma \int \frac{1}{S'}\frac{dN_M}{dt}dt = \frac{\Gamma}{S'}N_M,$$

where S' is the total resistance in the induction circuit. Hence, if the value of S' when the resonance oscillations are reduced to zero is S'_0, then the total angular momentum corresponding to M is given by

$$U = \frac{\Gamma}{S'_0}N_M. \qquad \ldots\ldots(144)$$

Γ may be obtained by passing a steady current i_0 through BB via the switch K to produce a small deflection ϕ given by

$$\Gamma i = C\phi, \qquad \ldots\ldots(145)$$

where C is the torsion constant of the suspension. Eliminating Γ and N_M, we have

$$U/M = \rho = 4\pi n C\phi/S'_0 i. \qquad \ldots\ldots(146)$$

The ratio can therefore be found from a knowledge of n, C, ϕ, S'_0 and i, all of which can be measured with precision.

The great advantage of the null-resonance method lies in the fact that the final expression for ρ does not contain the frequency of the alternating current, and, consequently, does not require its determination or its accurate maintenance. Measurements of the amplitude of resonance oscillations are not required, and if exact resonance is not attained the measurements suffer only through decrease in sensitivity and not through the introduction of error. Other advantages are that the magnetic moment and moment of inertia of the system need not be found.

Errors in Resonance Measurements. We shall not discuss in detail the various sources of error which occur in resonance measurements; many have already been discussed in the description of the direct method. It is very important to make the magnetic axis of the specimen exactly vertical. If, however, this is not done and the specimen possesses a permanent horizontal magnetic moment as well as one which changes with the current, it can easily be shown that the earth's horizontal field and that of the magnetising solenoid give rise to couples in quadrature with the gyromagnetic effect, and thereby cause a spurious increase in the resonance amplitude. The effects are eliminated by adjusting the compensating current for the earth's field until minimum resonance amplitude is observed. The field of the magnetising solenoid also gives rise to a couple in phase with the gyromagnetic effect which is eliminated by turning the torsion head through 180 degrees and repeating the measurement of S_0'. If a steady displacement of zero is observed it means that the magnetic axis is by no means vertical and must be adjusted.

The effects of stray fields on the needles N and on BB can readily be compensated. Correction should be made for leakage from the induction helix; Barnett avoided it either by winding the magnetising solenoid directly upon the specimen and surrounding the whole by an induction helix or by placing the magnetising solenoid coaxially within a long induction helix. Barnett considers that magnetostriction effects give rise to vertical motions of the specimen which through asymmetry are in part transformed into axial rotations. If the vertical component of the earth's field is not compensated the magnetostriction effects are bound to be asymmetrical. As the phase relation of these disturbances with respect to the gyromagnetic ratio is unknown, they ought to disappear from the mean of a series of results over a wide range of frequencies; in any case, they could not arise in the direct method, and the results from direct and resonance methods agree.

Controlled Resonance Methods. In the preceding discussion of resonance methods the gyromagnetic couple has been denoted by an expression of the form $G \cos(\omega t - \alpha)$, so that the various disturbances in quadrature might be written equal to $X \sin(\omega t - \alpha)$. The resultant couple would be of the form $R \cos(\omega t - \alpha - \gamma)$, i.e. the oscillations are given a phase displacement γ with respect to the gyromagnetic effect. If the damping is very small the amplitude of the oscillations is given by an equation $\theta = \theta_0 \sin(\omega t - \alpha - \gamma)$, with a phase difference γ with respect to the magnetism of the specimen, disappearing when the disturbances vanish. Coeterier and Scherrer* therefore concluded that if by artificial means γ could be made zero, then the presence of the quadrature disturbances would be immaterial, because they would be unable to affect the motion of the specimen. They saw that this could be done by using a "flat-topped" alternating current reversed exactly at the moment when the specimen was passing through its equilibrium position.

The specimen was suspended by a quartz fibre and carried a mirror which reflected a narrow beam of light upon a horizontal scale in which was cut a narrow slit coinciding with the equilibrium position of the reflected beam. Light passing through the slit fell upon a photoelectric cell and activated a system of relays which reversed the current supplied to the magnetising solenoid from a battery of accumulators. The inertia of the oscillating system was small and its period was about 5 sec., while the reversal of the current took place in about 5×10^{-4} sec.

This arrangement allowed much sharper resonance oscillations to be used than were possible in other work. It would seem, however, that the artificial elimination of γ results only in a first-order reduction of the effects of the quadrature disturbances, for the effect of the small phase difference α is not considered.

The method was mainly designed for measurements with pyrrhotite, and, to some extent, results with iron powder obtained in the course of the work were not regarded as important; the value for $\rho \cdot e/m$ was 1·01. The iron or pyrrhotite

* F. Coeterier and P. Scherrer, *Helv. Phys. Acta*, 5, 217, 1932; F. Coeterier, *ibid.* 6, 483, 1933.

powder was packed in a fine glass tube and suspended by a quartz fibre inside a highly evacuated vessel to keep damping and radiometric effects low. Many of the experimental details are similar to those used earlier by Sucksmith with paramagnetics and described on p. 223. The mean value of $\rho . e/m$ obtained with pyrrhotite was 3·17, giving $g = 0·63$. Inglis* has shown on the basis of a simple model in which the orbital momenta of the $3d$ electrons responsible for ferromagnetism are orientated antiparallel to their spins, so that each M_s is $-\frac{1}{2}$ and each M_l is 2, that

$$g = (2M_s + M_l)/(M_s + M_l) = \tfrac{2}{3} = 0·67.$$

In view of the peculiar magnetic behaviour of pyrrhotite, discussed on p. 152, the agreement between experiment and theory is very close, and better could not be expected with powdered material. See Note 8.

Spin and Orbital Effects. In the following table are collected the mean values of the gyromagnetic ratio for iron or substances containing iron as obtained by several workers.

TABLE XI. COLLECTED VALUES OF $\rho . e/m$ FOR IRON

Authors	Ratio	Substance	Method
Chattock and Bates	1·00±1%	Iron	Direct rotation by magnetisation
Sucksmith and Bates	1·00±1%	Iron	Modified resonance
Coeterier	1·00±(1%?)	Iron	Controlled resonance
Ray-Chaudhuri	1·00±1%	Iron oxides	Resonance
Barnett	1·038±1%	Iron	Magnetisation by rotation
Barnett	1·032±0·3%	Iron	Modified resonance
Coeterier	3·17±1·5%	Pyrrhotite	Controlled resonance

From a theoretical point of view it would be expected that the ratio would be greater than 1, i.e. the value of g would lie between 2 and 1, if ferromagnetism were not entirely due to electron spin, but partly to spin and partly to orbital motion. The above data leave little doubt upon this point, but to obtain further information Sucksmith† measured the ratio for

* D. R. Inglis, *Phys. Rev.* **45**, 118, 1934.
† W. Sucksmith, *Helv. Phys. Acta*, **8**, 205, 1935.

a series of nickel-copper alloys with low Curie points at temperatures above the latter, using the method described in the following section. Owing to iron impurities the effects of permanent magnetism in the specimens were considerable and difficult to eliminate, and experiments had to be confined to specimens with susceptibilities less than 180×10^{-6}, so that the gyromagnetic effect was always quite small. The value of g was found to be $1 \cdot 9 \pm 0 \cdot 1$, or 2 within the limits of experimental error. We may thus safely conclude that even above the Curie point the magnetic properties of these alloys are due to spin and not to orbital motion.

Values of the ratio for other ferromagnetics are given in the following table:

TABLE XII. COLLECTED VALUES OF $\rho.e/m$ FOR FERRO-MAGNETICS OTHER THAN IRON

Material	Ratio	Authors	Method
Nickel	$1 \cdot 002 \pm 0 \cdot 004$	Sucksmith and Bates	Modified resonance
	$1 \cdot 046 \pm 0 \cdot 004$	Barnett	,,
Heusler alloy	$1 \cdot 002 \pm 0 \cdot 006$	Sucksmith and Bates	,,
Cobalt	$1 \cdot 030 \pm 0 \cdot 064$	Sucksmith	,,
	$1 \cdot 078 \pm 0 \cdot 007$	Barnett	,,
Magnetite	$0 \cdot 990 \pm 0 \cdot 048$	Sucksmith	,,
Pyrrhotite	$3 \cdot 17$	Coeterier	Controlled resonance

Discrepancy exists between the results of Barnett and the English workers. In particular the values for cobalt invite comment. Sucksmith found no special difficulty in the determination of ρ for cobalt, although, as he used powder, the accuracy was not as high as usual, whereas Barnett encountered difficulties with torques believed to be of magneto-strictive origin. The difference of some 5 per cent between the iron and cobalt values is well outside the limits of experimental error. The high value for cobalt would mean that the orbital electron moment is not completely quenched. There is evidence of this with cobalt salts, of course, but it would seem strange that orbital motion should play such a prominent part in the metal.

Sucksmith's Experiments with Paramagnetics. The discussion of gyromagnetic experiments with ferromagnetic

222

substances has shown that even in the most favourable circumstances the angular momentum to be measured is very small. The difficulties of measurement are enormously increased when paramagnetic substances are used, but Sucksmith* managed to make a series of very interesting determinations with oxides of the rare earths using the simple resonance method with special technique. His apparatus is shown in Fig. 65b. The material under investigation was tightly packed in a very thin-walled glass tube, CD in Fig. 65a, about 0·5 to

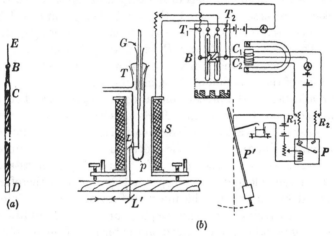

Fig. 65. Apparatus for gyromagnetic measurements with paramagnetics. (Sucksmith.)

0·6 mm. in internal diameter and about 8 cm. long, and plugged in position with a spicule of non-magnetic shellac. A small plane mirror was cemented at D and loops of glass fibre were used to attach the tube to a quartz fibre suspension, great care being taken to obtain as low a value of the moment of inertia as possible.

This system was mounted from the ground-glass joint G inside an evacuated glass vessel, and its motion was followed by light which passed to and from D via the lens L and mirror L'. The solenoid S was supplied with square wave current of

* W. Sucksmith, *Roy. Soc. Proc.* A, **128**, 276, 1930; A, **133**, 179, 1931; A, **135**, 276, 1932.

constant frequency, producing a maximum magnetic field of about 600 oersteds, by means of a battery of accumulators and a special reversing switch. The latter was controlled by an adjustible clock pendulum P', which was synchronised with the vibrations of the tube CD by projecting an image of P' on to the same scale as that which received the light reflected from D. The clock activated a series of relays which finally caused a commutator to reverse the current in the solenoid. On account of the heavy inductive load the commutator consisted of two deep channels of mercury and the wires on the rocker were cross-connected, the whole being immersed in paraffin oil to prevent fouling of the mercury surface by sparking. The resistances R_1 and R_2 regulated the time interval between "make" and "break", which was the same irrespective of the direction of the commutator.

In addition to the precautions already described earlier, it was necessary to avoid the effects of electrostatic charges produced on the walls of the specimen and the container during evacuation and retained on the specimen by the quartz fibre. Such charges were exposed to the alternating electrostatic fields of the solenoid. The suspended system was therefore screened by platinising the inner surface of the containing vessel and earthing it through the platinum wire p; in addition, screens of metal foil were placed on the former of the magnetising solenoid. Again, it was imperative to avoid mechanical disturbances by working at night, and metal supports were not used on account of the mechanical effects produced by eddy currents. By far the worse difficulties were those caused by traces of ferromagnetic impurity, which endowed the specimen with a horizontal magnetic moment whose magnitude varied with the hysteresis of the impurity. This horizontal moment was acted upon by any horizontal field of the solenoid and gave a couple, 90 degrees out of phase with the gyromagnetic effect, whose magnitude could be enormous in comparison with that of the latter. Extreme care was therefore exercised in handling and preparing the specimen and in the selection of shellac and mirror, the latter being cut with agate. Moreover, changes of temperature produced variations in the moment of inertia of the specimen, probably due to strains at the shellac joints; after a steady state was reached, the temperature was kept constant to within $0.25°$ C.

The special technique employed by Sucksmith depended upon the maintenance of the amplitude of the oscillating system with the correct phase, and a more complete mathematical discussion of the resonance method must now be given. In the case of a paramagnetic substance of mass m_0 and mass susceptibility χ placed in a field $\mathcal{H}$ the value of M is $m_0 \chi \mathcal{H}$. Now, $\mathcal{H}$ is produced by a battery current which is periodically reversed so quickly that the time taken to reverse it may be considered negligible. The magnetising current is therefore equivalent to an alternating current of square wave form. The magnetising field may therefore be written

$$\frac{4}{\pi} \mathcal{H}_m \left\{ \frac{\cos \omega t}{1} - \frac{\cos 3\omega t}{3} + \frac{\cos 5\omega t}{5} - \cdots \right\}$$

from Fourier's theorem, taking the integration limits from $+\pi/2$ to $+\pi$, where ω/π is equal to the number of times the field is reversed per second. Hence, neglecting the terms of higher frequency, equation (137) of p. 215 becomes

$$I \frac{d^2\theta}{dt^2} + k \frac{d\theta}{dt} + C\theta = -\left(\frac{1}{g} \frac{2m}{e}\right) \chi m_0 \frac{4}{\pi} \mathcal{H}_m \omega \sin \omega t. \quad \ldots\ldots(147)$$

Now, to the right-hand side of equation (147) must be added a term on account of disturbances in phase with $\mathcal{H}$; let this term be $D \cos \omega t$. Then the equation may be rewritten

$$I \frac{d^2\theta}{dt^2} + k \frac{d\theta}{dt} + C\theta = D \cos \omega t - \frac{2m}{e} \frac{4\chi m_0}{\pi g} \mathcal{H}_m \omega \sin \omega t,$$

or as $\qquad \dfrac{d^2\theta}{dt^2} + 2K \dfrac{d\theta}{dt} + p^2\theta = b \sin \omega t + d \cos \omega t$

$$= b \sin(\omega t + \gamma)/\cos \gamma, \quad \ldots\ldots(148)$$

where $\qquad 2K = k/I, \quad b = -\dfrac{8m}{e} \dfrac{\chi m_0 \mathcal{H}_m \omega}{\pi g I},$

$$d = D/I \quad \text{and} \quad \tan \gamma = d/b.$$

The solution of (148) is

$$\theta = \frac{b \sin \delta}{2K\omega \cos \gamma} \sin(\omega t + \gamma - \delta) + ae^{-Kt}\sin(\omega' t + \epsilon), \quad \ldots\ldots(149)$$

where $\tan \delta = 2K\omega/(p^2 - \omega^2)$, $\omega' = \sqrt{p^2 - K^2}$ and a and ϵ are constants. The second term on the right-hand side represents the growth of the amplitude in the early stages of the motion, and, therefore, when the oscillations have been properly established t may be considered infinitely great and the term be disregarded. Moreover, if perfect resonance is attained and no disturbing effects are present $\gamma = 0$ and $\tan \delta = \infty$, so that

$$\theta = \frac{-b \cos \omega t}{2K\omega} \cos \omega t \quad \text{and} \quad \theta_M = -b/2K\omega.$$

If, when disturbing effects are present, it is arranged that the phase of the oscillation is the same as that which would obtain at perfect resonance and $D = 0$, then

$$\sin(\omega t + \gamma - \delta) = \sin(\omega t - \pi/2), \text{ or } \delta = \pi/2 + \gamma,$$

whence

$$\theta = \frac{b \sin(\pi/2 + \gamma)}{2K\omega \cos \gamma} \sin\left(\omega t - \frac{\pi}{2}\right) = -\frac{b}{2K\omega} \cos \omega t, \quad \ldots\ldots(150)$$

which is the relation used in the simple resonance method. In other words, the true amplitude for perfect resonance may be measured even when disturbing effects are present if the oscillation is maintained in phase with the first harmonic of the magnetising current. The double amplitude $2\theta_M$ and g are given by the relations

$$2\theta_M = \frac{2b}{2K\omega} \quad \text{and} \quad g = \frac{2m}{e} \frac{2\chi m_0 T \mathscr{H}_m}{\pi \lambda I \theta_M},$$

since $K = 2\lambda/T$, where λ is the logarithmic decrement and T the period. We have already discussed how the phase difference γ was avoided in Coeterier's measurements, and we shall now describe Sucksmith's arrangements for getting rid of γ and δ.

The phases of the several effects treated above are shown in Fig. 66a, where in addition is shown the phase of the oscillations of a bifilar galvanometer connected to the secondary of a small anchor ring transformer whose primary was in series with the magnetising solenoid. With a period equal to about 2 sec. the gyromagnetic amplitude lagged some 16 degrees behind the magnetising current. The galvanometer mirror projected light on to the same scale as that used by the vibrating system and the pendulum controlling the magnetising current.

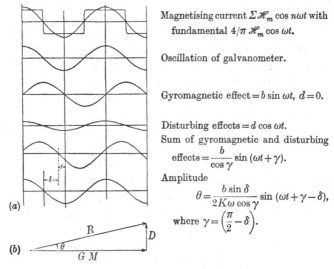

Magnetising current $\Sigma \mathscr{H}_m \cos n\omega t$ with fundamental $4/\pi \, \mathscr{H}_m \cos \omega t$.

Oscillation of galvanometer.

Gyromagnetic effect $= b \sin \omega t$, $d = 0$.

Disturbing effects $= d \cos \omega t$.
Sum of gyromagnetic and disturbing
effects $= \dfrac{b}{\cos \gamma} \sin (\omega t + \gamma)$.
Amplitude
$$\theta = \frac{b \sin \delta}{2K\omega \cos \gamma} \sin (\omega t + \gamma - \delta),$$
where $\gamma = \left(\dfrac{\pi}{2} - \delta\right)$.

Fig. 66. Phase diagram. (Sucksmith.)

Assuming that the time of response of the various relays is exceedingly small, then the pendulum is exactly in phase with the gyromagnetic amplitude, but if disturbing effects in quadrature are present then they will produce a marked phase change before a change in amplitude becomes perceptible. Thus, in Fig. 66 b, GM and D respectively represent the gyromagnetic and disturbing effects, and it is clear that while their resultant R may not be appreciably different from GM, the angle θ may be considerable. The galvanometer was used to provide the correct phase, the compensating current for the horizontal component of the earth's field being adjusted until the phase of the specimen's oscillation was the same as that of the galvanometer. The value $2\theta_M$ was then measured. Even when slight permanent magnetism due to impurity was present, its effects could be eliminated by using an incorrect value of the compensating current, which then introduced an effect equal and opposite to the effects of permanent magnetism. This was not possible when the accidental ferromagnetism of the specimen was large, and, therefore, variable, during the course of an experiment.

In the case of a tube containing dysprosium oxide Dy_2O_3 the value of $2\theta_M$ was between 2 and 4 mm. on a scale 60 cm. distant. In the expression for g the values of T and λ were found directly, and χ was determined in separate susceptibility measurements. I was found by the method of added inertia, rings cut from copper foil being used for the purpose. The limits of experimental error varied from 3 to 10 per cent.

In the last column of the following table are given some of the values of g obtained by Sucksmith, together with values of the splitting factor in the fourth column; the latter are calculated on the assumption that the state of the ion responsible for the paramagnetism is that given in the third column. The agreement between experiment and theory, in view of the great experimental difficulties, is extremely satisfactory.

TABLE XIII. g-VALUES FROM GYROMAGNETIC
EXPERIMENTS.—SUCKSMITH

Substance investigated	Para-magnetic ion	Nor-mal state	g theoretical				g experimental
			(1)	(2)	(3)	(4)	
Iron group:							
CrCl$_3$	Cr^{+++}	$^4F_{\frac{3}{2}}$	0·40	2·00	1·38	—	1·95
MnSO$_4$ MnCO$_3$	Mn^{++}	$^6S_{\frac{5}{2}}$	2·00	2·00	2·00	—	1·98
FeSO$_4$	Fe^{++}	$^5D_4$	1·50	2·00	1·67	—	1·89
CoSO$_4$ CoCl$_2$	Co^{++}	$^4F_{\frac{9}{2}}$	1·33	2·00	1·67	—	1·54
Rare earth group:							
Nd$_2$O$_3$	Nd^{+++}	$^4I_{\frac{9}{2}}$	0·73	—	—	0·76	0·78
Eu$_2$O$_3$	Eu^{+++}	$^7F_0$	—	—	—	6·56	>4·5
Gd$_2$O$_3$	Gd^{+++}	$^8S_7$	2·00	—	—	2·00	2·12
Dy$_2$O$_3$	Dy^{+++}	$^6H_{\frac{15}{2}}$	1·33	—	—	1·33	1·36

Under the theoretical values of g in section (1) are given the values calculated from the Landé splitting factor, while those under (2) are calculated on the view that the orbital momenta are completely quenched, when g must have the value for electron spin alone. The values under (3) are calculated on the assumption that the multiplet intervals are small, i.e. that L and S are independently quantised, when, as shown on p. 41, $p_{\text{eff}} = \sqrt{4S(S+1) + L(L+1)}$. Now, the experimental value of the magnetic moment per particle in unit field expressed in terms of the Bohr magneton is $\dfrac{N}{3RT}\{4S(S+1) + L(L+1)\}\mu_B$ while the corresponding value of the angular momentum is $\dfrac{N}{3RT}\{2S(S+1) + L(L+1)\}\dfrac{h}{2\pi}$, so that

$$g = \frac{4S(S+1) + L(L+1)}{2S(S+1) + L(L+1)}.$$

The experimental values for ions of the iron group suggest that the orbital moments are entirely quenched in the case of Cr^{+++} and Mn^{++}, but are only partly effected by the crystalline field in the case of Fe^{++} and Co^{++}. In the rare earth group the agreement with the Landé values under (1) is fairly satisfactory, but better agreement is given with the values under (4) from the more rigorous calculations of Van Vleck,* particularly in the case of Eu^{+++}. The main point is that there is a contribution to the magnetic moment provided by the temperature-constant paramagnetism described on p. 43, which is very important in the case of Eu^{+++} and Sm^{+++}, but there is no contribution to the measured angular momentum; hence the high value of g. See Note 9.

Experiments with Rotating Magnetic Fields. We have seen on p. 205 that when a bar of ferromagnetic material is placed in a transverse magnetic field which rotates about the axis of the bar, then the latter should show evidence of longitudinal magnetisation. This should be in the direction in which a left-handed screw would progress when rotated in the same sense as the transverse field, assuming that the motion of negative electricity is responsible for the magnetic changes. A priori it

* J. H. Van Vleck, *Electric and Magnetic Susceptibilities*, p. 255, 1932.

would be expected that such longitudinal magnetisation would be easily detected, for, assuming that the transverse field causes each elementary magnetic particle in the bar to precess with angular velocity Ω, then magnetisation equal to that acquired on the application of an external field $\mathscr{H} = \dfrac{2m}{e}\Omega$ should result in the case of orbital motion of the electrons. Accordingly, when the susceptibility of the magnetic material is k, the average value of the longitudinal magnetisation should be $\dfrac{2m}{e}\Omega k$, or, more generally, $\dfrac{2m}{ge}\Omega k$, when spin and orbital motions of the electrons are involved. So, by the use of modern high frequency technique and modern materials for which k is large, the magnetisation would be expected to reach high values. Unfortunately, the estimate just made is based on the assumption of perfectly free rotation of the magnetic carriers and must be reduced* by a factor of at least $\dfrac{3}{2}\dfrac{I_t}{I_\infty}$, where I_t is the transverse magnetisation of the bar, and I_∞ is the saturation value; I_t is very nearly equal to $\left(\dfrac{\mu-1}{\mu+1}\right)\dfrac{\mathscr{H}_t}{2\pi}$, or to $\dfrac{\mathscr{H}_t}{2\pi}$, where $\mathscr{H}_t$ is the strength of the applied transverse field. Hence, even under most favourable circumstances, the observed longitudinal magnetisation should not exceed $\dfrac{2}{g}\dfrac{m}{e}\Omega k\dfrac{3}{2}\dfrac{\mathscr{H}_t}{2\pi I_\infty}$, which is at most only a few per cent of the first estimate.

It was found by Fisher,† who was the first to carry out such experiments, and confirmed by Barnett,‡ that the longitudinal magnetisation was extremely small and often in the opposite direction to that predicted by the theory. Fisher used bars consisting of powdered iron or powdered magnetite pressed into glass tubes, while Barnett used bars built up from small discs of compressed permalloy dust or iron dust, to avoid the disturbing effects of eddy currents at the high frequencies employed.

---

* See S. J. Barnett, *Proc. Amer. Acad. Arts and Sci.* **68**, 230, 1933.

† J. W. Fisher, *Phys. Soc. Proc.* **34**, 177, 1922; *Roy. Soc. Proc.* A, **109**, 7, 1925.

‡ S. J. Barnett, *Proc. Amer. Acad. Arts and Sci.* **68**, 229, 1933.

The general experimental arrangements are as follows. A sensitive magnetometer, preferably of an astatic type with upper and lower needle systems at A and B, is encased in a series of concentric copper tubes to protect the needles from the action of high-frequency magnetic fields, while to avoid thermal effects the whole system is encased in felt. Two similar magnetic bars X and Y are mounted respectively east and west, with their centres equidistant from the lower needle system, and their axes horizontal and perpendicular to the meridian. Test coils are wound upon the bars, and flat magnetising coil systems L_1 and L_2 are used respectively to supply horizontal and vertical fields of equal intensity with a phase difference of $\pi/2$ at the near end of each bar and perpendicular to its axis. The change in the magnetometer deflection which occurs on reversing the direction of rotation of the resultant field is noted. Coils are used for testing the sensitivity of the magnetometer and control magnets. Errors due to permanent magnetism are eliminated by repeating the measurements with the bars interchanged and their directions reversed.

Fisher produced fields of the order of 100 oersteds rotating with frequencies of 2 to 5×10^4 cycles per second, using a valve oscillator, while Barnett used fields of about 15 oersteds at frequencies of 14,650 and 21,000 cycles per second. Barnett found a definite though small negative effect for permalloy in these fields. The negative results, however, do not in any way weaken the evidence for the view that the elementary magnetic particle is to be treated as a small gyrostat, for they can readily be explained, as Auwers* has shown on the basis of suggestions put forward by Bloch and Becker. The Weiss theory of ferromagnetism, p. 238, treats a piece of ferromagnetic material as made up of a large number of tiny regions or domains which are individually magnetised almost to saturation but whose magnetisation vectors are randomly orientated. Application of an external field is supposed to cause rotation of these vectors, but the conception of the rotation of whole regions in the substance in weak fields is not supported by experiment,

* O. v. Auwers, *Naturwissen.* **23**, 202, 1935.

for in very weak fields Barkhausen effects, p. 323, are not observed. In any case, Heaps* has found that magnetic hysteresis and viscosity can be present even when no Barkhausen effects are observed, which means that boundary changes can show time lags.

Bloch and Becker therefore suggested that changes in direction of the magnetisation vector can only take place at the boundary between two such domains, so that the domain whose vector is more or less in alignment with the applied field grows at the expense of its less well-aligned neighbour. The directions of such changes will be determined by the strong internal fields in the vicinity of the saturated domains, and need not directly follow the rotation of the applied field. The net contribution to the longitudinal magnetisation under these circumstances is practically zero; in other words, the Fisher experiment cannot succeed with weak fields, and the absence of longitudinal magnetisation can really be taken to lend support to the gyrostat picture, if these views are correct.

Measurements with Superconductors. The modern conception of a superconductor requires that its permeability in the superconducting state shall be zero, i.e. $B = 0$. When currents are induced in a superconductor by a magnetic field, they must be confined to its surface, and be sufficient to prevent lines of induction from passing through it. Are these currents ordinary electron currents or are they in some way connected with electron spin? I. K. Kikoin and S. V. Goobar†, who used a resonance method to measure the gyromagnetic ratio for superconducting lead in magnetic fields below the threshold value at which the resistivity of lead assumes a measurable value, found $g = 1$, which means that electron spin is not involved. Hence, we conclude that the large diamagnetism of superconductors is due to electrons moving in orbits in the crystal lattice as if they were "free" in the sense of having the ordinary values of e and m.

* C. W. Heaps, *Phys. Rev.* **54**, 288, 1938.
† I. K. Kikoin and S. V. Goobar, *Compt. Rend. Acad. Sci. U.S.S.R.* **19**, 249, 1938.

CHAPTER VIII

MAGNETIC SATURATION AND EQUATION OF STATE

Introduction to Theory of Ferromagnetism. In describing Langevin's theory of the magnetic behaviour of a substance which could be treated as a paramagnetic gas, we saw that if each gm. of the substance contains z elementary magnetic particles each of moment μ, then, when the substance is exposed to a magnetic field of intensity F, each particle behaves as if it possessed an average magnetic moment $\overline{m}$ given by

$$\frac{\overline{m}}{\mu} = \coth a - \frac{1}{a} = L(a) \qquad \ldots\ldots(151)$$

$$= \frac{a}{3} - \frac{a^3}{45} + \frac{2a^5}{945} - \ldots + \ldots,$$

where $\qquad\qquad a = \mu F/kT.$

Now, Weiss extended the theory to ferromagnetic substances by postulating the existence of an *internal field*, often termed the *Weiss molecular field*, $\nu\rho z\overline{m}$, where ν is a constant and ρ is the density of the substance, so that instead of $\mathscr{H} + \dfrac{4\pi}{3}\rho z\overline{m}$ we write

$$F = \mathscr{H} + \nu\rho z\overline{m}, \qquad \ldots\ldots(152)$$

where $\mathscr{H}$ is the applied or external field acting on the ferromagnetic, after allowance has been made for demagnetisation effects.

From equations (151) and (152) it follows that

$$\frac{\overline{m}}{\mu} = \frac{kT}{\nu\rho z\mu^2}a - \frac{\mathscr{H}}{\nu\rho z\mu}. \qquad \ldots\ldots(153)$$

Now, two cases are of special interest. First, we note that for high values of T equation (151) reduces to

$$\frac{\overline{m}}{\mu} = \frac{a}{3}, \qquad \ldots\ldots(154)$$

233

so that, eliminating a from (153) and (154), we have

$$\frac{\overline{m}}{\mu} = \frac{\mu\mathscr{H}}{3k}\frac{1}{T-\left(\dfrac{\nu\rho z\mu^2}{3k}\right)} = \frac{\mu\mathscr{H}}{3k}\frac{1}{T-\theta}, \quad \ldots\ldots(155)$$

where the parameter $\theta = \nu\rho z\mu^2/3k$ is termed the *Curie point*, or, more precisely, the *paramagnetic Curie point*, to which further reference is made on p. 256.

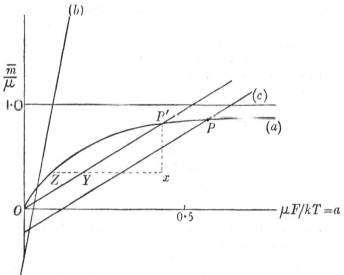

Fig. 67. Theory of ferromagnetism.

Since the molecular susceptibility of the material is given by $N.\overline{m}/\mathscr{H}$, we have from (155)

$$\chi_M = \frac{N^2\mu^2}{3R}\frac{1}{T-\theta} = \frac{C}{T-\theta}, \quad \ldots\ldots(156)$$

in other words, the substance obeys a Curie-Weiss law at temperatures above θ. By measuring the susceptibility over a range of temperatures $T > \theta$, we may therefore determine C and θ, which enables us to compute the values of μ and $\nu\rho$.

We may also consider this case graphically. In Fig. 67 curve (a) represents the Langevin function and the straight line (b) represents equation (153) when T is large. It is clear that the point of intersection of (a) and (b) changes its position considerably with the value of $\mathscr{H}$, and that when $\mathscr{H} = 0$, (a) and (b) intersect only at the origin where $\overline{m}/\mu = 0$, which shows that the substance is not magnetised in the absence of an applied field at high temperatures.

The second case arises when T is low, when, for example, (c) represents equation (153). The value of $\overline{m}/\mu$ at the point of intersection P now varies but little with the value of $\mathscr{H}$, or, at any rate, with such values of $\mathscr{H}$ as we can produce in the laboratory, and a considerable degree of magnetisation is acquired which depends markedly upon the slope of (c), i.e. upon T. The special case when $\mathscr{H} = 0$ is of the greatest interest, for, in this case the point of intersection P' shows that even when the applied field is removed magnetisation of the substance persists. It may be thought a priori that, as there is now a point of intersection at O as well, the latter represents the true state of affairs and the specimen is really unmagnetised. It is easy to show, however, that the point P' represents a stable condition whereas O does not. For, if we imagine that the magnetisation is decreased to the value at x, then this corresponds to a field appropriate to the point Z on curve (a); but we have a field corresponding to the point Y available, and therefore the magnetisation must go on increasing until the point P' is reached. In like manner we can see that if the magnetisation is supposed greater than that at P, it must decrease to the point P'; by a similar argument we can show that the point O represents an unstable condition. Of course, when the line OP' is tangential to curve (a) P' coincides with O, which is then the only stable point.

We see further that when $\mathscr{H} = 0$ equations (153) and (155) show that when $T = \nu\rho z\mu^2/3k$, (b) is tangential to (a) at the origin, so that we may say that ferromagnetism disappears at the *ferromagnetic Curie point*, θ_f. According to the present theory the ferromagnetic and paramagnetic Curie points coincide; we shall see on p. 256 that in practice they do not.

It follows, then, that in the second case the substance is *spontaneously magnetised* even when $\mathscr{H} = 0$. We denote the *intrinsic specific magnetisation* or the magnetic moment per gm.

which the substance possesses in the state required by P' by $\sigma_{0,T}$, and the value of the specific magnetisation corresponding to P by $\sigma_{\mathscr{H},T}$, while the *saturation specific magnetisation at absolute zero*, or the magnetic moment per gm. which the substance would possess in an infinitely strong field at absolute zero, when all the magnetic particles would be completely aligned, is denoted by $\sigma_{\infty,0}$. Equation (151) may now be rewritten when $\mathscr{H} = 0$ as

$$\frac{\overline{m}}{\mu} = \frac{z\overline{m}}{z\mu} = \frac{\sigma_{0,T}}{\sigma_{\infty,0}} = \coth a - \frac{1}{a}, \qquad \ldots\ldots(157)$$

and by a graphical method the relation between $\sigma_{0,T}$ and T can be found. It is clear that the ratio $\sigma_{0,T}/\sigma_{\infty,0}$ must depend only upon the ratio T/θ, for the slope of the line OP' is, from (153), given by

$$\frac{\overline{m}}{\mu} = \frac{T}{3\theta}a. \qquad \ldots\ldots(158)$$

We can thus either choose suitable values for the slope of OP' and hence draw up a table of T/θ against the corresponding values of $\overline{m}/\mu$ found by the intersection of the several lines with curve (a), or we can choose a particular value of a and by substitution in (157) we can find the corresponding value of $\overline{m}/\mu$ which together with the value of a can be substituted in (158) to obtain T/θ. Since the parameter a is thus eliminated the ratio $\sigma_{0,T}/\sigma_{\infty,0}$ is a function of T/θ alone, and the relation between them should be the same for all ferromagnetic substances, a statement which is sometimes termed the *law of corresponding states*. The experimental values of $\sigma_{0,T}/\sigma_{\infty,0}$ and T/θ are reproduced in Fig. 78, where the curve (1) represents the relation deduced from the above theory. The agreement between experiment and this theory is not good, particularly at low temperatures where $T \ll \theta$.

Now, the Langevin theory assumes that the magnetic particles may set with their axes in an infinite number of orientations with respect to the field, whereas the quantum theory permits certain discrete positions only. The right-hand side of equation (157) must therefore be replaced by an appropriate quantum expression, and when the total angular momentum of the particle is J and the splitting factor g, it follows from p. 42 that

$$\frac{\sigma_{0,T}}{\sigma_{\infty,0}} = \frac{1}{Jg\mu_B} \sum_{-J}^{+J} e^{\mu_B MgF/kT} \, Mg\mu_B \Big/ \sum_{-J}^{+J} e^{\mu_B MgF/kT}, \quad (159)$$

where $M = J, J-1, \ldots, -(J-1), -J$, and $F = \nu\rho\sigma_{0,T}$,

$$= \frac{2J+1}{2J} \coth\left(\frac{2J+1}{2J}\right) a - \frac{1}{2J} \coth\left(\frac{1}{2J}\right) a. \quad \ldots\ldots(160)$$

When $a \ll 1$ we may expand the last expression, and

$$\frac{\sigma_{0,T}}{\sigma_{\infty,0}} = \frac{J+1}{3J} a - \frac{J+1}{3J} \frac{2J^2+2J+1}{30J^2} a^3. \quad \ldots\ldots(161)$$

The slope of the tangent at the origin of the $(\sigma_{0,T}/\sigma_{\infty,0}, a)$ curve is thus equal to $(J+1)/3J$, whence we deduce that spontaneous magnetisation is possible only when $T < \theta$, where

$$\theta = \left(\frac{J+1}{3J}\right)(Jg\mu_B)^2 \frac{\nu\rho z}{k}.$$

When $T \gg \theta$ we have

$$\chi_M(T - \theta) = \frac{J+1}{3J} \frac{(NJg\mu_B)^2}{R}.$$

For the special case $J = \frac{1}{2}$, $g = 2$, $Jg = 1$, or the case when the magnetic carriers can set parallel or antiparallel to the field, equation (160) reduces to

$$\sigma_{0,T}/\sigma_{\infty,0} = \tanh(a) = \frac{T}{\theta} a,$$

since $\qquad\qquad \theta = \dfrac{\nu\rho z\mu_B^2}{k}. \qquad\qquad \ldots\ldots(162)$

The values of $\sigma_{0,T}/\sigma_{\infty,0}$ obtained from the intersection of $\sigma_{0,T}/\sigma_{\infty,0} = \tanh(a)$ and $\sigma_{0,T}/\sigma_{\infty,0} = (T/\theta) a$ when plotted against T/θ give the line (3) of Fig. 78. The agreement between experiment and theory is now much more satisfactory, although there are obvious discrepancies at low temperatures. There are other discrepancies at high temperatures which are discussed on p. 278.

Three points require emphasis. First, we have postulated that ν, or more strictly the product $\nu\rho$, is a definite constant which is unchanged by rise in temperature, and we have put forward no experimental proof for this somewhat unlikely suggestion. Secondly, the value of ν is of the order of 10^4, whereas the field which would act upon an isolated magnetic pole inside a small spherical cavity within a ferromagnetic magnetised to an intensity of magnetisation I is $\mathcal{H} + \dfrac{4\pi}{3} I$. In other words, the molecular field is some thousands of times as great as the field which can arise from forces of purely magnetic origin. It was therefore long ago concluded that the molecular field must be electrostatic in nature.

Thirdly, if a ferromagnetic is always spontaneously magnetised at room temperatures, why is it possible to demagnetise a piece of iron at these temperatures? The answer is that the Weiss theory really assumes that a ferromagnetic body is made up of a very large number of elementary *domains* or regions which are strongly magnetised even in the absence of an applied field. Therefore, at any temperature below 770° C. a piece of unmagnetised iron consists of such domains and fails to exhibit external magnetic effects, e.g. to influence a compass needle, merely because the magnetic vectors of the individual domains are so arranged or disorganised that the resultant moment of the spontaneous magnetisation vanishes. At low temperatures the effect of an applied field on a ferromagnetic specimen is simply to align the moments of the domains, but at higher temperatures, where $\sigma_{0\ T}$ is considerably less than $\sigma_{\infty,\ 0}$, the effect of the field is two-fold. It first produces an apparent increase in the bulk magnetisation of the specimen by reversal and turning processes described on p. 294 and then produces a real increase in the magnetisation of each domain. The latter increase in magnetisation is strictly reversible, and has nothing to do with the irreversible type of magnetisation associated with technical hysteresis phenomena observed in low fields. The Weiss theory provides a fairly reasonable explanation of technical hysteresis processes, but we shall not consider this aspect of the theory here.

Heisenberg's Theory of Ferromagnetism. The modern explanation of the large value of ν was provided by Heisenberg,[*] who discovered what are known as the *interchange interaction forces* or *exchange forces* of electrons in atoms, which depend upon the alignment of the electron spins in the atoms of a ferromagnetic, although the forces between the spins themselves are not responsible for ν. In the language of wave mechanics, the probability that an electron 1 in an atom a will exchange places with an electron 2 in an atom b, or the *exchange integral*, is given by

$$\int \psi_b(1)\,\psi_a(2)\left(\frac{1}{r_{ab}} - \frac{1}{r_{a_2}} - \frac{1}{r_{b_1}} + \frac{1}{r_{12}}\right)\psi_a(1)\,\psi_b(2)\,d\tau_1 d\tau_2.$$

In this integral $\psi_a(1)$ is the characteristic function of electron 1 when it is in atom a, $\psi_b(2)$ the characteristic function of electron (2) in atom b, $\psi_b(1)$ that when electron 1 is in atom b and so on; r_{12} is the distance apart of the electrons and r_{a_2}, r_{b_1} the distance between the nucleus of atom a and electron 2 and between the nucleus of atom b and electron 1. In general, this integral is negative, since the attractions between the nuclei and the electrons $(1/r_{a_2} + 1/r_{b_1})$ is greater than the repulsion between the nuclei and between the electrons $(1/r_{ab} + 1/r_{12})$. It is, however, positive when there exists a certain ratio of the distance D between neighbouring atoms in the metal crystal to the radius r of the energy shell in which the electron spins are uncompensated, a state of affairs found in the rare earths and in metals of the iron group. Slater[†] gives the following data:

Metal	Fe	Co	Ni	Cr	Mn	Gd
Ratio D/r	3·26	3·64	3·94	2·60	2·94	3·1

and concludes that for ferromagnetism to exist D/r must be greater than 1·5 but not much greater. This raises the interesting point that atoms with uncompensated electron spins which do not in the pure state exhibit ferromagnetism, because the value for D/r is not suitable, may combine with

[*] W. Heisenberg, *Zeit. für Phys.* **49**, 619, 1928.
[†] J. C. Slater, *Phys. Rev.* **36**, 57, 1930. See also N. F. Mott, *Magnetism*, p. 11, 1938, Inst. Physics.

other non-ferromagnetic elements to form crystals whose lattice constant or distance between neighbouring atoms permits a suitable value of D/r and thus give a ferromagnetic compound. In this way it is supposed that we may account for the ferromagnetism of compounds containing Mn, such as the Heusler alloys (CuAlMn), and MnAs, compounds of Cr* with H and Te, and AgF_2.† For example, metallic manganese which exhibits a paramagnetism more or less independent of temperature has a lattice constant of 2·58 Å, whereas MnAs and MnSb with 2·85 and 2·89 Å respectively are ferromagnetic. We thus gain the impression that ferromagnetism should be exhibited by far more substances than are known at present to show it.

Very simple derivations‡ have been given of the equation for $\sigma_{0,\,T}/\sigma_{\infty,\,0}$ based on the Heisenberg theory. For the exchange integral to be positive the spins must be directed parallel, and as the exchange forces fall off very rapidly with distance, account need only be taken of the interaction between neighbouring atoms. Let a particular atom be surrounded by z_1 neighbours, e.g. 8 for a body-centred and 12 for a face-centred cubic crystal. The spins will either be parallel or antiparallel. Let x be the number of parallel pairs formed with the electron of the central atom and y the number of antiparallel pairs.

Let $2E_0$ be the difference in the energy of interaction between two atoms when their spins are respectively parallel and antiparallel. Then, if we turn one atom from the parallel to the antiparallel position we increase the number of parallel pairs by $(y-x)$ and increase the number of antiparallel pairs by $(x-y)$, thereby expending an energy

$$E = 2E_0(x-y). \qquad \ldots\ldots(163)$$

Now the maximum moment due to the z_1 electrons is $z_1\mu_B$, and, if μ_r is the resultant magnetic moment of the system,

$$\frac{\mu_r}{\mu_B} = \frac{\bar{x}-\bar{y}}{z_1} = \frac{\sigma_{0,\,T}}{\sigma_{\infty,\,0}},$$

* L. F. Bates and A. Baqi, *Phys. Soc. Proc.* **48**, 781, 1936.

† W. Klemm, *Naturwissen.* **4**, 59, 1937.

‡ E. C. Stoner, *Proc. Leeds Phil. Soc.* **11**, 56, 1930; R. M. Bozorth, *Bell. Syst. Tech. Journ.* **19**, 26, 1940.

when we treat the whole crystal and take the average values $\bar{x}$ and $\bar{y}$ for x and y.

Now, from Boltzman's law, an atom will have the following probabilities P_p and P_a of being parallel and antiparallel, respectively,

$$P_p = [1 + e^{-E/kT}]^{-1}, \quad P_a = e^{-E/kT}[1 + e^{-E/kT}]^{-1}.$$

But, as all atoms behave in the same way, then, $\bar{x} = z_1 P_p$ and $\bar{y} = z_1 P_a$, so that

$$\sigma_{0,T}/\sigma_{\infty,0} = P_p - P_a = \tanh(E/2kT). \quad \dots\dots(164)$$

Now, on p. 237 we saw that the Weiss theory of the molecular field in conjunction with the Brillouin function for a paramagnetic gas gave equation (162):

$$\frac{\sigma_{0,T}}{\sigma_{\infty,0}} = \tanh a = \tanh \frac{\nu\rho z \bar{m} \mu_B}{kT} = \tanh \frac{\nu\rho\sigma_{0,T}}{kT}\mu_B. \quad \dots\dots(165)$$

Hence for agreement between (164) and (165) we must have

$$\frac{z_1 E_0}{kT}\frac{\sigma_{0,T}}{\sigma_{\infty,0}} = \frac{\nu\rho\sigma_{0,T}}{kT}\mu_B$$

or

$$(\nu\rho) = \frac{z_1 E_0}{\sigma_{\infty,0}}\frac{1}{\mu_B}$$

or

$$(\nu\rho)\sigma_{\infty,0} = \frac{z_1 E_0}{\mu_B}. \quad \dots\dots(166)$$

The quantity $(\nu\rho)$ has been bracketed to emphasise that ρ is the density of the crystal at T and not at absolute zero. The expression on the right-hand side of (166) represents the maximum value of the internal field.

We see, therefore, that in the case of a ferromagnetic we require to know the value of $\sigma_{\infty,0}$ and θ, and how the ratio $\sigma_{0,T}/\sigma_{\infty,0}$ depends upon the value of T/θ, whether agreement exists between the experimental results and the foregoing theories, and how far the law of corresponding states holds for the several ferromagnetics known to us. In the following sections will be described the experiments which have given us this information, and we will first discuss the measurement of $\sigma_{\infty,0}$.

The Approach to Saturation. In order to determine the quantity $\sigma_{\infty,0}$, and hence the atomic magnetic moment of any ferromagnetic substance, two sets of experiments are in general necessary. First, the way in which the specific magnetisation $\sigma_{\mathscr{H},\,T}$ in applied field $\mathscr{H}$ varies with $\mathscr{H}$, when T is constant, must be found. For high fields and low temperatures a relation of the form

$$\sigma_{\mathscr{H},\,T} = \sigma_{\infty,\,T}\left(1 - \frac{a}{\mathscr{H}}\right) \qquad \ldots\ldots(167)*$$

is obeyed. In equation (167), $\mathscr{H}$ is the resultant of the applied external field and the demagnetising field of the specimen used in the investigation, a is termed *the coefficient of magnetic hardness* and is usually a function of the temperature, and $\sigma_{\infty,\,T}$ is the specific magnetisation at T° K. which the specimen would acquire in an infinitely strong field. We see, then, that at least two measurements with different fields at a known temperature T are required in order to find $\sigma_{\infty,\,T}$.

Secondly, as we cannot make experiments at temperatures very close to absolute zero—perhaps it would be better to say, as such measurements have not yet been made—the quantity $\sigma_{\infty,\,0}$ must be found by extrapolation. Hence we need a series of values of $\sigma_{\infty,\,T}$ over a satisfactory range of temperatures as near to absolute zero as possible. It is found that

$$\sigma_{\mathscr{H},\,T} = \sigma_{\mathscr{H},\,0}(1 - AT^2 - BT^4 - \ldots), \qquad \ldots\ldots(168)$$

whence, for the particular case where a is not a function of the temperature, and, neglecting as usual terms of higher order than T^2, we obtain

$$\sigma_{\infty,\,T} = \sigma_{\infty,\,0}(1 - AT^2). \qquad \ldots\ldots(169)$$

When, however, this simplification is not possible, we must obtain a set of values of $\sigma_{\mathscr{H},\,T}$ for a series of different temperatures T and fixed values of $\mathscr{H}$, e.g. $\mathscr{H}_1, \mathscr{H}_2, \ldots$, and obtain by extrapolation a series of values of $\sigma_{\mathscr{H},\,0}$, e.g. $\sigma_{\mathscr{H}_1,\,0}, \sigma_{\mathscr{H}_2,\,0}, \ldots$, which, when plotted against $1/\mathscr{H}$, give a graph whose intercept at $1/\mathscr{H} = 0$ is the required value of $\sigma_{\infty,\,0}$.

Now, on the basis of the Heisenberg exchange theory of ferromagnetism Bloch† derived the equation

$$\sigma_{\infty,\,T} = \sigma_{\infty,\,0}\{1 - C(T)^{\frac{3}{2}}\}. \qquad \ldots\ldots(170)$$

* Cf. W. Steinhaus, A. Kussmann and E. Schoen, *Phys. Zeit.* **38**, 777, 1937.
† F. Bloch, *Zeit. für Phys.* **61**, 206, 1930.

It is clear that equation (170) is much more comprehensive than the others, and it is interesting that it is found particularly useful at low temperatures.

Accurate determinations of $\sigma_{\infty, T}$ for iron and nickel were first carried out by Weiss and Forrer,* and have recently been extended to lower temperatures by Fallot† in his work on the mean atomic moment of the iron atom in iron alloys. Cobalt was investigated by Allen and Constant.‡ In all cases an ellipsoid of the material under investigation was suddenly removed from an induction solenoid mounted between the pole pieces of an electromagnet, and the deflection of a ballistic galvanometer to which the solenoid was connected provided a measure of the intensity of magnetisation of the specimen.

Experiments of Weiss and Forrer. The arrangements used by Weiss and Forrer are shown in Fig. 68, where the specimen S is immersed in pentane P inside a copper vessel V, mounted inside a U-shaped Dewar flask. The arm of the flask outside the field was used to introduce liquid air to cool the copper vessel and its contents. A thermojunction T was placed close to the specimen. The latter was fixed in an ivory clamp on the handle t, provided with a spring—not shown—so that on releasing a catch the specimen was jerked upwards and out of the magnetic field. When, because of evaporation, the level of the liquid air fell below the lower extremity of V, the temperature of the pentane commenced to rise slowly. Series of magnetic measurements and thermocouple readings were made as the temperature rose, the rate of rise being increased if required by electrical heating.

The magnetic moment of the specimen was measured by means of two pairs of coils C_1 and C_2 wound as shown and connected in series opposition to a ballistic galvanometer. Each inner coil consisted of 600 turns and the outer of 200 turns. Their effective areas were arranged to be approximately equal so that the flux linked with each coil due to the above field was the same, although the total flux due to the specimen linked

* P. Weiss and R. Forrer, *Ann. de Phys.* 12, 297, 1929.
† M. Fallot, Thèse, Strasbourg, 1935.
‡ R. I. Allen and F. W. Constant, *Phys. Rev.* 44, 232, 1932.

16-2

with each coil was different; but spurious deflections of the galvanometer due to fluctuations in the magnet field were avoided. More complete elimination of such disturbances was obtained by rotating a movable coil which was also in series with C_1 and C_2.

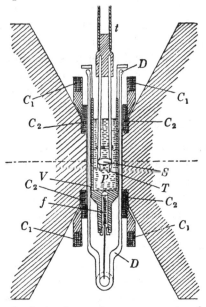

Fig. 68. Apparatus for low temperature measurement on ferromagnetics. (Weiss and Forrer.)

To standardise the galvanometer deflections a known current was reversed in a small water-cooled coil placed between the induction coils C_1 and C_2; this corresponded to measuring the deflection for a known change of magnetic moment, equal to twice that of the magnetic shell to which the calibrating coil was equivalent. Unfortunately, this was small compared with the magnetic moment of the specimen if coil and specimen had approximately the same dimensions. The standardisation was checked by measuring the magnetic moment acquired by a known quantity of manganese pyrophosphate contained in a thin-walled glass sphere of known susceptibility. The field was measured by a standardised search coil; the quantity $\mathscr{H}$ in the above equations is, of course, the resultant field obtained from the latter measurement after correction for the known demagnetisation of the ellipsoid.

Weiss and Forrer found the values given in the following table for the quantity a of equation (167) and for the specific magnetisation in a field of 17,000 oersteds for a specimen of vacuum-smelted electrolytic iron.

TABLE XIV

T	a	$\sigma_{17,000,\,T}$
288° K.	6·3	217·72
200	5·9	219·73
100	5·9	221·14
0	5·6*	221·23*

* Extrapolated.

At the same time the mean value of $\sigma_{\infty,\,0}/\sigma_{\infty,\,288}$ was found to be 1·0183, and the mean value of $\sigma_{\infty,\,288}$ for several determinations 217·79. Hence $\sigma_{\infty,\,0}$ was equal to $217·79 \times 1·0183 = 221·74$ e.m. units per gm., and the saturated magnetic moment per gm. atom at 0° absolute was 12382 e.m. units.

Fallot's Experiments. Fallot extended the Weiss and Forrer measurements to liquid hydrogen temperatures using the apparatus shown in Fig. 69 a, based on Ruhemann's† idea of liquefying hydrogen in the actual apparatus it is desired to cool. This idea has recently been extensively employed, as described in Chapter IX. To the ends of a pair of long annular spiral tubes S', forming a heat exchanger, was joined the chamber A, shown on a larger scale in Fig. 69 b, in which hydrogen was liquefied. Hydrogen under pressure entered the apparatus through the tube T_1 and passed through a spiral copper tube immersed in a vessel V containing liquid air. It proceeded through the inner portion of S' and entered A through a narrow opening O, where it suddenly expanded and liquefied. The unliquefied gas passed through the outer portion of S' and was evacuated through the german silver tube T_2 and a gas meter.

The chamber D, in which the specimen of ferromagnetic was placed, was soldered to the floor of A. A brass cone C was

† M. Ruhemann, *Zeit. für Phys.* 65, 67, 1930.

attached to V and formed a ground metal joint with a brass block attached to the outer brass container B. The cone C carried a side piece F which permitted the whole apparatus to be rotated rapidly through some 30 degrees about a horizontal

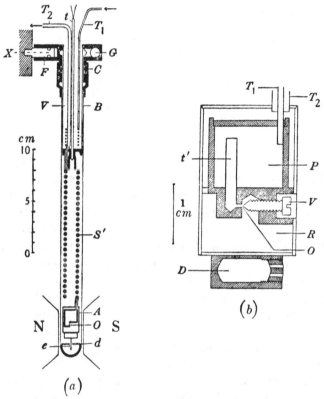

Fig. 69. Apparatus for low temperature measurements on ferromagnetics. (Fallot.)

axis X, and so to remove the specimen from the field between the pole tips NS. A brass pin e, fitting in a thin ivory disc attached to an ebonite support, prevented displacement of the inner components of the apparatus during the rotation. Through an orifice G in the block attached to C the space between V and A and the container B was highly evacuated. The leads of a thermojunction soldered to A passed through the tube t in which they were waxed, and allowed the temperature of A to be measured directly on a galvanometer. Enough liquid hydrogen could be stored in A to keep its temperature constant for three-quarters of an hour when a Dewar flask surrounded B. The temperature rose quite slowly when the hydrogen evaporated and measurements of the magnetic moment and the temperature were simultaneously made as in

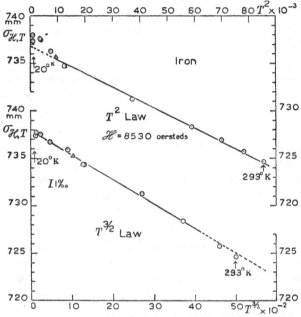

Fig. 70. Fallot's results for iron, $\sigma_{\mathscr{H}, T}$ as function of T^2 and $T^{\frac{3}{2}}$.

the Weiss and Forrer measurements. A number of very important and interesting corrections have to be applied to the measurements; these are discussed in the following section.

Representative results obtained by Fallot for iron are given in Fig. 70, in which the experimental values for $\sigma_{\mathscr{H}, T}$ are plotted as functions of $T^{\frac{3}{2}}$ and T^2, with the stated values of $\mathscr{H}$. The graphs show clearly that the $T^{\frac{3}{2}}$ law more satisfactorily represents the results at the lower temperatures, but that if $\sigma_{\mathscr{H}, 0}$ is to be found by extrapolation from results obtained

247

at the higher temperatures only, then the T^2 law is the more reliable; this was very clearly the case with nickel. This is an important point in measurements with alloys, for, when they have high Curie points Fallot uses the $T^{\frac{3}{2}}$ law, but when they have low Curie points, say between 100° and 500° C., he uses the T^2 law; when the Curie point is very low he uses the T^2 law and extrapolates from measurements over a somewhat restricted temperature range; the measurements with alloys are discussed on p. 252. Fallot found that $\sigma_{\infty, 0}/\sigma_{\infty, 288}$ was equal to 1·0196 in the case of iron. In the case of nickel his value for $\sigma_{\infty, 0}/\sigma_{\infty, 288}$ was 1·0570.

Image Effects. One of the most important corrections to be applied to the above measurements arises because of the existence of *magnetic images*. The magnetised specimen induces magnetic charges or images, analogous to those envisaged by Kelvin in dealing with electrostatic problems, in the pole tips between which it is placed. For example, if the permeability of the pole tips were infinite then the images would be equivalent to multiple "optical" images of the specimen's magnetism. Consequently, the deflection of the ballistic galvanometer is increased because of the additional lines of force between the specimen and its images. The deflection still measures the intensity of magnetisation of the specimen, of course, but the constant of proportionality is changed. Naturally, the effects produced are the greater the more permeable or less saturated the pole tips.

The effects were investigated by Weiss and Forrer who replaced the specimen by a special coil of many turns with the same volume as the specimen. The coil was wound on a copper former which was water cooled to enable it to carry a heavy current. The coil was then equivalent to a magnet and produced images in the pole tips. On reversing the current a galvanometer deflection was obtained; in the absence of image effects this would be independent of the field due to the electromagnet itself, i.e. of the saturation of the pole tips. With pole tips of high permeability producing fields up to 6000 oersteds, the correction to be applied for change in the factor of proportionality was very small and could be obtained with some precision with fields between 6000 and 11,000 oersteds, when the factor had to be increased by some few per cent. For fields above 11,000 oersteds it was large and uncertain, and with magnetically soft pole tips it is wise to limit intensity measurements to fields below 10,000 gauss. Allen and Constant found the image effect to be very large in their measurements on cobalt. Corrections for residual magnetism of the specimen producing effects after it has been removed from the field, and for the paramagnetism of the container used in Fallot's experiment, are also to be applied; they are, however, small compared with the image effect.

Domain Size Effect. An error to which practically no attention has been paid by any experimenter is bound to arise when a ferromagnetic contains a relatively small number of domains much smaller in volume than the average. Experiments on the Barkhausen effect, cf. p. 323, by Bozorth and Dillinger* showed that the discontinuities on the steeper parts of the ordinary technical hysteresis curve nearly all correspond to reversals of magnetisation in domains about 10^{-8} to 10^{-9} c.c. in volume and containing about 10^{15} to 10^{14} atoms in the ferromagnetic materials examined. However, domains containing much smaller numbers of atoms must exist. Stoner† first drew attention to some important magnetic effects which must arise if only a small fraction of a ferromagnetic is in the form of domains containing merely some 10^3 or 10^4 atoms, for these small domains will act as a trace of an impurity of extremely high magnetic moment. Stoner's argument, on account of the importance of the effect, will now be given *in extenso*.

Let $\sigma_{0, T}$ be the intrinsic specific magnetisation, μ_d the magnetic moment of a domain containing n_d atoms, and $\sigma'_{0, T}$ the apparent intrinsic magnetisation which we measure by experiment. Let A be the atomic weight and N the number of atoms in 1 gm. atom, then

$$\mu_d = \sigma_{0, T} A n_d / N, \qquad \ldots\ldots(171)$$

* R. Bozorth and J. Dillinger, *Phys. Rev.* **34**, 772, 1929 and **35**, 733, 1930.
† E. C. Stoner, *Roy. Soc. Phil. Trans.* A, **235**, 165, 1936.

whence, for nickel at room temperature we find

$$\mu_d = 5\cdot 3 \times 10^{-21} n_d. \qquad \text{......(172)}$$

When n_d is large so many orientations of μ_d in an applied field will be possible that the Langevin expression may be used to calculate the average moment $\bar{\mu}_d$ resolved parallel to the field, i.e.

$$\bar{\mu}_d/\mu_d = \coth a' - 1/a', \qquad \text{......(173)}$$

where

$$a' = (\mu_d/kT)\left(\mathscr{H} + \frac{4\pi}{3}\rho\sigma'_{\mathscr{H},T}\right) \approx \frac{\mu_d}{kT}\mathscr{H}$$

$$\approx 3\cdot 9 \times 10^{-5} n_d . \mathscr{H}/T, \qquad \text{......(174)}$$

since $\dfrac{4\pi}{3}\rho\sigma'_{\mathscr{H},T}$ may be neglected in comparison with $\mathscr{H}$ in the present discussion.

When $a' \gg 1$, equation (173) reduces to

$$\frac{\bar{\mu}_d}{\mu_d} = 1 - \frac{1}{a'} = \frac{\bar{\sigma}_{\mathscr{H},T}}{\sigma_{0,T}}, \qquad \text{......(175)}$$

where $\bar{\sigma}_{\mathscr{H},T}$ is the mean specific magnetisation corresponding to $\bar{\mu}_d$, so that from (174) we deduce that

$$n_d = \frac{T}{3\cdot 9 \times 10^{-5}\mathscr{H}}\left(\frac{\mu_d}{\mu_d - \bar{\mu}_d}\right) > \frac{T}{3\cdot 9 \times 10^{-5}\mathscr{H}}, \qquad \text{......(176)}$$

whence for $T = 300°$K. and $\mathscr{H} = 10^4$, $n_d > 10^3$. Now, if f_d represents $\sigma'_{\mathscr{H},T}/\bar{\sigma}_{\mathscr{H},T}$, then the *differential susceptibility* due to the domain size effect is given by

$$\frac{\partial\sigma'_{\mathscr{H},T}}{\partial\mathscr{H}} = \frac{\partial}{\partial\mathscr{H}}(f_d\bar{\sigma}_{\mathscr{H},T}) = f_d\frac{\sigma_{0,T}}{(a')^2}\frac{\partial a'}{\partial\mathscr{H}} = f_d . \sigma_{0,T}/a'\mathscr{H},$$

as long as (176) is true. Therefore, from (171) and (176)

$$\frac{\partial\sigma'_{\mathscr{H},T}}{\partial\mathscr{H}} \approx f_d\sigma_{0,T}\, RT/(\sigma_{0,T}An_d\mathscr{H}^2)$$

$$\approx 1\cdot 4 \times 10^6 f_d\, T/(n_d\mathscr{H}^2). \qquad \text{......(177)}$$

Hence, for

$$\frac{\partial\sigma'_{\mathscr{H},T}}{\partial\mathscr{H}} > 10^{-5},$$

$$\frac{n_d}{f_d} < \frac{1\cdot 4 \times 10^{11}T}{\mathscr{H}^2}, \qquad \text{......(178)}$$

which shows that when n_d is large f_d will be comparatively small, and the larger domains will make no appreciable contribution to $\partial\sigma'_{\mathscr{H},T}/\partial\mathscr{H}$, except for a real increase in the specific magnetisation, when $\mathscr{H}$ is large. Taking now the maximum possible value, namely $f_d = 1$, we find for $T = 300°$K. that $\partial\sigma'_{\mathscr{H},T}/\partial\mathscr{H}$ must be less than 10^{-6} if we choose $n_d > 10^7$ and $\mathscr{H} = 10^4$; in like manner $\partial\sigma'_{\mathscr{H},T}/\partial\mathscr{H}$ will again be less than 10^{-6} if we choose $n_d > 10^9$ and $\mathscr{H} = 10^3$. On the other hand, if we take the experimental value of $\partial\sigma'_{\mathscr{H},T}/\partial\mathscr{H}$ for nickel at $T = 300°$ K. and $\mathscr{H} = 10^4$ to be $1\cdot5 \times 10^{-5}$, then the whole of the differential susceptibility would be accounted for with $f_d = 0\cdot01$ and $n_d = 3 \times 10^3$.

Turning then to equation (167), $\sigma_{\mathscr{H},T} = \sigma_{\infty,T}(1 - a/\mathscr{H})$, we see that the domain size effect has an important bearing. So far, no experiments, with the possible exception of those of Gerloff, have been carried out with the domain size effect in mind. Until a method of eliminating this effect has been worked out, estimations of the way in which the specific magnetisation of a ferromagnetic varies with the applied field are bound to be subject to considerable error, although it is unlikely that extrapolated values of $\sigma_{0,T}$, obtained by producing back the graph of $\sigma_{\mathscr{H},T}$ for high values of $\mathscr{H}$ to cut the axis of ordinates, will be seriously incorrect.

Parasitic Magnetism. A kind of anomalous magnetisation whose origin is not fully understood has been recorded in the case of certain specimens of nickel and equation (167) of p. 242 has to be replaced by another of the form

$$\sigma_{\mathscr{H},T} = \sigma_{\infty,T}\left(1 - \frac{a}{\mathscr{H}} + b\mathscr{H}\right) \qquad \ldots\ldots(179)$$

in which, for ordinary temperatures $a = 20$, $b = 0\cdot2 \times 10^{-6}$. The term $b\mathscr{H}$ represents a magnetisation, due to some paramagnetic phenomenon, which is termed *parasitic magnetism* (*aimantation parasite*). Equation (179) can be true only over limited temperature ranges. The effect is important only when $\mathscr{H} > 20,000$ oersteds, but Fallot allowed for it in obtaining $\sigma_{\infty,0}$ for nickel. When we plot $\sigma_{\mathscr{H},T}$ as a function of $1/\mathscr{H}$, equation (179) gives a hyperbola to which equation (167) forms an asymptote; hence by drawing the asymptote to the experimental curve of $\sigma_{\mathscr{H},\,T}$, $1/\mathscr{H}$ the intercept $\sigma_{\infty,\,T}$ can be obtained without serious error. This procedure was also necessary with all the nickel alloys investigated by Sadron.* Fortunately, the parasitic magnetisation decreases rapidly with temperature (yet it is supposed to be of paramagnetic origin!), and hence extrapolation to $\sigma_{\infty,\,0}$ effectively takes account of the whole ferromagnetic moment of the specimen. In view, however, of the difficulty with which many alloys can be saturated, it is unwise to attach too much importance to the phenomenon at present.

Ferromagnetic Gadolinium. Metallic gadolinium has recently been isolated† from its chloride, some 0·12 gm. with an impurity, mainly silicon, of 0·73 per cent being available. The ferromagnetic Curie point of this specimen has been found‡ to be 16° C. Its specific magnetisation obeys equation (167) with a value of a greater than 1250, a value 200 times as great as that for iron. Yet in spite of its great magnetic hardness, gadolinium exhibits a greater specific magnetisation than iron in high fields. It also obeys equation (169), p. 242, the value of $\sigma_{\infty,\,0}$ being 253·5 e.m. units per gm.; hence its mean atomic moment is 3·2 times as great as that of iron. The differences between the gadolinium $\sigma_{\mathscr{H},\,T}$, $\mathscr{H}$ curves and those of other pure ferromagnetics are considerable and will, no doubt, receive further consideration. Between 90° and 360° C. it follows a Curie-Weiss law with $\theta = 29\cdot5°$ C.

Experiments with Ferromagnetic Alloys Experiments on the saturation of alloys at absolute zero may be considered as attempts to answer two questions, viz., whether the internal field in a ferromagnetic acts upon an atom of another metal dissolved in it in such a way that it exhibits a magnetic moment characteristic of the dissolved atom, and whether the atomic moment of a ferromagnetic is in any way affected by the presence of foreign atoms dissolved in it.

* C. Sadron, Thèse, Strasbourg, 1932.
† F. Trombe, *Compt. Rend.* **200**, 459, 1935; **201**, 656, 1935.
‡ G. Urbain, P. Weiss and F. Trombe, *Compt. Rend.* **200**, 2132, 1935.

The first question was tackled by Sadron* and the second by Fallot, using the methods already described. Sadron first found the mean atomic moment of an element dissolved in either nickel or cobalt, for series of elements in the same column of the periodic table, using the T^2 law of approach to saturation. Thus he found the mean atomic moments for a series of alloys of different concentrations of chromium, molybdenum and tungsten dissolved in nickel; his results for the chromium

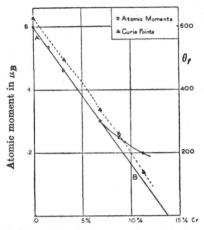

Fig. 71. Results with nickel-chromium alloys.

alloys are shown in Fig. 71. By producing the line AB until it cut the ordinate at a concentration of 100 per cent chromium he deduced that the chromium atom in this state possessed an atomic moment of $-(3 \cdot 8 \pm 0 \cdot 14)$ Bohr magnetons. In like manner he deduced that molybdenum and tungsten possessed $+3 \cdot 8$ Bohr magnetons within the limits of experimental error. Again, copper dissolved in nickel, and gold in cobalt, as if both non-ferromagnetic atoms possessed $0 \cdot 4$ Bohr magnetons; silicon dissolved in cobalt and tin in nickel with $3 \cdot 6$; vanadium and antimony in nickel with $4 \cdot 6$ and $4 \cdot 9$ respectively, values which were considered equal within the limits of experimental error. The case of manganese was interesting, for it dissolved in cobalt with a negative moment of -3 magnetons, in nickel

* C. Sadron, *Ann. de Phys.* **17**, 371, 1932; Thèse, Strasbourg, 1932.

with $+3$, and, apparently, in iron with $+0\cdot2$. The positive and negative signs attached to the moments of chromium and manganese were taken to show that these atoms set either parallel or antiparallel to the atomic field in the ferromagnetic. From these and other results was deduced the rule that the atomic moments of metals, in the same column of the periodic table, are equal when dissolved in nickel and cobalt.

Sadron also investigated the behaviour of dissolved elements belonging to the same horizontal row of the periodic table as nickel and cobalt, and his results are shown in Fig. 72, from

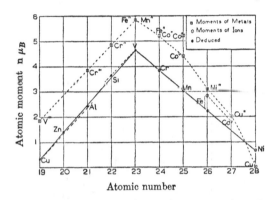

Fig. 72. Sadron's results for atomic moments of metals
dissolved in Co and Ni.

which it is seen that the atomic moments of the dissolved metals vary from element to element in a manner very similar to that in which the moments of the corresponding ions in aqueous solutions vary. The four results shown by the dots were deduced on the basis of the first rule. In discussing the validity of these rules it must be pointed out that the necessary extrapolations are often made with very dilute solutions of the metals, and it is easy to show that the smaller the concentration the greater is the uncertainty in the calculated value of the atomic moment.

Now, iron was found unsuitable as a solvent metal in Sadron's work, but it was used by Fallot in his examination of the second question. He measured the values of $\sigma_{\infty,0}$ for a series of iron alloys, using the T^2 or $T^{\frac{3}{2}}$ law according to the Curie point

254

found for the alloy. The mean atomic moments for a series of silicon-iron alloys are shown in Fig. 73, together with their Curie points. When the line AB is produced, it cuts the ordinate at 100 per cent silicon at zero atomic moment, thus showing that silicon combines with iron in the range 0 to 6 atoms per cent as if it possessed zero magnetic moment. Now, from the more rapid slopes of the portions BC, CD, etc. of the graph, it appears that the silicon atom possesses the property

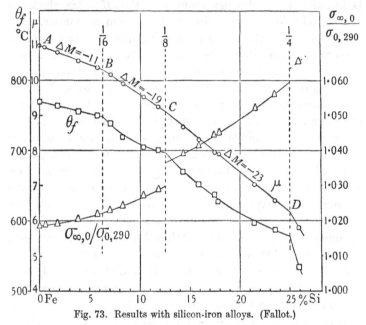

Fig. 73. Results with silicon-iron alloys. (Fallot.)

of reducing the magnetic effect of the iron atom in these alloys. Similar statements hold for aluminium and chromium, whereas gold and tin appear to increase the magnetic effect of the iron atom. Fallot considers that superstructures, such as Fe_3Si, Fe_7Si and $Fe_{15}Si$, occurring respectively at D, C and B, are easily made manifest on the curve of the ferromagnetic Curie point θ_f plotted against concentration, which is also shown in the figure. (μ is expressed in Weiss magnetons.)

Mott's Explanation of the Magnetic Properties of Alloys. The above results on alloys have recently been explained by

Mott* on the basis of a simple model in which is postulated a band of energy states corresponding to the 3d and another band corresponding to the 4s states of the atom. In the nickel atom these bands overlap so that there may be p electrons per atom in the s band and p gaps in the d band, where p is about 0·6 per atom. For cobalt and iron p is 0·7 and 0·2 respectively. Mott considers that the maximum binding energy is obtained in these alloys when the number of electrons in the 4s band is likewise about 0·6 per atom, and that if extra electrons are introduced by the addition of an atom of copper, zinc, aluminium or tin they must go into the 3d band as long as there is room for them. Here they can only take up states with spins oppositely directed to the direction of magnetisation, producing a diminution of the magnetic moment by 1, 2, 3 or 4 Bohr magnetons for the above metals respectively, in complete agreement with Sadron's results and some early work of Alder on copper-nickel alloys. Now, since the manganese atom has three less electrons than nickel, the addition of a manganese atom will decrease the number of electrons in the 3d band by 3 and so produce an increase of 3 magnetons in the magnetic moment. No explanation can however be given for the decrease found with cobalt, but the experimental results are meagre and it is not certain that the alloy specimens so far used have been solid solutions. Again, no change in the saturation moment occurs when palladium is added to nickel, presumably because both atoms have about 0·6 electron per atom in the s band. See Chapter XII.

Determination of Curie Points. The term Curie point is frequently not sufficiently precise,† and in practice we distinguish between the ferromagnetic Curie point, θ_f, and the paramagnetic Curie point, θ, of a substance; there is also some evidence that another temperature, θ_r, the temperature at which the coercive force of a ferromagnetic becomes zero, may be of importance; but it is not yet proven whether θ and θ_r do or do not coincide. The ferromagnetic Curie point is that

---

* N. F. Mott, *Phys. Soc. Proc.* **47**, 571, 1935.

† Cf. R. Forrer, *Journ. de Phys.* **1**, 49, 1930 and L. F. Bates, *Phys. Soc. Proc.* **43**, 87, 1931.

temperature at which the substance loses its spontaneous magnetisation or more precisely the point where the $\sigma_{0,\ T}^2$, T curve cuts the temperature axis, while the paramagnetic Curie point is the temperature, or parameter, which occurs in the Curie-Weiss equation representing the behaviour of the substance at temperatures well above θ_f; there is generally a difference of some $10°$ or $15°$ C. between the two points. θ is of course determined from susceptibility-temperature measurements of the kind described in Chapter III. The same apparatus, though in rather less sensitive form, may be used for the determination of θ_f, of which a knowledge is most important, as we have already seen. Thus Sadron, in his experiments on ferromagnetic alloys, employed the translation balance devised by Föex and Forrer with a non-uniform but feeble field of about 50 oersteds, obtaining the values of θ_f with an accuracy of about $\pm\,4°$ C.; the mean of the temperatures at which the ferromagnetism disappeared and reappeared was taken as the correct value of θ_f.

In computing θ it must be assumed that the graph of $1/\chi$ against T is linear. This is usually not the case unless $T \gg \theta_f$. For example, the graph for nickel is concave towards the $1/\chi$ axis below $500°$ C., but is linear between 500 and $850°$ C., giving $\theta = 377$, according to Sucksmith and Pearce*. These workers also found that iron 99·97 per cent pure, to which a little vanadium had been added to inhibit the formation of a γ phase between 920 and $1390°$ C., gave a linear graph above $900°$ C. with $\theta = 828°$ C. Cobalt 99·8 per cent pure gave a linear graph over the range 1230 to $1450°$ C. with θ lying between 1130 and $1155°$ C. according to the degree of annealing to which the specimen had been exposed. The relevant values of μ calculated for iron, nickel and cobalt from the slopes of these linear graphs were 3·15, 1·61 and $3·15\,\mu_B$. These magneton numbers are considerably greater than those found in low temperature measurements and shown in Table XV, but the discrepancy has been neatly explained by Stoner† on the

---

* W. Sucksmith and R. R. Pearce, *Roy. Soc. Proc.* A, **167**, 189, 1938.

† E. C. Stoner, *Roy. Soc. Proc.* A, **165**, 372, 1938; and *Leeds Philos. Soc. Proc.* **3**, 457, 1938.

basis of a collective electron treatment of ferromagnetism using Fermi-Dirac statistics.

In general, the remanent magnetism possessed by a ferromagnetic which has been placed in a strong field is sufficient to give satisfactory deflections of a sensitive astatic magnetometer system. The specimen can then be placed inside a copper block furnace heated by alternating current. The sudden changes in the magnetometer deflection when the temperature of the specimen is in the neighbourhood of θ_f are sufficient to enable the point to be fixed within 2° to 5° C. Ochsenfeld* used such an arrangement in his study of ferromagnetic alloys of chromium and tellurium and of compounds of manganese and nitrogen, and Forrer† used it in a study of the magnetic properties of specially treated nickel wires. The method has the disadvantage that only the temperature at which the magnetism disappears can be recorded, for, in the absence of a magnetic field, permanent magnetism does not reappear until the temperature has fallen well below θ_f in many cases. An astatic magnetometer is essential, as the measurements are usually made in institutions where electromagnets are likely to be in continual operation. It is, however, desirable to bear in mind that remanence is not necessarily to be associated with spontaneous magnetism, for a single crystal exhibits very little, and the last method may not be reliable. Its reliability must also come into question when the Curie point is so high that thermal diffusion occurs.

Bates, Gibbs and Reddi Pantulu‡ have described a simple oscillation method which gives results correct to $\pm 2°$ C. The specimen is fixed in an evacuated pyrex tube and suspended with its axis horizontal from a glass fibre attached to a fine phosphor bronze suspension, and allowed to oscillate in a field of some 120 oersteds. The period t is obtained for a series of different temperatures, and as $1/t^2$ is almost directly proportional to the magnetic moment of the specimen, marked changes in $1/t^2$ occur at θ_f.

* R. Ochsenfeld, *Ann. der Phys.* **12**, 353, 1932.

† R. Forrer, *Journ. de Phys.* **10**, 251, 1929. See also Chapter x, p. 324.

‡ L. F. Bates, R. E. Gibbs and D. Reddi Pantulu, *Phys. Soc. Proc.* **48**, 665, 1936.

With many alloys the ferromagnetism falls very rapidly at first but "tails off" before it completely disappears. This behaviour, see p. 279, is far more general than is commonly believed, even in the case of pure metals. The value of θ_f is then computed by producing the rapidly descending portion of the curve of magnetisation with temperature to cut the temperature axis; in fact, the procedure followed is like that used in finding the range of a group of α-particles in air.

All alloys used in magnetic work should be produced by fusing the necessary constituents together *in vacuo* in an induction furnace,* when, because of the convection currents set up in the molten material, the preparations are usually homogeneous. An examination of the θ_f of any alloy provides a fairly conclusive test of its homogeneity, for the presence of foreign or uncombined constituents is characterised by the occurrence of more than one θ_f for a specimen. Alloys should always be sealed in an evacuated tube and annealed for some hours in an ordinary furnace.

Magnetisation at High Temperatures. The experiments we have so far considered have given information concerning the behaviour of $\sigma_{0,\,T}$ in low temperature ranges, from 20° to about 300° K., and we must now discuss measurements at higher temperatures where T approaches θ_f. Such experiments have only been made with the elements iron and nickel. Weiss and Forrer† measured the intensity of magnetisation of a nickel sphere by the ballistic method, but instead of displacing it laterally from the gap of an electromagnet, they pulled the specimen and the furnace containing it through a hole into one of the pole pieces. Their apparatus is shown diagrammatically in Fig. 74. The magnet had large truncated pole tips pierced axially with holes 1·5 cm. in diameter. Yet, with a gap 2·5 cm. wide it was considered that the field over the whole region occupied by the nickel sphere was uniform to within 1 or 2 per cent.

The sphere Ni was heated in an electric furnace of special

* See *G.E.C. Journ.* **7**, No. 2, 1936.
† P. Weiss and R. Forrer, *Ann. de Phys.* **5**, 153, 1926. See also W. W. Stifler, *Phys. Rev.* **33**, 268, 1911.

design, made of meerschaum, wound with a non-inductive spiral of platinum wire, and water-jacketed. The brass tubes L and L' carried the heating current and through L' emerged the insulated wires of the thermocouple of which one junction was formed by forcing a silver and a constantan wire into conical holes in the nickel and soldering them in position. The whole furnace was pulled to the right by a cord attached to L' when it was desired to remove the specimen from the gap, and was restored to its normal position by a spring.

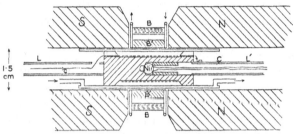

Fig. 74. Weiss and Forrer's apparatus for high temperature measurements.

The intensity of magnetisation was measured at different field strengths with the induction coils B and B' arranged as described on p. 243, the sensitivity of the ballistic galvanometer being varied over a wide range to permit accurate measurements as the magnetisation became progressively smaller at high temperatures. The relative accuracy of the measurements was about 1 part in 1000, but the absolute values might be wrong by as much as 1 per cent, owing to differences in the purity and thermal history of the specimens.

As the nickel specimen was not truly spherical the demagnetisation factor was found experimentally, using the knowledge that for very weak intensities of magnetisation the field inside the specimen was zero. With a pure specimen the uncorrected I, $\mathscr{H}$ curve was a straight line even up to half saturation values of I, and its slope was equal to the demagnetisation factor. Corrections were applied for the magnetic properties of the furnace and the image effect.

In this way Weiss and Forrer determined the isotherms of nickel, i.e. the $\sigma_{\mathscr{H}, T}$, $\mathscr{H}$ curves for a series of known temperatures between 20° and 630·4° C. Some of these isotherms are given in Fig. 75. There is no question of hysteresis effects in these experiments, for hysteresis phenomena are confined to such small field values that the hysteresis curves would practically coincide with the axis of ordinates. At the lower temperatures it is seen that $\sigma_{\mathcal{H}, T}$ increases proportionally to

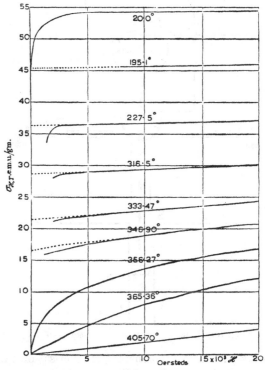

Fig. 75. $\sigma_{\mathcal{H}, T}$, $\mathcal{H}$ curves for nickel. (Weiss and Forrer.)

the field when the latter is strong. The curves show how difficult it is to determine the values of $\sigma_{0, T}$ with any certainty at temperatures in the neighbourhood of the Curie point, or, indeed, to discover from them the value of θ_f with an accuracy greater than 2° C. It may be argued that the isotherm at 346·90° C. shows the existence of spontaneous magnetisation as measured by extrapolation along the dotted line, while that at 356·27° C., being continuously curved, exhibits very much less, but this distinction is not sufficiently definite and really depends on preconceived ideas as to how $\sigma_{\mathscr{H}, T}$ ought to vary with $\mathscr{H}$. In fact, we conclude that the determination of $\sigma_{0, T}$ from $\sigma_{\mathscr{H}, T}$ curves is unreliable at temperatures much higher than $0 \cdot 9\theta_f$, and we shall see in the next chapter how reliable values can be obtained. As Weiss and Forrer showed,

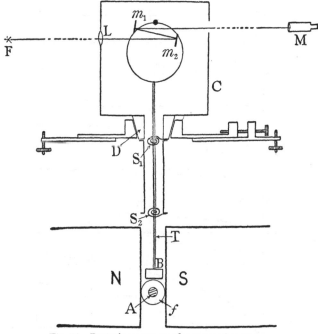

Fig. 76. Potter's apparatus for measuring $\sigma_{\mathscr{H}, T}$.

a series of $\mathscr{H}$, T curves for constant magnetisation can be obtained from the isothermals of Fig. 75, and, by extrapolation the temperature at which $\sigma_{0, T}$ has a specified value can be found; this no doubt provides more reliable $\sigma_{0, T}$, T data.

Potter's Experiments. The method of Weiss and Forrer could hardly be used with iron on account of the increased size of the furnace which would be necessitated by the high θ_f, but Potter has measured the $\sigma_{\mathscr{H}, T}$ curves for iron by an ingenious adaptation of the principle of the Kapitza and Webster method of measuring susceptibilities, p. 108, and the

use of a Sucksmith ring balance. The experimental arrangements are shown in Figs. 76 and 77. The iron sphere A was
mounted in a tube T' inside a non-magnetic furnace between
the plane pole tips of the electromagnet so that it distorted
what would otherwise be a very uniform magnetic field. A
coil B of about 150 turns was wound in three layers on an
ivory former and suspended by a quartz tube T from a
Sucksmith ring balance. B was placed with its axis strictly
parallel to the field, and when a current passed through it, the
force of attraction or repulsion between A and B was a direct
measure of the magnetic moment of the former. The force was

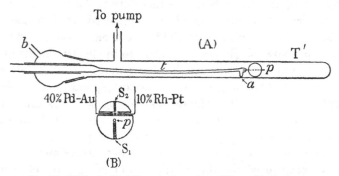

Fig. 77. Method of holding specimen in position. (Potter.)

measured by the displacement of an image in the tele-
microscope M when the current in B was reversed, the
sensitivity being determined by the current in B and the
distance between A and B. Lateral motion of B relative to the
case C was prevented by the two flat phosphor bronze springs
S_1 and S_2 mounted so that they turned in opposite directions.
The axis of B was carefully set by lateral and vertical motion of
C and by turning C about a vertical axis in the cone bearing D,
for, unless the axis of B was strictly parallel to the field a
torque was produced resulting in a big downward displacement
of B.

The method of holding A rigidly in position is shown in
Fig. 77. The sphere was mounted by a tiny set-screw S_1 on a
tapered tungsten pin sealed into the quartz tube t. The latter
was mounted rigidly in the head of the ground-quartz joint,

and it was curved as shown, with its end resting on two projections of which one is shown, so that the sphere was rigidly clamped against the tube T'. A 40 per cent palladium in gold —10 per cent rhodium in platinum thermojunction was inserted into the sphere.

Allowance was made for the pull of the field upon B, for it was impossible to make the field absolutely homogeneous; there was, however, little doubt about the accuracy of the correction. Allowance was also made for magnetic effects produced by the furnace, and for image effects. The Weiss and Forrer coil method was found inaccurate with Potter's apparatus, and the sphere was therefore replaced by an iron single crystal whose magnetisation showed an increase of about 0·5 per cent as the field rose from 7000 to 12,800 oersteds. Now, single crystals saturate in effective fields of a few hundred oersteds, hence the above increase could safely be attributed to the image effect and the corresponding correction applied to the results. The demagnetising field was taken to be $4\pi I/3$, the absolute value of I being obtained on the assumption that $I_{\infty, 290}$ was 1705 c.g.s. units per c.c.

The values of $\sigma_{\mathscr{H}, T}$ obtained from these experiments were not directly used for the extrapolation of $\sigma_{0, T}$ by plotting $\sigma_{\mathscr{H}, T}$ against $\mathscr{H}$, although they are probably more reliable for the purpose than others so used. Their use will be fully discussed in the next chapter.

Discussion of Results. It is convenient here to deal with the data for $\sigma_{0, T}/\sigma_{\infty, 0}$ obtained by the experiments so far described, but restricting our discussion to temperatures below $0·9\theta_f$. Tyler* has analysed the data, obtained by Weiss and Forrer for nickel, by Curie† for iron and by Bloch‡ for cobalt, and the values are shown plotted in Fig. 78. The data for iron and cobalt certainly require revision, although they are sufficiently accurate for our present purposes; Hegg's§ values for iron lie above those plotted in the figure and Allen and Constant's values for cobalt lie below.

* F. Tyler, *Phil. Mag.* **11**, 596, 1931.
† P. Curie, *Ann. de Phys. and Chem.* **5**, 289, 1895.
‡ O. Bloch, *Thèse*, Zürich, 1912.
§ T. Hegg, *Thèse*, Zürich, 1910.

In the figure curves (1), (2) and (3) respectively represent the theoretical variation of $\sigma_{0,T}/\sigma_{\infty,0}$ with T/θ_f on the classical Langevin theory, and on the quantum theory of p. 237 with $J = 1$ and $J = \frac{1}{2}$. Ignoring the data near $T/\theta_f = 1$, we see that the curve for $J = \frac{1}{2}$ fits the facts far better than the curve for $J = 1$, and as the curves for $J = 2, 3$, etc. would lie between curves (1) and (2), we could not expect better agreement

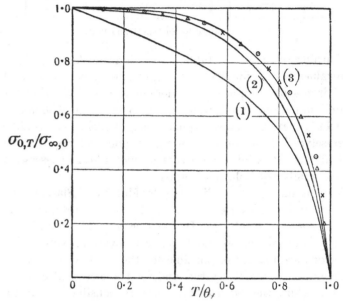

Fig. 78. Intrinsic magnetisation curves: (1) Classical curve (Langevin); (2) Quantum curve, $J=1$; (3) Quantum curve, $J=\frac{1}{2}$. ⊙ ⊙ ⊙ iron, × × × cobalt, △ △ △ nickel. ($\theta = \theta_f$, theoretically.)

between experiment and the present quantum theory. We conclude then that each magnetic carrier has a moment of $1\mu_B$, which we attribute to spin. In order to avoid confusion it is necessary to note that in many works the ordinates of Fig. 78 are labelled $\sigma_{\infty,T}/\sigma_{\infty,0}$, $\sigma_{\infty,T}$ being defined by equation (167). In these pages $\sigma_{0,T}$ denotes the intrinsic magnetisation of a Weiss domain in the absence of an applied field. Of course, $\sigma_{\infty,T}$ does not differ from $\sigma_{0,T}$ at low temperatures, but in view of experiments on the magnetocaloric effect it is desirable to distinguish between them at higher temperatures.

TABLE XV

Metal	Atomic weight	$\sigma_{\infty,0}$ e.m. units per gm.	Saturation atomic moment	θ_f ° C.	θ ° C.
Fe	55·84	221·8	2·221	770	828*
Co	58·97	163·9	1·716	1115	1138
Ni	58·68	57·6	0·606	358	376
Gd	157·30	253·5	7·10	16·0	29·5

* Value obtained by Sucksmith and Pearce.

Turning to the above table, in which are collected data of the values of $\sigma_{\infty,0}$ for the ferromagnetic elements, we see that each iron atom supplies 2·2 carriers on the average, while the gadolinium atom supplies over 7, and so on; we must therefore conclude that the electron spins in neighbouring atoms align themselves in a strong magnetic field independently of their neighbours. It is interesting that no metal atom possesses a whole number of Bohr magnetons.

This feature has been discussed by Mott* and Slater†, who explain it by reference to the number of vacant places in the 3d shells of electrons. In the ferromagnetic metals the energy bands of the 3d and 4s electrons overlap. Now, the 4s electrons are mainly responsible for the binding and lattice spacing in the metal crystals and for the electrical conductivity, while the 3d electrons are responsible for ferromagnetism. In general, there is on the average a fractional number of electrons in each band. For example, in the case of nickel, 10 electrons may be distributed among them, and we find, on the average, that 9·4 electrons are in the 3d band and 0·6 in the 4s band. Cobalt has 9 available electrons, 8·3 going to the 3d band and 0·7 to the 4s. Iron with 8 available electrons supplies 7·5 to the 3d and 0·5 to the 4s, the division being in all cases approximate only.

Now, if the 3d band contained a full complement of 10 electrons, then 5 would have their spins parallel and 5 anti-

* N. F. Mott, *Phys. Soc. Proc.* **47**, 571, 1935; N. F. Mott and H. Jones, *Properties of Metals and Alloys*, 222, 1936.
† J. C. Slater, *Journ. of Applied Phys.* **8**, 385, 1937.

parallel to an applied field. In the partially filled band we may still have 5 spins set parallel to the field when saturation is attained, but the remainder must set antiparallel. Thus the nickel atom gives us a net parallel spin of $5 - 4 \cdot 4 = 0 \cdot 6$, cobalt gives $5 - 3 \cdot 3 = 1 \cdot 7$, and iron gives $5 - 2 \cdot 5 = 2 \cdot 5$. We may further explain the behaviour of iron-cobalt alloys, where the combination results in the average number of electrons per atom of iron or cobalt lying between $7 \cdot 3$ and $8 \cdot 3$, the exact value depending on the composition. In fact Slater, by plotting the saturation magnetic moment against the average number of vacant spaces in the 3d shells, shows that the values for Ni-Cu, Ni-Zn, Ni-Co, Fe-Ni, Fe-Co, Fe-V and Fe-Cr alloys lie on a smooth curve, supporting the general rule that the saturation atomic moment is very nearly equal to the number of empty spaces in the 3d shell.

In iron-nickel alloys iron behaves as if it had $0 \cdot 8$ electron per atom in the 4s bands, and on the above views we should therefore expect a net spin of $5 - 2 \cdot 2 = 2 \cdot 8$, whereas we only observe $2 \cdot 2$ by direct experiment. Néel has provided an explanation based on the assumption that N electrons in the metal are coupled to give an angular momentum j units and a magnetic moment $2j\mu_B$. There are consequently $\frac{1}{2}Nj$ carriers spread over an electron energy band of approximately the parabolic form given by free electrons. If W is the maximum energy of the band and we define the coefficient of the internal molecular field by $\nu' = W/2N_B^2$, we find that we get saturation with each cell occupied by a single carrier of quantum number j when the actual coefficient ν of the internal field is greater than ν'. If, however, ν becomes smaller some cells become occupied by 2 carriers with quantum number j and $j-1$; as ν becomes smaller still, some cells become occupied by three carriers with quantum numbers j, $j-1$ and $j-2$, respectively. In each case, the saturation magnetisation is decreased, until finally, when the lowest cells are occupied by the maximum number $2j+1$ of carriers, the magnetic saturation is zero. Hence, there must always be a difference between the theoretical and the experimental saturation, unless $\nu \gg \nu'$, which depends upon the value of j.

CHAPTER IX

ENERGY CHANGES ASSOCIATED WITH REVERSIBLE MAGNETISATION

The Specific Heat of a Ferromagnetic. We have seen that the Weiss theory postulates the existence of a molecular field of magnitude $\nu\rho\sigma$ inside a ferromagnetic. Consequently, the magnetic energy associated with 1 c.c. of the material is

$$-\int_0^I F\, dI = -\int_0^\sigma \rho\nu\sigma(\rho\, d\sigma) = -\rho^2\frac{\nu\sigma^2}{2},$$

or the energy associated with 1 gm. is $-\rho\dfrac{\nu\sigma^2}{2}$, or, more precisely $-\dfrac{\rho\nu}{2}\sigma^2_{0,\,T}$ at a temperature $T°$ absolute and in the absence of an external field. The minus sign merely indicates that we have to supply heat in order to destroy the intrinsic magnetisation; in other words, the specific heat of a ferromagnetic is greater than that which would be exhibited by a normal metal under the same physical conditions. Hence the specific heat of a ferromagnetic may be written

$$C_f = C_n + C_m = C_n - \frac{\rho\nu}{2J}\frac{d(\sigma^2_{0,\,T})}{dT}, \qquad \ldots\ldots(180)$$

where C_n is the specific heat of a normal metal at the same temperature T, and J is the mechanical equivalent of heat. The last term in equation (180) is negative because $\sigma_{0,\,T}$ decreases with rise in temperature. Moreover, since $\sigma_{0,\,T}$ changes very rapidly in the neighbourhood of the ferromagnetic Curie point, the last term in equation (180) practically vanishes above θ_f. According to the simple Weiss theory, then, C_f should rise to a sharp maximum at θ_f, and a sudden drop in C_f should occur as the temperature rises above θ_f. If, however, we suppose that $\sigma_{0,\,T}$ does not fall sharply to zero at θ_f, but "tails off", then C_f will rise to a maximum at the temperature where $d\sigma^2_{0,\,T}/dT$ is a maximum, a temperature which is sometimes termed the *ferromagnetic critical point*,* and there will not be such a marked discontinuity in the specific heat. The investigation of equation (180) requires the simultaneous measurement of the specific heat and $\sigma_{0,\ T}$ of the same specimen at many temperatures over a wide range. The early experiments of Weiss and his collaborators were made by the method of mixtures, using a platinum coated iron cylinder. The method was unsuitable for a proper test, for it gave only average values of the specific heat over large temperature intervals, and did not permit magnetic and thermal data to be obtained at the same time.

Electrical methods, which give the *differential specific heat*, i.e. the average value of C_f over a very restricted temperature interval, have therefore been used by later workers. Thus, Sucksmith and Potter† used a nickel cylinder of which the middle portion was removed and replaced by a plug of the same sample. The plug had a screw thread in which a double-silk covered platinum wire was non-inductively wound and cemented, and the system formed a kind of Nernst calorimeter, the platinum being used both for supplying heat and for measuring the temperature of the specimen. The cylinder was suspended within an evacuated vessel surrounded by a furnace, mounted inside a solenoid provided with a secondary coil. The furnace and its contents were maintained at a steady temperature for some time, after which a known quantity of electrical energy was quickly supplied to the platinum wire. The rise in temperature of the cylinder was measured, correction being made for the temperature lag in the wire. The value of $\sigma_{0,\ T}$ was found by ballistic measurement with the solenoid.

Klinkhardt‡ employed a novel electron bombardment method for measuring the specific heat of iron at temperatures between 100° and 1000° C. Bates§ devised an electrical method for dealing with the specific heats of powdered ferromagnetics

---

* L. F. Bates, *Phys. Soc. Proc.* **42**, 441, 1930.
† W. Sucksmith and H. H. Potter, *Roy. Soc. Proc.* A, **112**, 157, 1926.
‡ H. Klinkhardt, *Ann. der Phys.* **84**, 167, 1927.
§ L. F. Bates, *Roy. Soc. Proc.* A, **112**, 157, 1926, and B. G. Whitmore, *Phil. Mag.* **7**, 125, 1929.

with low Curie points; this was used with manganese arsenide, MnAs, which has a ferromagnetic critical point at $42\cdot2°$ C., and whose specific heat rises from about $0\cdot1$ cal. per gm. per deg. C. at $25°$ C. to a maximum of nearly $0\cdot9$ at $42\cdot2°$ C., whereupon it falls almost to $0\cdot1$ again at about $45°$ C. There is no doubt that the maximum value of C_f occurs when $d\sigma_0^2 \, _T/dT$ is a maximum.

Pure nickel specimens in rods or wires some 2 mm. in diameter were used by Lapp* and by Grew,† the latter using metal $99\cdot97$ per cent pure. Grew followed Lapp's experimental arrangements fairly closely, except that he measured both the current supplied and the resistance of the nickel, while Lapp used a special type of wattmeter to measure the electrical energy supplied. Grew's arrangement is shown in Fig. 79, the nickel being mounted in an evacuated pyrex tube, and the electrical energy being supplied through the leads L, L'. The leads R, R', which were of the same material as the specimen, enabled the resistance of the portion between r and r' to be measured, while the temperature was measured by a thermocouple spot welded at a. The pyrex tube was enclosed in a furnace and the specific heat at chosen temperatures found as described above. The results obtained for pure nickel are shown in Fig. 80 and include those of Rodebush and Michaleck.‡ Magnetic data were in all cases obtained from separate experiments.

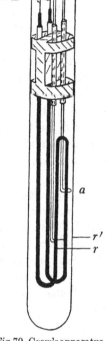

Fig. 79. Grew's apparatus for specific heat of nickel.

In a theoretical discussion it is convenient to regard the observed specific heat as made up of several terms, viz.

* Mme. Lapp, *Ann. de Phys.* **12**, 455, 1929.

† K. E. Grew, *Roy. Soc. Proc.* A, **145**, 509, 1934.

‡ W. H. Rodebush and J. C. Michaleck, *Journ. Amer. Chem. Soc.* **47**, 2117, 1925.

Observed Specific Heat $= C_q + (C_p - C_v) + C_e + C_1$, where C_q is the specific heat at constant volume of a normal metal given by the Debye theory, C_e represents the excess of the experimental value for a normal metal like copper over the Debye value, $C_p - C_v$ is the correction for thermal expansion and may be important near θ_f, and finally C_1 includes all the effects associated with ferromagnetism and any not separately mentioned in this analysis. A single determination of the specific heat at liquid-air temperatures gives the data for the calculation of C_q at that temperature, due allowance being made for the small magnetic and expansion contributions here. Hence the Debye characteristic temperature θ_D for the metal is known, and as C_q is a known function of T/θ_D it can readily be calculated for any value of T. We also have

$$(C_p - C_v) = \frac{T}{J} \frac{\alpha^2}{\rho\gamma},$$

where α and γ are the coefficients of cubical expansion and isothermal compressibility respectively. The quantity C_e is taken to be the same for nickel as for copper. Finally, the quantity $C_1' = -\frac{1}{2}\nu\rho\frac{d\sigma_{0,T}^2}{dT}$ can be calculated, $\nu\rho$ being found from measurements in the paramagnetic state which give $\theta = C\rho\nu$, where C is the Curie constant; it is assumed that $\nu\rho$ is the same above and below θ_f.

In Fig. 80 the curve C_1 represents the observed specific heat less $(C_q + C_p - C_v + C_e)$, while C_1' represents $-\frac{1}{2}\rho\nu\frac{d\sigma_{0,T}^2}{dT}$. There is obviously a difference between C_1 and C_1' which is of unknown origin. Moreover, the changes in specific heat are spread over a wider range of temperature around θ_f than the Weiss theory predicts, a feature confirmed by Ahrens,* who measured the differential specific heat of nickel by a comparison method. The differences between the results of the several experimenters are in part, and, perhaps, wholly due to differences in purity and mechanical treatment, for Grew

* E. Ahrens, *Ann. der Phys.* **21**, 169, 1934. See also C. Sykes and H. Wilkinson, *Proc. Phys. Soc.* **50**, 834, 1938, and L. Néel, *Compt. Rend.* **207**, 1384, 1938.

found that by adding copper to nickel θ_f is lowered, while the magnitude of the specific heat changes is reduced and they are extended over a wider temperature range. Zuithoff* considers that some irregularities in the specific heat of pure iron arise from the presence of occluded hydrogen.

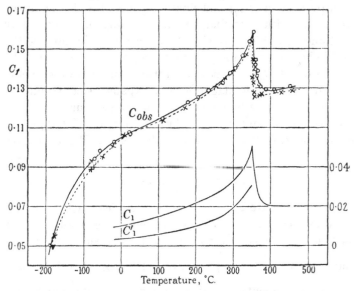

Fig. 80. Specific heat curves for pure nickel. C_{obs} —O— Grew's values of specific heat; —×— Lapp's values; —▼— Rodebush and Michaleck's values. C_1 excess specific heat; C'_1 magnetic specific heat.

The Magnetocaloric Effect. We have seen in the previous chapter that we cannot obtain accurate values of $\sigma_{0,\,T}$ from the curves of $\sigma_{\mathscr{H},\,T}$ against $\mathscr{H}$ when T is close to θ_f, and we shall now discuss experiments which enable us to obtain them. Let us suppose that a nickel sphere is suddenly introduced into a strong magnetic field, e.g. by the method of Weiss and Forrer described on p. 260, then the magnetisation of the sphere is increased. We shall neglect all hysteresis phenomena and consider solely the strictly reversible change in the magnetisation of the domains or *true magnetisation* of the nickel. An increase in this magnetisation means that the potential energy associated with the ferromagnetic domains is

* A. J. Zuithoff, *K. Akad. Amster. Proc.* **41**, 264, 1938.

decreased, and, if the energy thus set free cannot leave the sphere, a rise in temperature of the nickel must result. In like manner, if we suddenly remove the sphere from a strong magnetic field an adiabatic fall in temperature must result. This phenomenon is known as the *magnetocaloric effect*, and it is not apparent in the weak fields which are used for the investigation of hysteresis phenomena.

It is often helpful in all thermodynamical discussions in magnetism to consider $\mathscr{H}$ analogous to P and σ to V in the more familiar theory. We may calculate the observed changes in temperature as follows: Let U be the internal energy and S the entropy of 1 gm. of a ferromagnetic at a temperature T, and let $\sigma_{\mathscr{H},T}$ be the real or true specific magnetisation in an applied field $\mathscr{H}$; then, if dU is the change in internal energy, dQ is the quantity of heat absorbed, and dW the magnetic work done on the substance,

$$dU = J\,dQ + dW = T\,dS + \mathscr{H}\,d\sigma_{\mathscr{H},T}.$$

Hence,

$$dU - (\mathscr{H}\,d\sigma_{\mathscr{H},T} + \sigma_{\mathscr{H},T}\,d\mathscr{H})$$
$$= d(U - \mathscr{H}\sigma_{\mathscr{H},T}) = T\,dS - \sigma_{\mathscr{H},T}\,d\mathscr{H}.$$

But $d(U - \mathscr{H}\sigma_{\mathscr{H},T})$ is an exact differential, and, therefore,

$$\left(\frac{\partial T}{\partial \mathscr{H}}\right)_S = -\left(\frac{\partial \sigma_{\mathscr{H},T}}{\partial S}\right)_{\mathscr{H}}. \qquad \ldots\ldots(181)$$

Now, $(\partial S)_{\mathscr{H}} = JC_{\mathscr{H}}\,\partial T/T$, where $C_{\mathscr{H}}$ is the specific heat of the substance in a field $\mathscr{H}$, and $C_{\mathscr{H}}\,\Delta T$ is the heat supplied in establishing a field $\mathscr{H}$. Finally, on substituting the above value for ∂S in (181) we have, treating small finite changes, and assuming $C_{\mathscr{H}}$ constant,

$$\Delta T = -\frac{T}{C_{\mathscr{H}}}\left(\frac{\partial \sigma_{\mathscr{H},T}}{\partial T}\right)\frac{\Delta \mathscr{H}}{J}. \qquad \ldots\ldots(182)$$

When $\Delta\mathscr{H}$ is positive, ΔT is also positive, since $\partial\sigma_{\mathscr{H},T}/\partial T$ is negative.

We may, however, proceed in a different way, writing

$$J\,dQ = dU - dW = (JC_\sigma\,dT - \mathscr{H}_1\,d\sigma_{\mathscr{H},T}) - \mathscr{H}\,d\sigma_{\mathscr{H},T},$$

where C_σ is the specific heat of the substance when the magnetisation is constant, $\mathscr{H}_1$ is the internal field and $\mathscr{H}$ the applied
or external field. We may even look upon the last equation as
a definition of $\mathscr{H}_1$. On the Weiss theory, $\mathscr{H}_1 = \nu\rho\sigma_{\mathscr{H},T}$, so
that
$$J\,dQ = JC_\sigma\,dT - (\mathscr{H} + \nu\rho\sigma_{\mathscr{H},T})\,d\sigma_{\mathscr{H},T}. \quad \ldots\ldots(183)$$

For an adiabatic change, $dQ = 0$, and

$$dT = \frac{\mathscr{H} + \nu\rho\sigma_{\mathscr{H},T}}{JC_\sigma}\,d\sigma_{\mathscr{H},T}. \qquad \ldots\ldots(184)$$

Equation (184) can be integrated if $\nu\rho$ is independent of both
temperature and magnetisation. This is the assumption made
in the Weiss theory, and it is inadequate without experimental
proof. On the Heisenberg theory $\nu\rho$ depends both on temperature and magnetisation, but experiment also shows the
Heisenberg factor, p. 241, to be inadequate.

On the Weiss assumption, then, we write

$$\varDelta T = \frac{\nu\rho}{2JC_\sigma}(\sigma^2_{\mathscr{H},T} - \sigma^2_{0,T}), \qquad \ldots\ldots(185)$$

where C_σ is now expressed in cal. per gm. per deg. and $\mathscr{H}$ has
been neglected in comparison with the much greater internal
field. Equation (185) should therefore be valid at temperatures
below or even slightly above θ_f. At temperatures well above θ_f
we have $\sigma_{\mathscr{H},T} = \chi\mathscr{H}$, and if χ is independent of $\mathscr{H}$, on integration equation (185) becomes

$$\varDelta T = (\nu\rho + 1/\chi)\frac{\sigma^2_{\mathscr{H},T}}{2JC_\sigma} = \frac{A\sigma^2_{\mathscr{H},T}}{2JC_\sigma}, \quad \ldots\ldots(186)$$

where the constant A varies with T since χ decreases with rise
in T.

Equation (185) may be obtained by a simple argument. Let
the curves (1) and (2) of Fig. 81 a respectively represent the
variation of $\sigma_{0,T}$ and of $\sigma_{\mathscr{H},T}$ for a known value of $\mathscr{H}$, with
temperature. Let a specimen be initially in the state repre-
sented by the point A on curve (1) and let the field $\mathscr{H}$ be
switched on so that the temperature rises to that of B on

curve (2). As magnetocaloric changes are reversible we must get the same result if we proceed from A to B via C, i.e. if we first increase the temperature while keeping the magnetisation constant until C is reached and then increase the magnetisation while keeping the temperature constant from C to B. Neglecting the work done by the external field and assuming an adiabatic change from A to B, we have

$$J\Delta Q = 0 = \text{work done in going from } A \text{ to } C + \text{work done in going from } C \text{ to } B = JC_\sigma \Delta T - \tfrac{1}{2}\rho\nu(\sigma^2_{\mathcal{H},T} - \sigma^2_{0,T}).$$

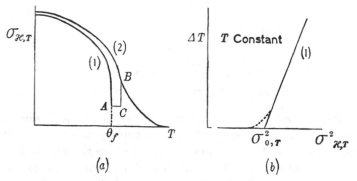

Fig. 81. Magnetocaloric phenomena.

Hence, on rearranging the terms we have equation (185). We see, then, that on plotting ΔT against $\sigma^2_{\mathcal{H},T}$ for a chosen value of T we must obtain a straight line such as (1) in Fig. 81 b, whose intercept on the magnetisation axis gives $\sigma^2_{0,T}$. In practice, however, we find graphs with "flat feet", indicated by the dotted line, the reasons for which we shall discuss later.

It follows that we may carry out two sets of experiments. First, we may place a ferromagnetic substance in a strong magnetic field and verify equation (182). Secondly, we may measure the changes in temperature corresponding to the establishment of measured values of $\sigma_{\mathcal{H},T}$ and find $\nu\rho$ graphically by using equation (185).

Experimental Investigation of the Magnetocaloric Effect. Equation (182) has been confirmed with nickel by Weiss and

Forrer,* and with iron by Potter.† Weiss and Forrer used the apparatus described on p. 260. A galvanometer in the thermo-couple circuit was initially balanced on a potentiometer and its deflection was observed when the furnace containing the nickel was pulled aside into a pole piece; this deflection was a measure of the adiabatic change in temperature. Rapid determinations were made by using a critically damped galvanometer of short period. The way in which ΔT for a series of known fields $\Delta\mathscr{H}$ depends upon the temperature is shown in Fig. 82.

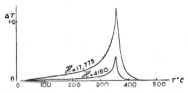

Fig. 82. Magnetocaloric effect in nickel.

Potter used the apparatus described on p. 262 with similar galvanometer arrangements for measuring the changes in temperature of the specimen, except that, as the field had to be switched on and off, a galvanometer of somewhat longer period was needed. By taking readings both when the field is switched on and switched off, eddy current effects may be eliminated, although these are small in any case. For, on exciting the field we observe $+\Delta T_1 = +\Delta T + \Delta T_e$, and on switching off we observe $-\Delta T_2 = -\Delta T + \Delta T_e$. Thus a series of values of ΔT for known changes $\Delta\mathscr{H}$ can be obtained at chosen temperatures. On the whole, we may say that experiment proves a linear relation to exist between ΔT and $\Delta\mathscr{H}$ in temperature regions where relation (182) would be expected to hold, and any slight concavity towards the $\Delta\mathscr{H}$ axis of the lines for the higher temperatures is explained by the known decrease of $\partial\sigma_{\mathscr{H},T}/\partial T$ with increase in $\mathscr{H}$, and a slight convexity of the lines for the lower temperatures by error in the value of the demagnetising factor used in calculating the effective value of $\Delta\mathscr{H}$.

* P. Weiss and R. Forrer, *Ann. de Phys.* 5, 153, 1926.
† H. H. Potter, *Roy. Soc. Proc.* A, **146**, 362, 1934.

Now from the slopes of these lines of ΔT against $\Delta \mathscr{H}$ taken at the origin we obtain $\Delta \mathscr{H}/\Delta T$ for $\mathscr{H} = 0$, which, combined with the appropriate values of $\partial \sigma_{\mathscr{H}, T}/\partial T$, obtained as described below, enables us to calculate $C_{\mathscr{H}}$ by substitution in equation (182). The values of $C_{\mathscr{H}}$ so obtained agree as well as could be expected with those from direct determinations, in view of the fact that the calculated values depend on the slopes of two separate graphs. The calculated values of $C_{\mathscr{H}}$, as we realise from Fig. 82, show a sharp maximum as θ_f is approached.

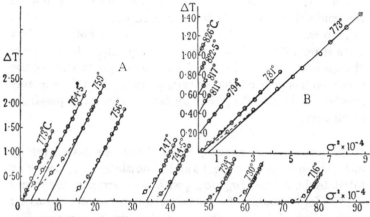

Fig. 83. Rise of temperature as a function of $\sigma^2_{\mathscr{H}, T}$. (Potter.)

Magnetocaloric Determinations of the Weiss Constant. We have seen that in Weiss and Forrer's experiments the magnetocaloric change in temperature ΔT and the value of $\sigma_{\mathscr{H}, T}$ were measured at the same time, when the specimen of nickel was pulled inside a pole piece. In Potter's experiments, when a field $\mathscr{H}$ was switched on, ΔT was measured as described above, and the value $\sigma_{\mathscr{H}, T}$ was found as indicated on p. 262, except that when T was greater than θ_f the values of $\sigma_{\mathscr{H}, T}$ had to be measured separately, to determine by extrapolation the true value of $\sigma_{\mathscr{H}, T}$ to which a known rise in temperature ΔT corresponded.

The curves obtained were in good agreement with theory. Some of Potter's curves showing ΔT as a function of $\sigma^2_{\mathscr{H}, T}$ are reproduced in Fig. 83. Apart from the distinct "flat feet", there is no doubt that at temperatures below θ_f and with strong fields we get a series of parallel lines, whose common slope should give $\nu\rho/2JC_\sigma$. At temperatures above but close to θ_f the slope of the individual line should be $\dfrac{1}{2JC_\sigma}(\nu\rho + 1/\chi)$ and the lines should remain almost parallel. The lines showed increasing slope with increase in $(T - \theta_f)$, but, as we shall see on p. 282, this does not necessarily mean that these experiments prove that $\nu\rho$ is a function of temperature when $T > \theta_f$.

The "flat feet" of the curves of Fig. 83 can be explained if the true value of $\sigma_{\mathscr{H}, T}$ is not measured in all circumstances, or if in weaker fields the domain vectors are not accurately aligned parallel to $\mathscr{H}$, so that a magnetocaloric temperature change is recorded before the apparent magnetisation of the specimen has reached the value $\sigma_{0, T}$. In a way, this is merely another mode of stating that hysteresis effects are playing a small part.

The Law of Corresponding States. From the data given by curves such as those of Fig. 83 and a knowledge of $\sigma_{\infty, 0}$ we may tabulate the values of $\sigma_{0, T}/\sigma_{\infty, 0}$ with the corresponding values of T/θ_f for a number of ferromagnetics and so obtain further information of the correctness of the law of corresponding states, p. 236. The new data are particularly important when T approaches θ_f, for we saw on p. 261 that in this region the values of $\sigma_{0, T}$ obtained by extrapolation from $\sigma_{\mathscr{H}, T}$ are somewhat uncertain. The collected results are shown in Fig. 84, where the broken curve represents the theoretical curve, already discussed on p. 265, for magnetic carriers with $J = \frac{1}{2}$.

If the experimental results are correct, then it is clear that the law of corresponding states is not obeyed, since there is such a large difference between the curves for iron and nickel. It is doubtful, however, whether on the present data we are justified in forming this conclusion; it would appear desirable to repeat the measurements with a nickel single crystal using Potter's method. The discrepancy between the experimental results and theory is now somewhat greater than was represented on p. 265. In particular, the enlarged picture of the behaviour of $\sigma_{0,T}/\sigma_{\infty,0}$ in the region of θ_f given in Fig. 84b shows that the experimental curve does not cut the temperature axis at a definite angle as predicted by the modified Weiss theory, but approaches it gradually. The "foot" so formed varies from specimen to specimen, but as one is found even with a single iron crystal, we may conclude that it is not due to impurities, or to the several domains possessing slightly different Curie points. There is, however, one point which has never been mentioned in this work, as far as the author is aware, and that is whether any part is played by *temperature*

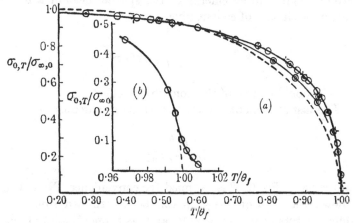

Fig. 84. Law of corresponding states. - - - theoretical; ⊙ nickel, Weiss and Forrer; ○ Heraeus electrolytic iron, ⊙ single crystal iron, + Armco iron, Potter.

hysteresis, i.e. the ferromagnetic loses its ferromagnetism at θ_f on heating and regains it at a temperature below θ_f on cooling, a phenomenon which occurs in some ferromagnetic compounds to a very marked degree. A very slight temperature hysteresis in the neighbourhood of θ_f would have a profound effect on the measurements of ΔT.

The Weiss Constant and Domain Size. We have seen on p. 251 that by direct experiment we cannot measure the true specific magnetisation $\sigma_{\mathscr{H},T}$ of each domain but only some apparent value, which we will now denote by $\sigma'_{\mathscr{H},T}$, because of the effects of domain size. We have seen, too, that although $\sigma'_{\mathscr{H}, T}$ is taken to be an exact measure of $\sigma_{\mathscr{H}, T}$ when $\mathscr{H}$ is greater than 5000 oersteds, we have no proof that this assumption is correct; indeed, we shall shortly review evidence which shows that it is sometimes seriously in error. Values of $\sigma'_{0, T}$ are obtained from $\sigma'_{\mathscr{H}, T}$ curves by extrapolation to the value of the specific magnetisation in infinitely small fields when T is well below the Curie point; again, we have no direct proof that $\sigma'_{0, T} = \sigma_{0, T}$. At higher temperatures, values of $\sigma'_{0, T}$ are obtained from magnetocaloric measurements, definite extrapolations being made from values of $\sigma'^2_{\mathscr{H}, T}$ obtained in strong fields. Measurements of the specific heat of a ferromagnetic should, of course, give

$$dE/dT - \frac{\nu}{2}\frac{\rho}{J}\frac{d}{dT}(\sigma_{0, T})^2,$$

which should be free of the domain size effect.

The last statement shows that $\nu\rho$ may be defined as

$$-\frac{J}{\sigma}\left(\frac{\partial E}{\partial \sigma}\right)_T,$$

where the suffix 0, T is dropped for convenience. Now, it is known that the quantity $\nu\rho$ can be determined from four different sets of measurements, and, as Stoner* pointed out, the several values ought to agree. The value of $\nu\rho$ for nickel obtained from measurements of the paramagnetic susceptibility above θ_f, based on equation (155), p. 234, was $11\cdot7 \times 10^4$, that from specific heat measurements was $11\cdot5 \times 10^4$, that from magnetocaloric data varied from $4\cdot3$ to $9\cdot8 \times 10^4$ over a wide temperature range, while the value from saturation atomic moment data was 16×10^4.

Stoner discussed in considerable detail the thermodynamical formulae which enable $-\frac{1}{\sigma}\left(\frac{\partial E}{\partial \sigma}\right)_T$ to be calculated from the various sets of experimental data. In particular, he showed

* E. C. Stoner, *Roy. Soc. Phil. Trans.* A, **235**, 165, 1936.

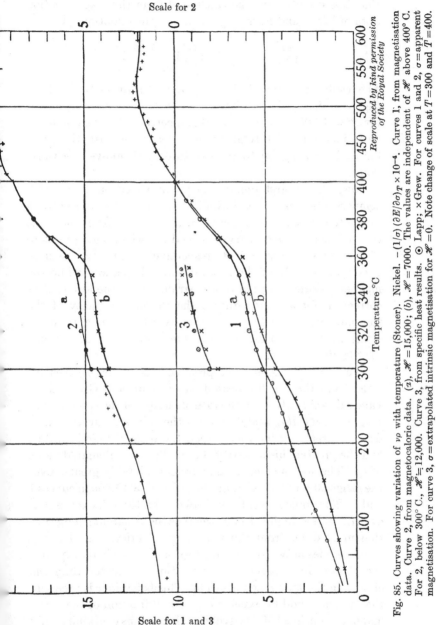

Fig. 85. Curves showing variation of $\nu\rho$ with temperature (Stoner). Nickel. $-(1/\sigma)(\partial E/\partial\sigma)_T \times 10^{-4}$. Curve 1, from magnetisation data. Curve 2, from magnetocaloric data. (a), $\mathscr{H}=15{,}000$; (b), $\mathscr{H}=7000$. The values are independent of $\mathscr{H}$ above 400° C. For 2, below 300° C., $\mathscr{H}=12{,}000$. Curve 3, from specific heat results. $\odot$ Lapp; $\times$ Grew. For curves 1 and 2, $\sigma=$ apparent magnetisation. For curve 3, $\sigma=$ extrapolated intrinsic magnetisation for $\mathscr{H}=0$. Note change of scale at $T=300$ and $T=400$.

Reproduced by kind permission of the Royal Society

Scale for 2

Scale for 1 and 3

Temperature °C

that this quantity can be calculated from the magnetisation data of Weiss and Forrer by the use of the equation

$$\left(\frac{\partial E}{\partial \sigma}\right)_T = \mathscr{H} + T\left(\frac{\partial \sigma}{\partial T}\right)_{\mathscr{H}} \bigg/ \left(\frac{\partial \sigma}{\partial \mathscr{H}}\right)_T .$$

The results of his calculations are shown in curves $1a$ and $1b$ of Fig. 85 which are based on the $\sigma'_{\mathscr{H}, T}$ data of Weiss and Forrer for $\mathscr{H} = 15{,}000$ and $\mathscr{H} = 7000$ respectively. A very marked variation of $\nu\rho$ with temperature appears to exist, the value for $\nu\rho$ increasing up to temperatures well above the Curie point.

Now, Potter first remarked that a variation of $\nu\rho$ with temperature appeared to be indicated from Weiss and Forrer's magnetocaloric measurements on nickel, and from the same data Stoner obtained curves $2a$ and $2b$, which again refer to $\mathscr{H} = 15{,}000$ and $\mathscr{H} = 7000$ respectively. The two sets of curves are clearly of the same general shape, and while the magnetocaloric values of $\nu\rho$ appear definitely the lower, after correction for an error due to the thermal capacity of the thermocouple, the values from the two sets of data agree quite closely. In both cases, however, the values of $\nu\rho$ are greater for the higher fields.

In curve 3 of Fig. 85 are reproduced the values of $\nu\rho$ calculated from the specific heat data of Lapp and Grew; these values depend only upon the rate of change of the spontaneous magnetisation $d(\sigma_{0, T})/dT$, whereas the values of curves 1 and 2 depend directly or indirectly upon the quantity $\partial(\sigma'_{\mathscr{H}, T})/\partial\mathscr{H}$, and the greater this quantity the smaller the calculated value of $\nu\rho$. The specific heat results are considerably greater over the range $300°$ to $350°$ C. than those for $\mathscr{H} = 15{,}000$ in curves 1 and 2. The conclusions to be drawn from these facts are that even in fields as high as 15,000 oersteds the apparent magnetisation must differ from the true magnetisation and that the discrepancies between the three sets of curves will disappear if the change in apparent magnetisation is greater than the change in the true magnetisation. This behaviour is precisely that which would be expected if a small fraction of the ferromagnetic consisted of relatively small domains containing only

10^3 or 10^4 atoms, which would behave as a small impurity of high atomic moment embedded in the substance; this has been discussed on p. 250. The effect is unimportant in the determination of $\sigma'_{\mathscr{H}, T}$ with $\mathscr{H}$ between 5000 and 20,000. The direct estimates of $\sigma_{0, T}$ from $\sigma'_{\mathscr{H}, T}$ curves are not likely to be seriously in error, although $(\partial\sigma'/\partial\mathscr{H})$ may differ considerably from $(\partial\sigma/\partial\mathscr{H})$, the difference between them being the larger the smaller the field $\mathscr{H}$, and relatively smaller the higher the temperature T. It is important to realise that, although the quantity $(\partial\sigma/\partial\mathscr{H})$ has now been proved to be so important, no experiments, with the possible exception of those of Gerloff,[*] have been made with the set purpose of recording this particular quantity. We still have therefore little precise information concerning the variation of $\nu\rho$ with temperature.

The Magnetocaloric Effect with Paramagnetics. By evaporating liquid helium under reduced pressure, temperatures down to $0.7°$ K. may be obtained, but if we wish to produce still lower temperatures a new method and a new technique must be used. Debye[†] and Giauque[‡] pointed out that this might be provided by the magnetocaloric effect with a paramagnetic as working substance, and such remarkable progress has recently been made that it is now possible to use it in lecture-room demonstrations of superconductivity phenomena.[§]

Turning to equation (182) of p. 273 we have, since $\sigma_{\mathscr{H}, T}$ for a paramagnetic is equal to $C\mathscr{H}/T$, where C is the Curie constant,

$$dT = -\frac{T}{C_{\mathscr{H}}}\frac{d}{dT}\left(\frac{C\mathscr{H}}{T}\right)_{\mathscr{H}}\frac{d\mathscr{H}}{J} = +\frac{C}{C_{\mathscr{H}}}\frac{\mathscr{H}}{T}\frac{d\mathscr{H}}{J}$$

for the rise or fall in temperature dT of a paramagnetic suddenly placed in or removed from a field $\mathscr{H}$. Now, it is obvious that when T is very small $C_{\mathscr{H}}$ will be a function of T, and from an experimental point of view it is sometimes preferable to measure the quantity of heat which has to be

[*] G. Gerloff, *Zeit. für Phys.* **99**, 585, 1936.
[†] P. Debye, *Ann. der Phys.* **81**, 1154, 1926.
[‡] W. F. Giauque, *Journ. Amer. Chem. Soc.* **49**, 1864, 1927.
[§] F. Simon, *Royal Institution Proc.* 1935, Feb. 1st Meeting.

absorbed to enable an adiabatically cooled substance to regain its original temperature. This is given by

$$\int_0^{\mathscr{H}} C_{\mathscr{H}}\, dT = \int_0^{\mathscr{H}} \frac{C}{JT}\mathscr{H}\, d\mathscr{H},$$

or, if we prefer to consider changes in entropy, we have

$$\left[\varDelta S\right]_0^{\mathscr{H}} = -\int_0^{\mathscr{H}} \frac{C_{\mathscr{H}}}{T}\, dT = -\int_0^{\mathscr{H}} \frac{C}{JT^2}\mathscr{H}\, d\mathscr{H} = -\frac{C}{2J}\left(\frac{\mathscr{H}}{T}\right)^2,$$

$$\dots\dots(187)$$

representing the entropy which 1 gm. of substance loses if the field $\mathscr{H}$ is established while T remains constant.

Now, no substance strictly obeys the Curie law at all temperatures; for, in the absence of an external magnetic field the carriers would have to set with their axes randomly orientated in all directions, and so, even at the absolute zero of temperature would still be in a state of disorder, i.e. still possess entropy; this is contrary to the Nernst heat theorem. Thus, the Gd^{+++} ion in gadolinium sulphate, being in a $^8S_{\frac{7}{2}}$ state, would possess a zero point entropy of

$$R\log_e(2J+1) = R\log_e 8 . \text{ergs per gm.-ion.}$$

Moreover, there must always be some tendency of the carriers to set in definite directions because of the interaction between neighbouring carriers. If the interaction energy is denoted by U, then, in the region of a characteristic temperature θ_m, defined by $k\theta_m = U$, deviations from the simple Curie law would be expected. Kürti and Simon* emphasise the importance of θ_m, for, the lower the value of θ_m, the lower is the temperature obtainable by the magnetocaloric method. θ_m must be lower than $1°$K. for the substance to be of practical use. Fortunately, its value can be found from specific heat measurements made at temperatures above θ_m.

Simon and Kürti made the tentative assumption that in a field $\mathscr{H}$ the normal $(2J+1)$ positions of the ion are separated by equal energy increments $g\mu_B\mathscr{H}$, so that, bearing in mind the

* N. Kürti and F. Simon, *Roy. Soc. Proc.* A, **149**, 152, 1935.

Boltzmann expression, the distribution of the ions in the
different levels is a function of $g\mu_B\mathscr{H}/kT_i$ only, as long as the
initial temperature $T_i \gg \theta_m$ and $\mu_B\mathscr{H} \gg k\theta_m$. On the removal
of the field $\mathscr{H}$ we may assume that the levels are separated by
$k\theta_m$ and the distribution of the ions in these levels is then a
function of $k\theta_m/kT_f$ alone, where T_f is the final temperature
reached by the adiabatically cooled substance. Hence,

$$T_f/T_i = k\theta_m/g\mu_B\mathscr{H}. \qquad \ldots\ldots(188)$$

Taking $g = 2$ for salts of the iron group, $T_f = 7\cdot5\theta_m T_i/\mathscr{H}$, so
that the final temperature is directly proportional to θ_m and
is the lower the lower the initial temperature and the greater
$\mathscr{H}$. See Chapter XV.

In applying this simple expression it must be remembered
that the lattice vibrations are important at low temperatures,
and that in cases where it is desired to cool non-paramagnetic
substances by mixing them with paramagnetic ones, the
entropy of the added substance may be important, particularly when high fields and very low temperatures are employed.
The specific heat of the paramagnetic substance is also important, since it practically determines the rate at which the
temperature will rise after cooling. In the region of θ_m the
specific heat shows a marked anomaly because of the energy
required to transfer the ions to the higher energy levels. Thus
for iron ammonium alum it is calculated that $\theta_m = 0\cdot061$ and
the maximum value of the specific heat is $1\cdot7$ cal. per gm.-ion
per deg. in the region of $0\cdot08°$ K. Hence, 1 c.c., containing
$1/258$ gm.-ion of this alum at about $0\cdot05°$ K., has a thermal
capacity of $0\cdot006$ cal. per deg., which is about equal to that of
16 tons of lead at the same temperature or that of 2 kilograms
of lead at $1°$ K. Such an enormous thermal capacity means
that the rate of rise of temperature after cooling must be small,
and shows how essential it is to choose as paramagnetic substance one whose θ_m lies in the actual temperature region in
which experiments are to be made.

Experimental Details. The design of apparatus is greatly
conditioned by the size of the magnet used to produce the field
$\mathscr{H}$, and by the method by which the very low temperature of

the paramagnetic substance is measured. For example, when the magnetic susceptibility is measured for the latter purpose, the design is much simplified. Now, the working substance must first be cooled by contact with liquid helium, in such a manner that the contact may readily be removed. These conditions are well satisfied when the cooling is done by surrounding the substance with helium gas. For, if we start with a substance at a temperature of 1° K. and cool it adiabatically to 0·3° K., the pressure of the helium vapour in contact with it falls from 0·15 to 7×10^{-10} mm., and the gas therefore condenses on the surface of the substance and so isolates it; the substance thus acts very efficiently as its own pump. In spite of direct contact between the loosely packed substance and its container, it is found that the latter does not cool, so that it may be kept in permanent contact with liquid helium. (Soft solder is superconducting at these temperatures and must not be used.)

Kürti* found that equation (187) was obeyed in the case of gadolinium sulphate over ranges of temperature starting from 4·4° and 2·6° K. and with fields up to 10,000 oersteds, and he found its specific heat over the range 1·6° to 20° K. The apparatus used was similar to that of Ruhemann and its *modus operandi* has been described on p. 246. A thin germansilver calorimeter containing about 0·5 gm. of gadolinium sulphate was suspended by a thin-walled capillary tube inside an evacuated chamber surrounded by liquid helium. The calorimeter contained helium gas and acted as its own gas thermometer, the capillary tube being connected to a membrane manometer.† It was also provided with a manganin heating coil. The whole of the above apparatus was contained in an evacuated vessel and surrounded by liquid air. The magnetic field was provided by a coil system S immersed in the liquid air. The magnetocaloric changes in temperature produced by switching the field on and off were measured by the manometer. The corresponding changes in entropy were found by measuring the thermal capacity of the calorimeter

* N. Kürti, *Zeit. für Phys. Chem.* **20** B, 305, 1933.
† G. Kornfeld and E. Klinger, *Zeit. für Phys. Chem.* **4** B, 37, 1929.

system over the same range of temperature using the heating coil as in a Nernst calorimeter experiment. In addition to verifying equation (187) Kürti measured the specific heat in a field of 1800 oersteds and showed that it obeyed the relation

$$C_{\mathscr{H}, T} - C_{0, T} = \frac{1}{J} \left(\frac{\partial (\mathscr{H} \sigma_{\mathscr{H}, T})}{\partial T} \right)_{\mathscr{H}} = -\frac{\mathscr{H}}{J} \left(\frac{\partial \sigma_{\mathscr{H}, T}}{\partial T} \right)_{\mathscr{H}}$$

$$= -\frac{\mathscr{H}}{J} \frac{\partial}{\partial T} \left(\frac{C \mathscr{H}}{T} \right)_{\mathscr{H}} = +\frac{C}{J} \left(\frac{\mathscr{H}}{T} \right)^2 ,$$

and the results agreed with the assumption that $\theta_m = 0.26°$ K. and showed that the specific heat did not obey the Debye T^3 law at these low temperatures, but, after reaching a minimum at 5° K., increased rapidly with further decrease in temperature.

The Production of Very Low Temperatures. The magneto-caloric effects have been used by many workers for the production of very low temperatures. The relevant portions of the apparatus used by Kürti and Simon* are shown in Fig. 86. The temperature of the liquid helium container V is lowered to 1·6° K. by reducing the pressure above the liquid. Liquid helium is also condensed in the thin-walled german-silver cryostat C by means of the tube G. The metal shield S is soldered to V, so that heat losses from C are so small that its temperature can be reduced to 1° K. by connecting a diffusion pump to G. About 0·3 to 1 gm. of the paramag-

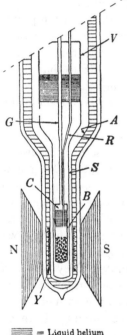

= Liquid helium
= ,, hydrogen

Fig. 86. Apparatus for low temperature investigations. (Kürti and Simon.)

netic is loosely packed in the vessel B which is attached to C

* N. Kürti and F. Simon, *Roy. Soc. Proc.* A, **149**, 152, 1935. See also *Very Low Temperatures*, Book Three, Science Museum, 1937.

by ground-metal surfaces smeared with tap grease, and helium gas added to it via the tube R. The portions of the apparatus so far described are enclosed in the evacuated vessel A and surrounded by liquid hydrogen.

The susceptibility of the paramagnetic is measured by a mutual inductance method. The primary and secondary of one pair of coils are wound on the brass bobbin Y and these are respectively joined in series with the primary and secondary coils of a compensating mutual inductance. The deflection of a ballistic galvanometer in the secondary circuit when a current is sent through the primary measures the susceptibility, which at these low temperatures is high. When the susceptibility as a function of temperature is known, low temperatures can easily be read from a graph. Such temperatures are denoted by T^* to indicate that they are not referred to the absolute scale of temperature.

To produce a lower temperature, the electromagnet is first switched on; there results a considerable rise in temperature of the substance, but the heat is speedily carried away by the release of condensed gas, which may take a few seconds to complete. The magnet is then switched off, the substance cools, and, as explained on p. 286, the thermal insulation becomes extremely good.

With the arrangements of de Haas, Wiersma and Kramers,[†] thermal insulation was much less satisfactory because of the weighing method used for measuring the change in susceptibility. They placed the finely powdered paramagnetic in a vessel suspended by a quartz fibre from one arm of a vacuum balance with electromagnetic control. The susceptibility was measured by the magnetic pull of a non-uniform field in which the substance was placed. Thus with a sample of CeF_3 weighing 90·4 mgm. the magnetic pull in a non-uniform field of 2250 oersteds was 41·1 units at the final temperature of 0·19° K., caused by a reduction of the field from 27,600 oersteds, and 11·9 units when the initial temperature 1·3° K. was regained.

The following table gives a few of the results of measurements so far recorded with various paramagnetics:

† W. J. de Haas, E. C. Wiersma and H. A. Kramers, *Physica*, **13**, 175, 1933.

TABLE XVI

Substance	T_i^*	H_i	H_f	T_f^*	Mean θ_m	Experimenters
$Fe.NH_4(SO_4)_2.12H_2O$	1·23	14,100	0	0·038	0·061	Kürti and Simon
$MnSO_4(NH_4)_2.SO_4.6H_2O$	1·23	8,000	0	0·09	0·11	,,
$Gd.SO_4.8H_2O$	1·15	5,400	0	0·35	0·21	,,
,,	1·50	8,000	0	0·25	—	Giauque and MacDougall
$K.Cr(SO_4)_2.12H_2O$	1·14	5,500	0	0·16	0·095	Kürti and Simon
,,	1·30	19,500	350	0·05	—	de Haas and Wiersma
CeF_3	1·30	27,600	850	0·13	—	,,

Applications of Magnetic Cooling. As applications of the method of magnetic cooling are likely to be of importance in the near future, some description of work already performed is advisable here. For example, the arrangements and technique of Kürti and Simon have been used in the search for new superconducting metals. By using a pill of powdered salt and powdered metal both are cooled by demagnetising the salt, and tests for superconductivity can be applied. The contribution of the metal to the magnetic susceptibility of the pill is negligible until a superconducting state is established, when the persistent currents induced in the metal cause it to behave as a strongly diamagnetic substance, producing a change in the susceptibility of the pill from a positive to a negative value.

Now, when a superconducting metal is heated the temperature at which its distinctive properties disappear is determined by the magnitude of the field. At every temperature there is a certain minimum or threshold value of the magnetic field which renders superconduction impossible. Hence, two separate sets of experiments are carried out with the pill. First, the temperature of the cooled pill is allowed to rise while magnetic measurements are made with a moderate field, and, secondly, the procedure is repeated with a weak field. In the latter case superconductivity is free to show itself and is detected by the change in susceptibility. Unfortunately, the thermal effects associated with the passage of the metal from one conductivity state to the other, and the existence of a somewhat indeterminate field owing to the presence of the salt in the pill, make quantitative measurements uncertain. However, use can be made of the residual magnetic moment which persists in a superconductor on removal of a field greater than the threshold value, for it vanishes at the normal transition point at which superconduction ceases in zero field. When a small known field is switched on, the galvanometer deflection depends upon whether or no the field coincides in direction with the residual moment, so that galvanometer deflections obtained by a first and a second switching on of the field differ at temperatures below the transition point. In the last experiments a packed tube rather than a pill should be used, as very pure superconductors do not retain residual moments when in a highly compact form. By such means cadmium, zirconium and hafnium were found to be superconductors.

Another important application consists in the measurement of the energy changes associated with radioactive transformations, which, on account of their minuteness, have been measured directly in but few cases. The apparatus of Kürti and Simon provides an extremely sensitive calorimeter. For example, the thermal capacity of a calorimeter made of tungsten, whose Debye characteristic temperature is above $330°$ K., is about 30,000 times less at $1·3°$ K. than its value at room temperature, and its thermal insulation at still lower temperatures is ideal. Hence it should be possible to use a composite pill of paramagnetic salt and tungsten when cooled by the magnetocaloric method to measure the energy of γ-rays received from a radon source of strength $0·1$ millicurie placed 3 cm. from the pill. Alternatively, the energy of the absorbed γ-rays could be used to examine the variation of the specific heat of a substance at low temperatures. So far, experiments have only been made with the working substance, iron-ammonium sulphate, without the addition of other material like tungsten.

The Temperature Scale below $1°$ *K.* We have seen how values of low temperatures computed on the assumption that the Curie law holds below $1°$ K. are denoted by T^*, because the law cannot be expected to hold in this region; for, the splitting produced by the crystalline field is here of the same magnitude as kT, and, in addition, the purely magnetic coupling of the ions plays an important part. The expression $\mathscr{H} + \dfrac{4\pi}{3} I$ for the field within a spherical cavity is inadequate when we consider magnetic dipoles, and Onsager† and Van Vleck‡ have shown that it must be replaced by a more complicated expression. When a dipole is placed in the spherical cavity it polarises the surrounding medium, and sets up a *reaction field* parallel to the dipole axis and not, in general, parallel to the lines of induction outside. Hebb and Purcell§ distinguish between three types of salts: viz. (1) salts like $Gd\,(C_6H_4NO_2\,.\,SO_3)_3\,.\,7H_2O$ and $Gd\,(PMo_{12}O_{40})\,.\,30H_2O$ in which the magnetic ions are so far apart that only splitting due to crystalline fields need be considered; (2) salts like $Cs\,Ti(SO_4)_2\,.\,12H_2O$ where crystalline splitting is very small, as the Ti^{+++} ion is in a ^2S state with no orbital angular momentum upon which the field can act, so that only magnetic or spin-spin coupling need be considered; (3) salts like $Gd_2\,(SO_4)_3\,.\,8H_2O$ in which the ions are close enough to require both crystalline splitting and magnetic coupling to be considered. Hebb and Purcell examined all these cases and calculated the departures from the Curie law at temperatures below $1°$ K., and compared the T^* temperatures with the Kelvin scale.

The experimental data are at present very meagre. Kürti, Lainé and Simon‖ worked with $Fe\,.\,NH_4(SO_4)_2\,.\,12H_2O$ and measured the rate of change of its entropy and the rate of change of its total heat Q with respect to T^* using γ-rays as a source of heat, as described on the opposite page. This enabled them to determine the absolute temperature from the equation $T = (\partial Q/\partial T^*)_{\mathscr{H}}/(\partial S/\partial T^*)_{\mathscr{H}}$. They found, in addition to the specific heat anomaly described on p. 285, that a much sharper maximum of 20 cal. per gm. ion per deg. occurred at $0\cdot03°$ K. The former anomaly is attributed to crystalline splitting and the latter to magnetic coupling, and it is suggested that it

† L. Onsager, *Journ. Amer. Chem. Soc.* **58**, 1486, 1936.
‡ J. H. Van Vleck, *Journ. Chem. Phys.* **5**, 320, 1938.
§ M. H. Hebb and E. M. Purcell, *Journ. Chem. Phys.* **5**, 338, 1938.
‖ N. Kürti, P. Lainé and F. Simon, *Compt. Rend.* **204**, 675 and 754, 1938.

portrays a kind of ferromagnetism. Giauque and MacDougall[†] found the susceptibility of Gd $(PMo_{12}O_{40}).30H_2O$ by several methods, notably by self-inductance measurements with an Anderson bridge, and considered that their T^* value was only some $0.03°$ less than the absolute value at $0.15°$ K. The results of Casimir, de Haas and de Klerk for $KCr(SO_4)_2.12H_2O$ showed that above $0.07°$ K. the relation between T and T^* was reliable to within $0.01°$ K. The lowest temperature on record is $T^* = 0.0043$ or $T = 0.0034$, obtained[‡] with a mixture of potassium chrome alum and potassium aluminium alum.

The Electrical Resistance of a Ferromagnetic. The changes of resistance which a normal metal exhibits when placed in a magnetic field, usually with the direction of the current in the metal parallel to the field, have been very thoroughly investigated by several workers with fields up to 50,000 oersteds, and by Kapitza[§] with fields up to 300,000. Leaving aside abnormal metals, bismuth, antimony, arsenic and the ferromagnetic metals, the results for different elements are very similar. If R is the resistance of the metal in zero field at a given temperature and ΔR the increase caused by the application of a magnetic field $\mathcal{H}$, it is found that the ratio $\Delta R/R$ increases proportionally to $\mathcal{H}^2$ for comparatively low values, but proportionally to $\mathcal{H}$ for high values of the field. Sometimes the ratio $\Delta R/R_0$ is used, where R_0 is the resistance in zero field at $0°$ C., for some workers consider that the latter ratio is less dependent upon the purity of the metal specimen than the former.

Now it has been known for some years that the resistance of nickel in the absence of a magnetic field increases in a peculiar manner with rise in temperature. Thus, Cabrera and Torroja[||] and, later, Gerlach and Schneiderhan[¶] showed that in addition to the change which comprises the more or less linear increase of resistance of a normal metal with rise in temperature there is

† W. F. Giauque and D. P. MacDougall, *Journ. Amer. Chem. Soc.* **60**, 376, 1938.

‡ W. J. de Haas and E. C. Wiersma, *Physica*, **2**, 335 and 438, 1935.

§ P. Kapitza, *Roy. Soc. Proc.* A, **119**, 358 and 401, 1928; **123**, 292, 1929.

|| B. Cabrera and J. Torroja, *Ann. de la Soc. Espan.* **11**, 443 and 525, 1913.

¶ W. Gerlach and K. Schneiderhan, *Ann. der Phys.* **6**, 772, 1930.

a change which is occasioned by the magnetic state of the nickel, i.e. by its spontaneous magnetisation. For example, at very high and very low temperatures where the rate of change of the spontaneous magnetisation with temperature is small, there is almost a normal linear change of resistance with temperature, but, in the neighbourhood of the ferromagnetic Curie point, where the spontaneous magnetisation varies rapidly, marked changes occur. Borelius* has shown that these changes may be represented by the equation

$$R_T = (R_N)_T + c(I_{\infty,0}^2 - I_{0,T}^2),$$

where R_T is the observed value of the resistivity at $T°$ K., $(R_N)_T$ the resistance of a normal metal at that temperature, $I_{\infty,0}$ and $I_{0,T}$ the spontaneous magnetisation (per c.c.) at 0 and $T°$ K. respectively, and c is a constant. It follows that the temperature coefficient of resistance of a ferromagnetic may be considered as made up of the normal temperature coefficient and a coefficient proportional to the rate of change of the energy of spontaneous magnetisation per c.c., but other effects depending on the presence of impurities will undoubtedly complicate matters.

As $I_{\mathscr{H},T}$ is increased when the ferromagnetic is placed in a strong magnetic field, particularly in the region of the ferromagnetic Curie point, we should expect corresponding changes of resistance. This phenomenon has been investigated by Gerlach, Englert,† Potter,‡ Scharff§ and others. Iron and nickel at room temperatures are found to show an increase in resistance in a longitudinal field and a decrease in a transverse field. In the neighbourhood of the ferromagnetic Curie point decreases are observed in both longitudinal and transverse fields. Heusler alloy shows decreases in both fields even at room temperature. Gerlach explains the difference in behaviour at high and low temperatures of iron and nickel on the assumption that two distinct processes contribute to the change in $I_{\mathscr{H},T}$ caused by the field, namely reversals and

---

* G. Borelius, *Ann. der Phys.* **8**, 262, 1931.
† E. Englert, *Ann. der Phys.* **14**, 589, 1932.
‡ H. H. Potter, *Roy. Soc. Proc.* A, **132**, 560, 1931; *Phil. Mag.* **13**, 233, 1932.
§ Gertrud Scharff, *Zeit. für Phys.* **97**, 73, 1935.

orientation processes. A contribution produced by reversal of the magnetic vector in a magnetic domain through 180° is comparatively small and causes a decrease, while the effect of orientation or turning of the vector (or extension of its boundary on the Becker hypothesis) is rather large and causes an increase in resistance in longitudinal fields. Orientation processes are very important at low temperatures. Fortunately, the phenomena are much simplified at temperatures in the immediate vicinity of the ferromagnetic Curie point, where, for the substances so far investigated, the same decrease in resistance is obtained for a given field applied longitudinally or transversely.

Potter first recognised that the resistance changes near the Curie point could be correlated with magnetocaloric effects, and he measured the ratio $(\Delta R/R)_T$ and the quantity of heat $C_{\mathscr{H}}\Delta T$ corresponding to the establishment or removal of a field $\mathscr{H}$ for iron, nickel and Heusler alloy. The wire was wound on a thin quartz sheet and its free ends used as potential leads, while current was supplied through two stout copper leads pinched on to the wire; these leads also acted as supports. A calibrated thermojunction was used to measure the temperature of the wire system, which, to prevent oxidation and maintain steady conditions was placed inside an evacuated quartz tube. The latter was mounted inside a long electric furnace with thick copper core, and set vertically between the poles of an electromagnet. A ground joint in the tube allowed the wire system to be adjusted with the wire parallel or perpendicular to the field.

Clearly, the whole of the wire in such a winding cannot be set parallel to the field, but correction can readily be applied for those portions perpendicular thereto, and, in any case, with experiments at temperatures near θ_f it can be neglected. The main source of error lies in the value assumed for the demagnetisation factor of the specimen, for the effect is considerable even at temperatures near θ_f. The magnetocaloric changes ΔT were not measured with the wire specimen but with a solid specimen; this is unfortunate, for the purity of the individual specimens has great effect on the results. The values of C_σ were obtained from Klinkhardt's work, p. 269. On plotting $\Delta R/R$ and $C_\sigma \Delta T$ against the corresponding values of T/θ_f for a field of 7000 oersteds, the two curves being made to coincide at their maxima, very good agreement between the resistance and magnetocaloric curves was obtained over the range $T/\theta_f = 0\cdot99$ to $1\cdot07$ in the case of iron and Heusler alloy, and over an even wider range in the case of nickel. The maximum values of $\Delta R/R$, $C_\sigma \Delta T$ and their ratio with $\mathscr{H} = 7000$ were found for iron, nickel and Heusler alloy; the ratio differs considerably for the three substances, but it increases as θ_f increases.

An interesting point arises when a very close study of the resistance changes in pure nickel with rise in temperature is made in zero field. Svensson* has shown that at θ_f a very sharp discontinuity is recorded in the resistance with temperature curve, a finding confirmed by Potter.† Yet we know from specific-heat measurements that the energy associated with ferromagnetism does not suddenly disappear at θ_f but "tails off". This can readily be explained as another domain size effect. The mean free path of an electron at the Curie point is about 20 times the interatomic distance, so that with a breakdown of the larger domains, the resistance which depends on *long-distance order* or superlattice structure of the atoms shows a sudden change. If, when the larger domains have broken down, spin interaction still persists in the smaller domains or regions of short-distance order, then magnetic energy still remains and we get the "tail" shown in specific-heat and $\sigma_{0,\,T}$, T/θ_f curves. Bragg and Williams‡ in their studies of atomic arrangement in alloys emphasise that the average orientation of electron spins in a ferromagnetic corresponds to the degree of ordered arrangement in an alloy, and that the critical temperature above which the atoms in an alloy are arranged in a completely random manner, and below which order or superlattice exists, corresponds to θ_f.

The Thermoelectric Properties of Ferromagnetics. It has

* B. Svensson, *Ann. der Phys.* **22**, 97, 1935.
† H. H. Potter, *Nature*, **139**, 411, 1937.
‡ W. L. Bragg and E. J. Williams, *Roy. Soc. Proc.* **145**, 699, 1934; **151**, 540, 1935; **152**, 231, 1935.

long been known that the thermoelectric properties of some
ferromagnetics exhibit anomalies in the regions of their
ferromagnetic Curie points, and they excited much interest
when Dorfman and Jaanus* suggested that the variation of
the Thomson coefficient of a ferromagnetic represented the
behaviour of its specific heat in zero external field, provided
that the ferromagnetism was mainly, if not entirely, due to
free or conduction electrons. This view, however, cannot now
be maintained. Dorfman and Jaanus measured the thermoelectric power of nickel between 0 and 400° C., and found that
in the neighbourhood of θ_f there was a sudden change in the
value of $eT(d^2E/dT^2)$. The latter expression represents the heat
absorbed when an electron with charge e is taken from one
point in the metal to another at a temperature 1° higher; the
expression $T(d^2E/dT^2)$ represents, of course, the Thomson
coefficient.

The observed effects were shown to be of the order of
magnitude predicted by the theory, but they were opposite in
sign. A similar result was obtained† in the case of iron, and
Bates‡ showed that in the case of manganese arsenide the sign
of the observed effect depended on the amount of free arsenic
or impurity present, and that pure specimens appeared to give
the sign required by theory. The experimental procedure consisted in measuring the thermoelectric E.M.F., ΔE, between
two ends of a metal rod or wire which were maintained at
temperatures T and $T+\Delta T$ respectively, a neutral metal
such as copper being used to complete the thermoelectric
circuit. In this way were obtained data for the graph of
$\Delta E/\Delta T$ against T, from which values of $T(d^2E/dT^2)$ could be
found. In view, however, of the fact that such experiments
have not, so far, given results of theoretical importance, and
appear to be markedly influenced by secondary phenomena,
further details will not be given here. An example of secondary
phenomena is provided by Foster's§ measurements of the

* J. Dorfman and R. Jaanus, *Zeit. für Phys.* **54**, 277, 1929.

† J. Dorfman, R. Jaanus, K. Grigorow and M. Czernichowski, *Zeit. für Phys.* **70**, 796, 1931.

‡ L. F. Bates, *Phil. Mag.* **13**, 393, 1932.

§ A. W. Foster, *Phil. Mag.* **18**, 470, 1934.

thermoelectric power of nickel and of two alloys of nickel containing 1 and 2 per cent of chromium. The Curie point changes in the Thomson coefficient shown by these alloys are respectively 30 and 10 per cent of those exhibited by pure nickel.

CHAPTER X

MAGNETOSTRICTION

Introduction. It has long been known that ferromagnetics exhibit *magnetostriction* or changes in dimensions when exposed to a magnetic field, but real progress in the interpretation of the experimental data has only been made within the last few years. The more important phenomena are usually, for convenience, treated under two headings, viz. the production of mechanical strains by magnetic fields and the production of magnetic changes by mechanical stresses. Under the first heading we treat the longitudinal and transverse *Joule effects* or the changes in the linear dimensions of a ferromagnetic which result when a magnetic field is applied along a specified direction, e.g. along the axis of a rod. We also treat the *Guillemin effect*, or the bending which a ferromagnetic exhibits when exposed to a magnetic field. In addition, we have the twisting of a ferromagnetic under the combined action of circular and longitudinal fields, known as the *Wiedemann effect*, and, further, the volume changes and the changes in Young's modulus and in the rigidity and bulk moduli which accompany magnetisation.

Under the second heading we deal with the changes in magnetic induction caused by longitudinal and transverse mechanical stresses of mechanical origin, known respectively as the longitudinal and transverse *Villari effects*. Here, too, we consider the changes in magnetic induction produced by bending, torsion or hydrostatic pressure. Besides the phenomena listed under the above headings there are others which must be studied in conjunction with them, as, for example, the discontinuous changes in magnetisation known as the Barkhausen effect.

The Joule Effect. It is well to realise at the outset that precision measurements in magnetostriction are only possible with pure materials, carefully annealed and free from the effects of cold working. The maximum changes of length which accompany technical saturation of ferromagnetics are merely some 5×10^{-5} times the original length, and, consequently, can only be measured if adequately magnified by mechanical or optical devices. Special difficulties accompany the measurements, and these may be appreciated by considering the interferometer method of measurement used by Lochner* and by Stevens† for the longitudinal Joule effect.

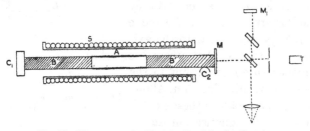

Fig. 87. Measurement of longitudinal Joule effect by
Michelson interferometer.

The rod specimen A lies wholly within the magnetising solenoid S, Fig. 87, brass extension pieces BB' enabling the system to be clamped at C_1 and supported at C_2. The mirror M of the Michelson interferometer is mounted at the end of B', so that when A is magnetised and it expands or contracts, the interference fringe system moves across the field of view and the change in dimensions is thus known from the fringe shift. The specimen must be in a uniform field and the demagnetising factor must be small; hence S should be as long as possible. The temperature of the specimen and extension pieces must be kept constant, and S must therefore be wound on a water-cooled former. If A is not placed symmetrically with respect to the solenoid field, magnetic forces will tend to displace it parallel to the axis and thus cause it to be strained. If the horizontal component of the earth's field is not neutralised S must be set perpendicular to the magnetic meridian.

Optical levers‡ are frequently employed in conjunction with mechanical levers. In a form used by Williams,§ shown in Fig. 88, the specimen is mounted vertically, and as it expands

* S. J. Lochner, *Phil. Mag.* **36**, 498, 1893.
† J. S. Stevens, *Phys. Rev.* **7**, 19, 1898.
‡ See A. F. Pollard, *Journ. Scient. Instr.* **15**, 37, 1938, and B. H. C. Mathews, *ibid.* **16**, 124, 1939.
§ S. R. Williams, *Journ. Opt. Soc. Amer.* **14**, 385, 1927.

or contracts it causes a lever L to fall or rise respectively. To the far end of L is attached a fine phosphor bronze strip, which is wound upon a narrow roller and weighted. The longer L and the narrower the roller, the greater is the motion of the strip and the greater the rotation of the mirror m. When levers are employed they should be arranged, as in the above example, perpendicular to the axis of the specimen, so that errors due to temperature fluctuations in the lever system are a minimum. Allowance has to be made in these experiments for demagnetisation due to

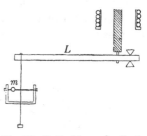

Fig. 88. Optical lever for Joule effect. (Williams.)

free poles. This can of course be avoided by using a ring specimen and measuring the magnetostrictive change in diameter of the ring by an optical lever system.*

Lloyd† has described a high frequency method of measurement which has given satisfactory results with short polycrystalline materials, and which could clearly be used with single crystals. Handley‡ has adapted the method for use with cold drawn wires, and Fricke§ has used a similar method for measurements of the transverse effect. The magnetostriction change of dimensions is used to produce a change in the capacity of a parallel plate condenser forming part of an oscillatory circuit, as in the Whiddington‖ ultramicrometer. The combination note formed with a standard oscillator whose frequency can be accurately adjusted is picked up by an amplifier-detector unit and compared with the frequency of a valve-maintained tuning-fork. Changes in capacity in the first circuit can thus be readily detected and measured accurately, the calibration being made by displacing one condenser plate through a known distance.

* Cf. Shelford Bidwell, *Roy. Soc. Phil. Trans.* A, p. 205, 1888.
† H. Lloyd, *Journ. Scient. Instr.* **6**, 81, 1929. See also W. Alexander and J. Swaffield, *Beama Journ.* **41**, 99, 1937.
‡ J. H. Handley, *Journ. Scient. Instr.* **6**, 84, 1929.
§ W. Fricke, *Zeit. für Phys.* **80**, 324, 1933.
‖ R. Whiddington, *Phil. Mag.* **40**, 634, 1920.

The Joule Effect with Single Crystals. Measurements with single crystals are complicated by the smallness of the specimens normally available. The apparatus used by Honda and Masiyama* is shown in Fig. 89 a, b, dealing respectively with the longitudinal and the transverse effects. In the former the ellipsoidal-shaped crystal C is placed upon a support A and placed inside a solenoid S. A quartz tube T passes over a roller R, and to its ends are connected the metal attachments

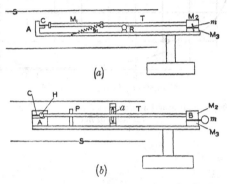

Fig. 89. Honda and Masiyama's apparatus for Joule effects with single crystals.

M_1 and M_2, and two springs S_1 cause M_1 to press the crystal against its support. Between M_2 and the metal surface M_3 is a very narrow roller to which the mirror m is attached. As C expands or contracts, m rotates and deflects a beam of light whose displacement is measured with a micrometer eyepiece. In investigating transverse effects, i.e. the changes in dimensions which take place at right angles to the applied magnetic field, a spring at P presses a brass attachment H against the crystal C. When the latter expands or contracts laterally, H is forced outwards and the metal block B moves sideways, causing a fine roller carrying the mirror m to rotate.

The variation of the longitudinal Joule effects of typical polycrystalline specimens with the applied field is shown in Fig. 90, while typical results for single crystal specimens of iron are shown in Fig. 91. The effect is termed positive when

* K. Honda and Y. Masiyama, *Sci. Reports Tôhoku*, **15**, 755, 1926 and **20**, 574, 1931. See also D. Kirkham, *Phys. Rev.* **52**, 1162, 1937.

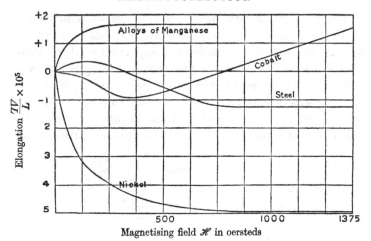

Fig. 90. Longitudinal Joule effect for polycrystalline metals as function of field.

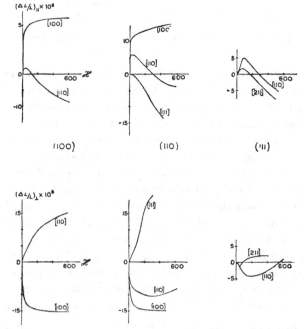

Fig. 91. Joule effects in iron single crystals. (Honda and Masiyama.)

302

an increase in dimensions results on the application of the field. In the latter cases, the longitudinal strains $(\Delta L/L)_{\parallel}$ and the transverse strains $(\Delta L/L)_{\perp}$ are of considerable interest in discussing the changes of volume produced by magnetisation, p. 308. With single crystals of nickel $(\Delta L/L)_{\parallel}$ is negative along all directions, the contraction being greatest along the [100] and least along the [111] directions, while $(\Delta L/L)_{\perp}$ is always positive, being greatest in the [100] and least along the [111] directions.

The Joule Effect with Non-ferromagnetics. The existence of a Joule effect in non-ferromagnetic substances was first proved by Kapitza* using the magnetic balance described on p. 112 with intense magnetic fields. The diaphragm of the balance was attached by means of a quartz connecting rod to the top of the metal specimen under examination. The lower end of the specimen was attached by a quartz tube surrounding the quartz rod to a massive ring piston which moved with considerable friction in oil. Consequently, slow changes of temperature did not produce serious distortions of the balance diaphragm, and yet the rapidity with which the magnetic field was applied meant that the lower end of the specimen was rigidly fixed for the purposes of the experiment. The extensometer so constructed was capable of measuring changes in length down to 10^{-7} cm. in a short interval of time; it had to be calibrated by a special method. Kapitza definitely established the Joule effect in bismuth, antimony, graphite and gallium and considered that it probably existed in the cases of tin and tungsten.

Wolf and Goetz† were able to make measurements with pure and alloyed bismuth crystals, using a steady magnetic field of 28,000 oersteds produced by a 500 K-watt solenoid consisting of two coaxial coils of heavy copper tape. The actual apparatus was of complicated design, but the simple principle upon which it worked may be seen from Fig. 92. The bismuth crystal Bi changed its length relatively to that of a neutral material N and so rotated the mirror M, causing a beam of light to fall

* P. Kapitza, *Roy. Soc. Proc.* A, **135**, 556, 1932.

† A. Wolf and A. Goetz, *Phys. Rev.* **46**, 1095, 1934.

upon a Moll thermo-relay *Th*, which is a differential thermo-couple of constantan C between copper ribbons Cu of low thermal capacity, mounted inside an evacuated glass bulb. The bulb was mounted on a micrometer stage and adjusted so that in the zero position the light from *M* heated each junction equally and no current flowed through the galvanometer *G*; but when *M* was rotated slightly the junctions were unequally heated and a current flowed. A motion of the spot of light by

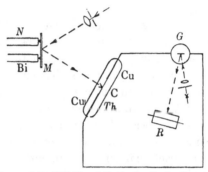

Fig. 92. Apparatus of Wolf and Goetz for Joule effect in bismuth.

about 0·0007 mm. gave a galvanometer deflection of 1 mm. The deflections were recorded photographically, and were calibrated by raising the temperature of the crystal about 0·2° C.

The need for such large magnification arises because the magnetostriction of bismuth in a field of 25,000 oersteds is still about 100 times smaller than that of nickel in a moderate field. Wolf and Goetz obtained results in good agreement with those obtained by Kapitza for pure crystals of bismuth. It was found that the magnetostriction was positive in a direction parallel to the trigonal axis of bismuth and negative at right angles thereto, the magnetostriction equations being respectively $(\varDelta l/l) = \text{const.} \times \mathcal{H}^{1·6}$ and $(\varDelta l/l) = \text{const.} \times \mathcal{H}^{2·3}$, the constant depending on the crystal orientation. The effects of certain impurities were considerable; e.g. about 0·4 per cent of tin or 2 per cent of lead reduced the magnetostriction perpendicular to the trigonal axis to zero, and greater quantities of these substances reversed its sign.

Becker's Theory of Magnetostriction Processes. Becker[*]
has given an interesting picture of the behaviour of a ferromagnetic in the region of technical saturation. He supposes
that the Weiss domains are magnetised almost to saturation
when $T \ll \theta_f$. Then, assuming the total energy associated with
a distorted dipole lattice in the presence of an applied field to
be made up of the internal magnetic energy of the lattice, the
energy stored in the elastic deformation and the potential
energy of the domains in the applied field, the $I, \mathcal{H}$ curve can
be obtained from the condition that the total energy shall be a
minimum, and fairly satisfactory, though qualitative, pictures
of technical hysteresis and saturation processes are thus
provided. Certain aspects of the theory were considered in
Chapter IV.

From the fact that when a nickel wire is magnetised
longitudinally it invariably shortens, Becker deduced that
under the simultaneous influence of a severe longitudinal stress
F and a longitudinal field $\mathcal{H}$ it should be magnetised with the
magnetic vectors making an angle of average value θ with the
axis of the wire given by

$$\cos \theta = \frac{I_s}{3\lambda_{0,s} F} \mathcal{H}, \qquad \dots\dots(189)$$

where I_s is the technical saturation intensity of magnetisation
and $\lambda_{0,s}$ is the experimental value of the contraction per unit
length when saturated. Under these conditions the intensity of
magnetisation parallel to the axis of the wire is

$$I_\parallel = \frac{I_s^2}{3\lambda_{0,s} F} \mathcal{H} = C\frac{\mathcal{H}}{F}, \qquad \dots\dots(190)$$

where C is a constant. It follows from (189) that under very
severe tension a nickel wire must be magnetised perpendicularly to its axis since $\cos \theta = 0$, and from (190) that technical
saturation should be obtained in a field

$$\mathcal{H}_s = \frac{3\lambda_{0,s} F}{I_s}.$$

[*] R. Becker, *Zeit. für Phys.* **62**, 253, 1930; **64**, 660, 1930.

Becker and Kersten investigated the magnetisation of a nickel wire under tension by a method used by Preisach.* The wire was mounted vertically inside a long solenoid. Its middle portion was surrounded by an induction solenoid connected to a ballistic galvanometer. The ends of the wire were held in brass clamps, one being directly attached to a strong aluminium frame, and the other attached via a calibrated spring balance to the usual spindle and nut stretching device. The longitudinal magnetisation was measured by the ballistic method, and equation (190) was verified.

Kirchner† investigated the effects of tension, pressure and torsion on the longitudinal magnetostriction of a nickel wire using a comparative ultramicrometer method. He proved that for nickel the magnetostrictive contraction per unit length under a tension F, and in fields of sufficient intensity, λ_F obeyed a law of the form $\lambda_F = C'(I^2 - I_F^2)$, where C is a constant. Under increasing pressure the magnetostriction of nickel decreased slowly towards zero; and under increasing torsion it reached a limiting value of about $0·72\,\lambda_{0,s}$, and, again, for large torques obeyed a law $\lambda_t = C'(I^2 - I_t^2)$.

From the experimental verification of equation (190) it follows from thermodynamic arguments that the saturation magnetostrictive contraction $\lambda_{F,s}$ of a nickel wire under longitudinal tension F and field $\mathscr{H}_s$ is given by

$$\lambda_{F,s} = C\mathscr{H}_s^2/2F. \qquad \ldots\ldots(191)$$

Equation (191) may be rewritten as $C = I_s^2/2\lambda_{F,s}$. Now, Becker and Kersten found that $C = I_s^2/3\lambda_{0,s}$; hence it follows that $\lambda_{F,s} = \frac{3}{2}\lambda_{0,s}$, or, the saturation magnetostrictive contraction of a nickel wire under severe tension is exactly 50 per cent greater than that of an unstretched wire, a prediction, verified by Becker and Kersten and by Kirchner, which it would be difficult to comprehend except in terms of the Weiss domain theory.

We see that for the unmagnetised, unstretched wire $\cos^2\theta_0 = \frac{1}{3}$, since the magnetic vectors are orientated in all

* F. Preisach, *Ann. der Phys.* **3**, 737, 1929.
† H. Kirchner, *Ann. der Phys.* **27**, 49, 1936.

possible directions, while for saturation $\cos^2 \theta_s = 1$. Now, from Kirchner's experimental results, $\lambda_{F,s}$ is proportional to $(I_s^2 - I_F^2)$, and we have

$$\lambda_{F,s} = \alpha(\cos^2 \theta_s - \cos^2 \theta_F) = \alpha(1 - 0) = \alpha,$$

$$\lambda_{0,s} = \alpha(\cos^2 \theta_s - \cos^2 \theta_0) = \alpha(1 - \tfrac{1}{3}) = \tfrac{2}{3}\alpha,$$

where α is a constant, so that therefore $\lambda_{F,s}/\lambda_{0,s} = \tfrac{3}{2}$, and we have a ready interpretation of the relation deduced by Becker, much on the lines of the explanation of the changes of resistance of nickel in a magnetic field. In the case of torsion we must assume that the vectors set at an angle of 45° to the axis, so that $\cos^2 \theta_t = \tfrac{1}{2}$; we have then

$$\lambda_{t,s} = \alpha(\cos^2 \theta_s - \cos^2 \theta_t) = \alpha(1 - \tfrac{1}{2}) = \alpha/2,$$

$$\lambda_{0,s} = \tfrac{2}{3}\alpha,$$

whence $\lambda_{t,s} = \tfrac{3}{4}\lambda_{0,s}$, while $0 \cdot 72 \lambda_{0,s}$ is found by experiment. Under heavy pressure we have $\lambda_{p,s} = \alpha(\cos^2 \theta_s - \cos^2 0) = 0$, i.e. the magnetostriction disappears.

Effects of Shape on Magnetostriction. It has hitherto been assumed that magnetostriction effects are independent of the shape of the specimen, but this is not correct. Becker[*] and Hayasi[†] have shown that shape is important. It is found, for example, that the longitudinal Joule effect for an ellipsoid should theoretically be greater by a term proportional to $I^2 . D$ than for an infinitely long and thin rod, where I is the intensity of magnetisation and D the demagnetising factor of the ellipsoid. Moreover, the volume effect and the transverse Joule effect should each be greater for the ellipsoid by a term proportional to $I^2 . D$.

Kornetzki[‡] made measurements with a series of iron ellipsoids to check these predictions in the case of the longitudinal effect. Disturbances due to translational forces in non-uniform fields were difficult to avoid, and fields of 10,000 oersteds were produced by currents of short duration in a solenoid which was not water-cooled. The limits of experimental error were

[*] R. Becker, *Zeit. für Phys.* **87**, 547, 1933.
[†] T. Hayasi, *Zeit. für Phys.* **72**, 177, 1931.
[‡] M. Kornetzki, *Zeit. für Phys.* **87**, 560, 1933

rather large but the theoretical predictions could be regarded as confirmed, the value of $(\Delta l/l)_{\parallel}$ being $-3\cdot15 \times 10^{-6} \pm 5$ per cent for a sphere instead of -7×10^{-6} as found for an infinitely long rod.

More comprehensive theoretical discussion shows that in weak fields, where, as we saw on p. 62, $I = \mathcal{H}/D$, the *volume magnetostriction* $\omega = (\delta V/V)$ is equal to $\frac{1}{2}\mathcal{H}^2/k'D$, where k' is the compressibility of the material, which is a parabolic relation between ω and $\mathcal{H}$, while in high fields ω becomes a linear function of $\mathcal{H}$. There is, of course, a transition region in the ω, $\mathcal{H}$ curve between these two cases for fields of moderate intensity.

Volume Changes in Magnetostriction. Volume changes accompanying magnetisation have been measured by Honda and Nagaoka,* Masiyama| and Kornetzki.‡ The specimen was in all cases an ellipsoid and was mounted inside a dilatometer containing boiled water; the changes in volume of the ellipsoid were manifested by the motion of the meniscus in a capillary tube. The dilatometer was placed within a magnetising solenoid provided with an internal water jacket, and in some cases was also surrounded by a copper tube.

Kornetzki's work showed that the ω, $\mathcal{H}$ curve fell into three parts—an initial parabolic portion depending only upon the shape of the specimen, a linear portion for fields above that required for technical saturation, and a transition portion between them. With a small value of D, ω may appear to decrease with $\mathcal{H}$ at first, but with a large value of D, ω always increased with $\mathcal{H}$, in the case of iron specimens.

Auwers§ investigated the ω, $\mathcal{H}$ curves of single crystals of iron-nickel alloys and of polycrystalline specimens of iron-nickel-cobalt alloys. In the case of the single crystals there was practically no change in volume until technical saturation was exceeded when a linear increase in volume occurred, the value of $d\omega/d\mathcal{H}$ being independent of the orientation of the crystal

---

* K. Honda and Nagaoka, *Phil. Mag.* 4, 56, 1902.

† Y. Masiyama, *Sci. Rep. Tôhoku*, 22, 338, 1933.

‡ M. Kornetzki, *Zeit. für Phys.* 87, 560, 1933; 98, 289, 1935.

§ O. von Auwers, *Phys. Zeit.* 34, 824, 1934.

lattice with respect to the field, although the intercept which the ω, $\mathscr{H}$ curve, when produced, made with the $\mathscr{H}$ axis became smaller as the direction of easiest magnetisation was approached. The value of $d\omega/d\mathscr{H}$ however depended considerably on the constitution of the crystal.

The Weiss Constant and Magnetostriction. It is clear that a ferromagnetic substance which exhibits volume magnetostriction must change its magnetisation when it is compressed in the presence of an applied field, and that this change must be related to the change in volume or in lattice constant produced by the compression. The question therefore arises as to what effect a change in volume may have on the quantity $\nu\rho$ of the Weiss theory of ferromagnetism. If the product $\nu\rho$ is dependent upon the distance apart of the carriers in a material, i.e. upon the density ρ, then the spontaneous magnetisation at a given temperature must depend upon the volume, and, consequently, the paramagnetic Curie point, defined by $\theta = \dfrac{\nu\rho z\mu^2}{3k}$, must change with pressure. Indeed, Kornetzki suggests that this explains Steinberger's* observation that a 30 per cent nickel-iron alloy loses its ferromagnetic properties when exposed to a pressure of 10,000 atmospheres at room temperature.

Now, in Chapter VIII the Curie point θ_f of a substance was always measured under atmospheric pressure and no attempt was made to maintain the test specimen at constant volume. If, however, we deliberately keep the volume of a specimen constant at some chosen value, say V_T, the volume normally found under atmospheric pressure at a temperature T, the substance would be expected to lose its ferromagnetism at a different temperature to θ_f, say $(\theta_f)_{V_T}$. As we are interested in $\nu\rho$ it will be more convenient to speak of $(\theta)_V$ and $(\theta)_{V_T}$ for obvious theoretical reasons. We may write

$$\frac{1}{(\nu\rho)_{V_T}}\frac{\partial(\nu\rho)_{V_T}}{\partial T} = \frac{1}{(\theta)_{V_T}}\frac{(\partial\theta)_{V_T}}{\partial T} = 2\alpha\frac{1}{(\theta)_{V_T}}\frac{(\partial\theta)_{V_T}}{\partial\omega},$$

* R. L. Steinberger, *Physics*, 4, 153, 1933.

which Kornetzki* deduces is equal to

$$-\frac{3\alpha}{T}\frac{\dfrac{\partial\omega}{\partial\mathscr{H}}\cdot\dfrac{1}{k'}}{\partial\sigma_{\mathscr{H},T}/\partial T-3\alpha\cdot\dfrac{\partial\omega}{\partial\mathscr{H}}\cdot\dfrac{1}{k'}},$$

very approximately, where α is the coefficient of linear expansion and k' is the compressibility. All the quantities in the last expression can be measured and we can therefore find how $\dfrac{1}{\nu\rho}\dfrac{\partial(\nu\rho)}{\partial T}$ varies with temperature if we determine a series of ω, $\mathscr{H}$ curves for different values of T.

Kornetzki carried out measurements of the volume magnetostriction of several ellipsoids of iron and iron-nickel alloys over a temperature range 20° to 95° C. The dilatometer was placed inside an electrical furnace within a Dewar flask enclosed by the magnetising solenoid, elaborate temperature control being avoided by the use of an automatic recording apparatus which simultaneously recorded the volume changes and the applied fields. The experimental error was of the order of 10 per cent, but in the case of iron $\dfrac{1}{(\nu\rho)_{V_T}}\dfrac{\partial(\nu\rho)_{V_T}}{\partial T}$ was found practically independent of the temperature and equal to 5×10^{-4} per degree, which means that the Curie point would be about 400° C. if the iron were demagnetised under the pressure necessary to maintain the volume at 400° C. the same as that under atmospheric pressure at room temperature; unfortunately the necessary 60,000 atmospheres is not obtainable.

The Villari Effect. We have seen that when a ferromagnetic body is magnetised it is strained. Matteucci† showed that, conversely, when a magnetised bar is strained its magnetisation is changed. Thus, when a weakly magnetised iron bar is stretched the magnetisation increases. Villari‡ later showed that the magnetisation of a strongly magnetised iron bar is decreased by tension, and showed that for a certain intensity

* M. Kornetzki, *Zeit. für Phys.* **98**, 289, 1935. Cf. R. Smoluchowski, *Phys. Rev.* **59**, 309, 1941.   † C. Matteucci, *Compt. Rend.* **24**, 301, 1847.
‡ E. Villari, *Ann. der Phys.* **126**, 87, 1865.

of magnetisation of the iron there is no change on applying tension. The change in magnetisation or, more exactly, magnetic induction, which accompanies longitudinal stress is known as the *Villari effect* and the reversal of the sign of the effect is termed the *Villari reversal*, the point on the $I, \mathscr{H}$ curve where the application of longitudinal stress does not change I being called the *Villari reversal point*. The effects were studied by Kelvin* and Ewing†.

It is very instructive to discuss the reciprocal relation which exists between the longitudinal Joule and Villari effects. We may define the constant λ of the former by the equation $p = \lambda . \mathscr{H}$, where p is the magnetostrictive stress (in dynes per sq. cm.) produced by a field $\mathscr{H}$. The mechanical energy stored in the material is then $p^2/2E$ ergs per c.c., where E is Young's modulus. Now this is balanced by the change in magnetic energy $B^2/8\pi\mu$. Therefore

$$B^2/8\pi\mu = p^2/2E = \lambda^2 \mathscr{H}^2/2E = \frac{\lambda^2 B\mathscr{H}}{2\mu E};$$

consequently,

$$B/4\pi = \frac{\lambda^2 \mathscr{H}}{E} = \lambda p/E = \lambda \times \text{(extension per}$$

unit length produced by p),

or $\quad B = 4\pi\lambda \times$ (extension per unit length produced by p).

We see then that the change in induction B which accompanies a given longitudinal stress is connected with the strain by the same constant λ which connects the stress with the applied field $\mathscr{H}$. We shall make use of this result on p. 316.

The Wiedemann Effect. When a longitudinal and a circular magnetic field are simultaneously applied to a long rod of ferromagnetic material, the resultant lines of force form helices about the axis of the rod, and any change in dimensions of the material along the helices causes the rod to twist; this phenomenon is called the *Wiedemann effect*. Wiedemann‡ suspended a vertical iron wire and caused it to be acted upon by a small vertical magnetic field while at the same time an

---

* Lord Kelvin, *Published Papers*, **11**, 332.
† J. A. Ewing, *Magnetic Induction in Iron and other Metals*, p. 197, 1910.
‡ W. Wiedemann, *Elektrizität*, **3**, 680, 1883.

electric current passed along the wire. He found that the free end of the iron wire twisted in a clockwise direction when viewed by an observer looking along the wire in the direction of the current flow and in the direction of the lines of force of the solenoid; the twist was then said to be positive. On reversal of either the current or the longitudinal field the direction of the twist was reversed. He further found that if the longitudinal field was gradually increased while the current in the wire was maintained constant the twist reached a maximum value in fields of some 15 to 36 oersteds, and gradually decreased in stronger fields until the twist became zero and eventually changed sign.

Clearly, the use of a wire or rod is open to the objection that the circular magnetic field produced by the current at any point in it varies directly as the distance of the point from the axis of the specimen, so that it becomes difficult to compare experimental results with those expected on theoretical grounds. Williams* surmounted this difficulty to some extent by using tubular specimens, but it is not always possible to obtain ferromagnetic specimens in this form. The apparatus used by Pidgeon† is shown in Fig. 93.

The solenoid S was attached to a rigid beam P_1 and the specimen WW with its extension pieces was suspended on knife-edge supports from a similar but independent beam. The vertical component of the earth's magnetic field was compensated by an additional solenoid wound on S, and the specimen was surrounded by a water-jacket J. Electrical connections were made through the mercury cups V_1 and V_2 and the wire specimen was kept taut by the weight F immersed in

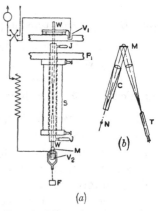

Fig. 93. Pidgeon's apparatus for the Wiedemann effect.

* S. R. Williams, *Amer. Journ. Sci.* **36**, 555, 1913.
† H. A. Pidgeon, *Phys. Rev.* **13**, 209, 1919.

oil. The optical arrangements are also shown in Fig. 93 b. Light from a Nernst filament N fell upon the collimator C, in whose slit a fine glass fibre was mounted lengthwise, virtually to form a double slit. The parallel beam issuing from the collimator was reflected by a plane mirror M. The reflected beam was viewed in the travelling microscope T, and the diffraction pattern produced by the double slit served to provide accuracy in setting.

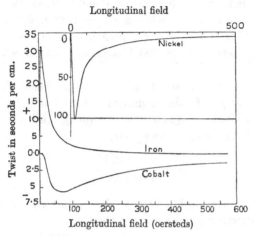

Fig. 94. Pidgeon's results for the Wiedemann effect; circular field constant.

Some of Pidgeon's results are shown in Fig. 94; they have been corrected for any slight residual twist observed when, after careful demagnetisation, either a longitudinal or a circular field alone was applied. Errors may arise due to imperfect initial demagnetisation, changes in temperature produced by the heating effects of the current, and lack of homogeneity in the specimen. In addition to the curves shown, other instructive sets may be obtained by plotting the observed twist against the current in the specimen for a series of fixed values of the longitudinal field.

Initial Permeability and Magnetostriction. We saw on p. 79 that satisfactory permanent magnet materials were those in which severe internal strains were present, and that substances of high permeability had to be as strain-free as possible. Now, if a substance exhibits magnetostriction, then it is bound to be strained when placed in a magnetic field and the strain will tend to oppose the magnetisation, i.e. the initial permeability of the material will be lowered. While thorough annealing will relieve strains produced by working, phase changes, etc., it will not relieve magnetostriction strains. In many cases, however, these can be relieved by annealing while the material is in a magnetic field. Thus Bozorth and Dillinger* showed that when a 65 per cent nickel iron alloy was annealed at 650° C. in a field of only 10 oersteds, the maximum permeability was increased from 20,000 to over 600,000.

Kersten† has obtained a relation between the initial permeability μ_i, I_0 the technical saturation intensity, the saturation Joule magnetostriction $(dl/l)_s$ and the average value of the internal stress p_i, viz.

$$\mu_i = \frac{0 \cdot 018\, I_s^2}{16\pi^2\, (dl/l)_s\, p_i}.$$

When the annealing has been perfectly carried out and only magnetostriction strains remain, then

$$p_i = E(dl/l)_s, \quad \text{and} \quad \mu_i = \frac{0 \cdot 018\, I_s^2}{16\pi^2\, (dl/l)^2 E},$$

which gives us a theoretical upper limit for the value of μ_i. It is an important fact that the permalloy of highest initial permeability is also that for which the magnetostriction is zero, or practically zero, and for which μ_i is theoretically infinite or extremely large.

Magnetostriction Oscillators. Since the application of a magnetic field causes changes in the dimensions of ferro-magnetic bodies, it is easy to see that, by applying an alternating magnetic field of appropriate frequency, a rod of ferro-magnetic material may be set into longitudinal vibration in

---

* Joy Dillinger and R. M. Bozorth, *Physics*, 6, 279 and 285, 1935 and R. M. Bozorth, *Bell Syst. Tech. Journ.* 15, 63, 1936.

† M. Kersten, *Zeit. für tech. Phys.* 12, 665, 1931.

unison with the current producing the field when resonance magnetostriction oscillations of considerable amplitude may be produced. Magnetostriction oscillators were first studied by Pierce,* who investigated their possibilities as standards of frequency. The oscillators† were generally in the form of thin rods. Each rod was coupled to an electrical circuit by placing it in a solenoid carrying a small alternating current superimposed on a steady magnetising current, the rod being set into longitudinal resonant vibration in its fundamental mode with a node at its centre and antinodes at its ends. In addition to the demagnetising effects of the poles induced in such oscillators, disturbing effects arise because the mechanical stress is greatest at the centre of a rod and zero at its ends. Hence, the modern oscillator now takes the form of a closed circular ring wound with a uniform toroidal winding. As the ring vibrates, all parts of it move radially with the same amplitude and phase, and when the vibration is excited by the passage of an alternating current of suitable frequency, the magnetisation and the mechanical stress do not vary along the circumference, although these quantities must vary over the cross-section of the ring, because of eddy currents induced in it.

Let us suppose that a ring oscillator is rigidly clamped so that no motion is possible and that a current of pulsatance ω is supplied to the winding. Then a mechanical stress

$$p = \lambda \cdot \mathcal{H}_0 \sin \omega t$$

will be set up, where λ is a constant representing the *dynamical Joule magnetostriction effect*. If we now allow the ring to vibrate, this stress may be treated as a motive force and the equation of motion takes the conventional form for a system vibrating with one degree of freedom. If the ring is of mean density ρ, area of cross-section A and of material whose Young's modulus is E, then the radial force on an element subtending an angle $\delta\theta$ at the centre of the ring is $\lambda A \mathcal{H}_0 \sin \omega t \, \delta\theta$. Hence, the circumferential stress due to magnetostriction is

* G. W. Pierce, *Proc. Amer. Acad. Arts and Sci.* **63**, 1, 1928.
† L. H. Vincent, *Proc. Phys. Soc.* **41**, 476, 1929 and **43**, 157, 1931.

equivalent to a total force tending to expand the ring equal to $2\pi A\lambda\mathscr{H}_0 \sin \omega t$, and the equation of motion is

$$2\pi r A\rho\ddot{x} + k_1\dot{x} + 2\pi A Ex/r = 2\pi A\lambda\mathscr{H}_0 \sin \omega t, \quad ......(192)$$

where $\ddot{x}$, $\dot{x}$ and x respectively represent the radial acceleration, velocity and displacement at any instant t, k is a damping constant and the last term on the left-hand side represents the effect of resistance to extension. For steady vibration the solution of equation (192) is

$$x = \frac{2\pi A\lambda\mathscr{H}_0}{\omega}\left\{1\bigg/\sqrt{k_1^2 + \left(2\pi r A\rho\omega - \frac{2\pi A E}{\omega r}\right)^2}\right\}\sin(\omega t - \epsilon),$$
$$......(193)$$

and when resonance sets in at $\omega = \omega_0$, then $\epsilon = 0$ and

$$2\pi r A\rho\omega_0 = \frac{2\pi A E}{\omega_0 r}$$

or $$\omega_0 = \frac{1}{r}\sqrt{E/\rho} = \frac{1}{r}V_s, \quad ......(194)$$

i.e. $$2\pi r n_0 = V_s,$$

where V_s is the velocity of sound in the ring, and n_0 is the frequency of the alternating current for resonance.

By differentiating (193) we see that the maximum radial velocity $\dot{x}_r$ at resonance is $2\pi A\lambda\mathscr{H}_0/k_1$. Now the electromotive force e induced in the winding of N turns by the Villari effect is equal to

$$-NA\frac{dB}{dt} = -NA\frac{d}{dt}\left(\frac{4\pi\lambda x}{r}\right) = -4\pi NA\frac{d}{dt}\left(\frac{\lambda x}{r}\right),$$

from the relation between I and x discussed on p. 311. Hence

$$e = -4\pi NA\lambda\frac{\dot{x}}{r}$$

$$= -\frac{4\pi NA\lambda}{r}(2\pi A\lambda)\mathscr{H}_0\frac{\cos(\omega t - \epsilon)}{\sqrt{k_1^2 + \left(2\pi r A\rho\omega - \frac{2\pi A E}{\omega r}\right)^2}},$$
$$......(195)$$

so that at resonance the maximum e.m.f. is

$$e_r = -\frac{8\pi^2 N A^2 \lambda^2 \mathscr{H}_0}{r}\frac{}{k_1} = -\frac{8\pi^2 N A^2 \lambda^2}{rk_1}\left(\frac{2Ni_0}{r}\right)$$

$$= -\frac{16\pi^2 N^2 A^2 \lambda^2}{r^2 k_1} i_0, \text{ since } \mathscr{H}_0 = \frac{2Ni_0}{r},$$

where i_0 is the maximum value of the current supplied to the coil.

Now from equation (142), p. 216, we see that the maximum displacement will be half that at resonance at two frequencies $\omega_0 \pm \delta\omega$, where

$$k_1 = 2 \cdot \delta\omega \cdot \rho r A \sqrt{1/3}. \qquad \ldots\ldots(196)$$

Hence, when the resonance is very sharp, we may make a series of determinations of the reactance and resistance of the system over a range of known frequencies and so determine $\delta\omega$ and e_r, whence λ can be found.

Smith* measured the reactance of a ring oscillator consisting of stampings of 40 per cent nickel-iron alloy, coated with paraffin wax and separated by paper rings to prevent eddy currents. The ring rested in a wooden channel over which two toroidal coils were wound, one to carry a direct current and the other to form the arm X of a Wheatstone bridge system. Alternating current of adjustable frequency was supplied by a valve oscillator, the detector was a sensitive thermionic voltmeter connected to the bridge through a tuned transformer, and a variable inductance L and resistance R were used to obtain balance. $\delta\omega$ was found to correspond to about 25 cycles per sec. for a resonance frequency of about 15,100 cycles per sec.

In the above theory eddy current and hysteresis effects have been ignored. These, however, play an important part, and the reader is referred to a paper by Butterworth and Smith† for further information. The more complete theory shows that a high degree of resonance may be obtained with a ring showing small magnetostriction changes provided that the permeability and resistivity of the material are large.

* F. D. Smith, *Phys. Soc. Proc.* **42**, 181, 1930.
† S. Butterworth and F. D. Smith, *Phys. Soc. Proc.* **43**, 166, 1931.

The Magnetostriction Echo Depth Recorder. Wood, Smith and McGeachy* have recently used magnetostriction oscillators in marine echo-sounding devices or depth recorders. The

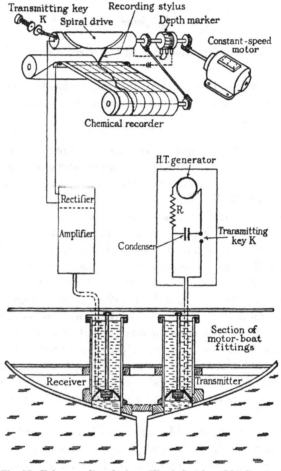

Fig. 95. Echo-sounding device. (Wood, Smith and McGeachy.)

general arrangement is shown in Fig. 95. Two magneto-striction oscillators, a transmitter and a receiver are fitted inside reflectors in separate compartments quite close to one

* A. B. Wood, F. D. Smith and J. A. McGeachy, *Journ. Inst. Elect. Eng.* **76**, 550, 1935.

another in a ship. Theoretically, the same oscillator could be used as transmitter and receiver, but there are obvious advantages in using separate oscillators for the two purposes. The magnetostriction oscillations of the transmitter are communicated by the reflector to the water in the transmitting compartment and pass through the hull of the ship to the sea-bed in the form of longitudinal compressional waves, whence they are reflected and are picked up by the receiver. The changes in pressure caused by their impact on the receiver produce changes in the magnetisation of the ferromagnetic; these are amplified, and a rectified current is fed to a chemical recorder as shown in the figure.

The chemical recorder is worked by a spiral drive, and the commencement of the motion is synchronised with the transmission of a signal, a small current causing the stylus to make a mark on the sensitised paper. A straight line or band corresponding to the series of transmission signals is traced on the paper as shown on the left of the diagram. The reflected signal is received when the stylus has moved some short distance across the paper, and a mark is produced as before. Hence for each passage of the stylus a mark is made corresponding to the reception of the reflected signal. In this way is traced the wavy band shown in the figure, and the distance between the two bands is a measure of the depth of the sea-bed beneath the ship at the time of record; this distance depends, of course, on the known rate at which the stylus is made to traverse the paper. A relatively simple type of damped impulse high-frequency transmission can be used, as indicated in the figure, or continuous-wave transmission can be employed. In the figure is also shown an auxiliary commutator which produces a series of equidistant dots in the paper, representing known depth intervals.

The magnetostriction oscillator most frequently used in these devices is made of nickel of ordinary commercial purity, chosen because of the large magnetostriction changes in relatively weak magnetic fields, simple composition and good mechanical properties, resistance to corrosion by water, and because it can be obtained in the form of suitable stampings.

A ring is made of stampings insulated from each other by films of oxide or of insulating varnish to oppose the formation of eddy currents. The ring is magnetised by current supplied to a toroidal winding. A nickel ring of mean diameter 10 cm., weighing about 3 K gm., gives a frequency of about 15,900 cycles per sec. The mean diameter fixes the frequency, and the radiation damping is settled by the thickness of the oscillator.

The receiver should possess some permanent magnetism and this is easily impressed by a large steady current of short duration. The method of mounting the transmitter is shown in the figure, which also shows the type of reflector used to impress directional properties upon the radiation. On account of the high frequencies employed diffraction effects are quite small, the transmitter and receiver require no elaborate screening and can be fitted quite close together on the ship, giving the great advantage that soundings can be taken directly beneath the ship. The form described above is capable of recording depths of 0 to 30 fathoms, which are the depths really important in navigation, correct to about 1 ft. of water.

Surface Fields on Ferromagnetics. Perhaps the most striking evidence for the domain theory of ferromagnetism is provided by the deposits formed by magnetic powders on the polished surface of a ferromagnetic. Suspensions of rouge, (Fe_2O_3), particles about 1μ in diameter in ethyl acetate were used by Bitter,* who obtained the first pictures of the complicated patterns into which the particles settled upon the ferromagnetic surface, whether it was magnetised or unmagnetised. Akulov and Degtiar† obtained very clear patterns with an alcohol suspension of Fe_2O_3, while Becker and Freundlich‡ used an almost colloidal suspension in propyl alcohol. McKeehan and Elmore§ and Soller|| used true sols of Fe_2O_3 in very weak KOH solution; these sols did not precipitate on iron, but swarms of particles collected in regions where the

* F. Bitter, *Phys. Rev.* **41**, 507, 1932.
† N. Akulov and M. Degtiar, *Ann. der Phys.* **15**, 750, 1932.
‡ R. Becker and H. F. W. Freundlich, *Zeit. für Phys.* **80**, 292, 1932.
§ L. W. McKeehan and W. C. Elmore, *Phys. Rev.* **46**, 226, 1934.
|| Th. Soller, *Zeit. für Phys.* **106**, 485, 1937.

field was non-uniform and enabled sharp micro-photographs to be made with about 1 min. exposure.

The patterns are not formed on non-ferromagnetics and they show conclusively that there exist stray magnetic fields between adjacent elements of the surface, with lines of force coming out of one element and entering another. McKeehan and Elmore used a single crystal silicon iron disc which they magnetised perpendicularly to the surface, and their results supported the view that the surface was made up of roughly cubical blocks, of edge about 2μ, magnetised parallel to the surface either along a [100] or along a [110] direction, giving place to lines 4μ apart when the field is on. Other workers, who used fields parallel to the ferromagnetic surface, obtained lines with spacings of the order of 40 to 100μ. The spacing appears to be more or less accidental. Elmore* sprinkled iron filings on a simple model of hardened steel cubes packed close together to form a square surface and showed that a typical maze pattern was obtained. Polishing appears to play an important part in determining the pattern, for unpolished strips of iron give straight-line patterns. Soller showed that these lines appear when a specimen originally magnetised parallel to its surface is stretched in a direction perpendicular to the magnetisation and simultaneously exposed to a field perpendicular to the surface.

The Barkhausen Effect. Further direct evidence supporting the Weiss domain theory of ferromagnetism has been obtained from the experiments now to be described. In Fig. 96 AB represents a ferromagnetic wire which may be magnetised by the solenoid S. Upon the wire is mounted an induction helix or search coil s_1 consisting of a very large number of turns, say, 10,000, which is joined in series opposition with an identical coil s_2, which does not surround the wire and placed some distance from s_1, and an amplifier and telephone receiver. The coil s_2 acts as a compensator for the lines of force through s_1 from the magnetising field alone, and could be replaced by the secondary of a mutual inductance whose primary is in the magnetising current circuit.

* W. C. Elmore, *Phys. Rev.* **51**, 982, 1937.

If the changes in the magnetisation of the rod were continuous and there were no disturbing effects, there would be no sounds heard in the telephone when the magnetising current changed slowly. As, however, the magnetisation processes are discontinuous and independent of each other, a series of "clicks" or "noises" corresponding to abrupt changes in different parts of the wire are heard. Sometimes the discontinuities are so pronounced that the noises are heard when the coils s_1 and s_2 are joined direct to the telephone. Such noises

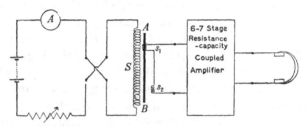

Fig. 96. Apparatus for investigation of individual
longitudinal Barkhausen effects.

were first investigated by Barkhausen* and the phenomenon is known as the *Barkhausen effect*. It has been described as the magnetic analogue of the "cry of tin",† and it is intimately connected with magnetostriction processes.

The effect has been studied quantitatively by Bozorth and Dillinger,‡ Pfaffenberger§ and Preisach.|| In many cases the specimen wire was 60 cm. long and 1 mm. or less in diameter. Bozorth and Dillinger replaced the telephone of Fig. 96 by a transformer in whose secondary was a thermojunction connected to a critically damped Moll galvanometer. The galvanometer deflections were recorded photographically and provided a measure of the average value of the square of the current through the secondary. Preisach connected s_1 and the secondary of a mutual inductance direct to a ballistic galvanometer to obtain the hysteresis cycle of the wire, and to a string

* H. Barkhausen, *Phys. Zeit.* **20**, 401, 1919.
† Cf. E. N. da C. Andrade and B. Chalmers, *Nature*, **29**, 650, 1932.
‡ R. M. Bozorth and Joy Dillinger, *Phys. Rev.* **35**, 733, 1930.
§ J. Pfaffenberger, *Ann. der Phys.* **87**, 737, 1928.
|| F. Preisach, *Ann. der Phys.* **3**, 737, 1929.

galvanometer when the Barkhausen discontinuities were investigated. It must be emphasised that s_1 and s_2 must be some distance apart or the induction in the wire in s_1 may produce a flux in s_2; they may then both surround the wire, since the Barkhausen changes are random in magnitude and time of occurrence. Mechanical vibrations must, of course, be avoided.

Nature of the Barkhausen Discontinuities. It is clear from the above experiments that if we plotted the $I, \mathscr{H}$ curve for a ferromagnetic on an extremely large scale, we should find it made up of a series of steps. Van der Pol* examined the magnitude of these steps in the several portions of the hysteresis cycle, and found that they were pronounced only in the steep portions of the curve, where I changes rapidly, and were extremely small in the region of saturation. Preisach concluded that the total contribution of the discontinuous induction processes depended on the type of hysteresis cycle, but that there was no simple relation between the slope of the $I, \mathscr{H}$ curve and the Barkhausen effects. Moreover, he found that the magnitude of the individual steps varied over a wide range and could be given no precise physical significance. Barkhausen effects are certainly found in the region of saturation but high amplification is necessary for their detection.

Estimates of the size of the Barkhausen domains, or of the number of atoms involved in a discontinuity, have been made by van der Pol,* Tyndall† and Bozorth and Dillinger.‡ The estimate of the latter workers appears to be the most reliable, but it is based upon a series of rather complicated experiments, to say nothing of theoretical assumptions, and it will not be outlined here. They found that the average volume of the ferromagnetic in which the magnetisation changed as a unit was about 10^{-8} to 10^{-9} c.c. in the case of vacuum-annealed armco iron, iron single crystals, vacuum-annealed nickel and permalloys of different compositions and thermal treatment, and the extreme values were $1 \cdot 2 \times 10^{-9}$ c.c. for iron and

---

* B. van der Pol, *Proc. Amsterdam Acad.* **23**, 637, 1920.

† E. P. Tyndall, *Phys. Rev.* **24**, 439, 1924.

‡ R. M. Bozorth and Joy Dillinger, *Phys. Rev.* **35**, 733, 1930.

21-2

45×10^{-9} c.c. for a 50 per cent nickel-iron alloy. These volumes are much greater than the volumes of the individual crystals found, for example, in a specimen of hard worked iron, and the average domain thus appears to contain about 10^{15} atoms. It may safely be assumed that Barkhausen discontinuities are independent of crystal size and lattice, and that the number and sizes of the discontinuities in a given specimen are constant.

It is possible to make very large Barkhausen discontinuities by using a wire with positive magnetostriction under sufficient tension to ensure that all the domain vectors are aligned parallel to its axis. This means that the hysteresis cycle can virtually be made a rectangle, as Forrer* first proved.

Time of Duration of a Barkhausen Reversal. Barkhausen† attempted to find the average time of duration of a discontinuity by making use of the principle that an oscillatory circuit can only be vigorously excited by an impulse which lasts for less than half the period of one oscillation, and with his amplifier he showed that the duration was less than 3.3×10^{-6} sec. Preisach used a receiver consisting of two high-frequency and one low-frequency stages, and found the Barkhausen effects to persist even with the shortest waves possible with the arrangement, 200 m. He accordingly extended his work by using nickel-iron specimens whose hysteresis cycles were practically rectangles and whose Barkhausen impulses amounted to single impulses only. By exposing such a specimen to an alternating field the impulses could be repeated regularly and made to give a periodic current in a high-frequency circuit. For, if the hysteresis cycle has the approximate shape of a parallelogram, the flux ϕ changes sign in a short interval of time τ and then remains constant for a time $T - \tau$ if the frequency of the applied field is $1/T$. The Fourier expansion for the induced E.M.F. is given by

$$E = \frac{8\phi}{\pi\tau} \sum_{k=1}^{\infty} \frac{\sin(k\omega\tau/2)}{k} \sin k\omega t,$$

* R. Forrer, *Journ. de Phys.* **1**, 109, 1926.
† H. Barkhausen, *Zeit. für tech. Phys.* **5**, 518, 1924.

the appropriate expression for a series of sharply defined potentials, ω being the pulsatance of the magnetising current. The amplitude of the kth harmonic is therefore equal to

$$E_k = \frac{8\phi}{T} \frac{\sin (k\pi\tau/T)}{k\pi\tau/T}.$$

Hence the potential curve will contain a considerable number of equally pronounced harmonics as long as their time of duration is large compared with the duration of the discontinuous change, or $T/k > \tau$, in which case

$$\frac{\sin (k\pi\tau/T)}{k\pi\tau/T} \approx 1.$$

Thus the amplitude only decreases to a marked degree in the neighbourhood of a frequency whose half period is equal to that of the duration of the discontinuous process, or when $T/k \approx 2\tau$.

On coupling the helix s_1 and its compensator s_2 of Fig. 96 directly to the antenna coil of a short-wave receiver which was made to oscillate, the high-frequency components supplied by the coil s_1 produced combination frequencies with that of the receiver. The frequency of the magnetising current was 6000 cycles per sec., and when the condenser in the oscillatory circuit of the receiver was gradually rotated a succession of beat notes was generated as one after another the successive harmonics supplied by s_1 came into combination with the fundamental frequency of the receiver. In this way harmonics with wave-lengths down to 30 m. were recognised, but with shorter wave-lengths the combination tone gradually faded away. Hence the average time of duration of a Barkhausen process must have been less than 10^{-8} sec.

The Transverse Barkhausen Effect. Bozorth and Dillinger[*] suggested that, as the magnetic vectors of domains would not all be set parallel to the direction of the applied field during the investigation of the longitudinal effects, transverse effects would be registered in coils set with their axes at right angles

[*] R. M. Bozorth and Joy Dillinger, *Phys. Rev.* **38**, 192, 1931; **39**, 353, 1932; **41**, 345, 1932.

21-3

to the field. They used tubular specimens 60 cm. long and about 5 mm. in diameter. By passing a heavy current through a copper wire along the axis of the tube the specimen was magnetised transversely while a current in a solenoid S surrounding it produced longitudinal magnetisation. Two search coils s_1 and s_2 were wound on the tube and joined in series opposition to the amplifier and thermocouple detector described on p. 322. The magnetisation perpendicular to the specimen's axis was measured by a few turns of wire threaded along the axis and down the outside of the tube and connected to a ballistic galvanometer; the search coil so formed could also be connected to an amplifier and thermocouple detector to record the transverse discontinuities, and it was arranged that both transverse and longitudinal discontinuities were recorded on the same drum. The deflection of the string galvanometer for longitudinal changes was invariably in the same direction, but the transverse changes gave deflections half in one direction and half in the other, proving that the average change in magnetisation perpendicular to the field was zero.

On measuring the ratio of average transverse to the average longitudinal effect as a function of the induction B, it was found that it was less than unity until B exceeded 15,000 after which it rose rapidly, values obtained from the virgin B, $\mathscr{H}$ and the hysteresis loop curves falling on the same graph of the ratio against B. The results definitely support the view that in low fields on the average all the magnetic vectors change by very nearly 180°, whereas in high fields rotation of the vectors through 90° is common.

Velocity of Propagation of Barkhausen Discontinuities. We have seen on p. 324 how Preisach made use of the large discontinuities which are observed in the magnetisation of highly strained wires. It is difficult to account for the sudden reversal of magnetisation in the whole length of the wire unless, as Langmuir first suggested, the magnetisation is initially reversed in a limited region or nucleus in the wire, from which the discontinuity spreads outwards with a finite velocity.

Experiments will now be described which prove the correctness of this suggestion and tell us how the velocity of propagation depends upon the applied magnetic field and upon the elastic strain which the wire undergoes.

In the arrangement used by Sixtus and Tonks* shown in Fig. 97 the initial discontinuity or nucleus in the nickel-iron wire AB was produced by passing a current through the small coil a, whilst the wire as a whole was magnetised to a chosen extent by a current in the main magnetising coil S. The short

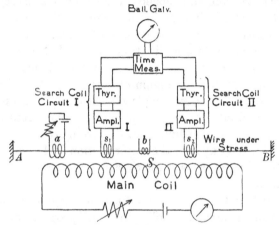

Fig. 97. Sixtus and Tonks's apparatus for investigation of velocity of Barkhausen discontinuities.

5000 turn search coils s_1 and s_2 were connected to amplifiers and thyratron circuits in order to measure the time a discontinuity took to pass from s_1 to s_2. The wire AB was subjected to tension, or torsion, or both. To protect the wire, and to permit adjustment of the positions of s_1 and s_2, it was enclosed in a glass capillary tube. In most cases the distance between s_1 and s_2 was 20 cm., the centre of each being 10 cm. from the mid-point of AB, the coil a being placed 24 cm. from the mid-point.

The time of passage was measured by an amplifier circuit devised by Turner,† thyratrons being used to give constant

* K. J. Sixtus and L. Tonks, *Phys. Rev.* **37**, 930, 1931.
† L. B. Turner, *Radio Rev.* **1**, 317, 1920.

impulses to a timing circuit whatever the voltages induced in s_1 and s_2. The effect was to send a current through a ballistic galvanometer for the short time between the impulse delivered by s_1 and that from s_2. For a fixed value of the tension the velocity was, in general, a linear function of the main magnetising field $\mathscr{H}$, of the form $v = A(\mathscr{H} - \mathscr{H}_0)$, where A is a constant of the order of magnitude 25,000 cm. per sec. per oersted, and $\mathscr{H}_0$ may be considered to be that limiting field in which the velocity of propagation would be zero. $\mathscr{H}_0$ decreases with increase in tension, and depends on the previous history of the wire. A appears to increase with rise in temperature.

If discontinuities are started when the field in a restricted region exceeds a certain limiting value, then it should be possible to arrest the propagation of the discontinuity by establishing a restricted region of very low field in its path. Sixtus and Tonks* therefore placed a stopping coil b between s_1 and s_2 and adjusted the stopping current therein until s_2 recorded no changes in flux. They then found that a demagnetising field was set up over the major portion of the wire inside b which exactly neutralised the excess of the applied field over the limiting field.

Sixtus† found it possible to adjust the magnitude or duration of a single impulse supplied to a until propagation started and to impart a succession of "flat-topped" impulses of equal magnitude and duration for this purpose. In both cases he found $t \times (\mathscr{H}_a - \mathscr{H}_0) = $ constant, where t was the total time the added field $\mathscr{H}_a$ lasted. From this he concluded that for propagation a nucleus had to grow to sufficient size. By moving a short search coil along the wire a flux distribution curve was obtained, whose height measured the size of the nucleus; there was a definite relation between the height of this curve and the main field in which the nucleus could start propagation. Assuming the value of the demagnetisation constant of the nucleus and knowing its length, its diameter could be estimated. The length of a nucleus was found in the case of a 15 per cent nickel-iron wire to be a few mm. and its diameter about 10^{-4} to

* K. J. Sixtus and L. Tonks, *Phys. Rev.* **43**, 931, 1933.
† K. J. Sixtus, *Phys. Rev.* **48**, 425, 1935.

10^{-3} cm., giving a volume of about 10^{-8} c.c. for the size of a Barkhausen domain.

Sixtus and Tonks have given a theory of propagation of Barkhausen effects based on the suppositions that the penetration of a discontinuity into the wire is determined by eddy currents in the material, while the propagation along the wire depends on surface conditions. They consider that the propagation into the wire requires a time δt which they call the penetration time, and they calculate this time on the basis that magnetisation is reversed at any point in the discontinuity where the total effective field $\mathscr{H}_m$ exceeds the critical field $\mathscr{H}_0$, and that reversal occurs so rapidly that the eddy currents generated reduce $\mathscr{H}_m$ to $\mathscr{H}_0$ at every instant. In this way they find

$$\delta t = 3 \cdot 94 \times 10^{-8} \, a^2 \, \Delta I / \rho (\mathscr{H} - \mathscr{H}_0),$$

where a is the radius of the wire, ΔI the change in intensity of magnetisation associated with the discontinuity, ρ the resistivity and $\mathscr{H}$ the superimposed main magnetising field. Unfortunately, to obtain agreement with the experimental results it is necessary to multiply the right-hand side of the above equation by a numerical factor, $0 \cdot 035$, of unknown significance. An interesting feature, however, of this work is that it appears to account satisfactorily for large discontinuities, over 100 cm. long, found experimentally.

Döring* has given a new theory of propagation in which account is taken of the energy in the transition layer† between two Weiss domains. He pictures a small nucleus in the wire as growing by displacing its boundary, and likens the process to the growth of drops within a supersaturated vapour. He thus shows that the experiments on large Barkhausen discontinuities lead to a value for the energy stored in each square cm. of the transition boundary of $2 \cdot 7$ ergs, which is in satisfactory agreement with theory.

* W. Döring, *Zeit. für Phys.* **108**, 137, 1938; and *Probleme der technischen Magnetisierungskurve*, p. 26, 1938. Cf. H. Haake, *Zeit. für Phys.* **113**, 218, 1939.
† F. Bloch, *Zeit. für Phys.* **74**, 295, 1932.

CHAPTER XI

INTERPRETATION OF THE
HYSTERESIS CYCLE

Introduction. We discussed the several ways in which the magnetic properties of ferromagnetic materials may be investigated in Chapter II, where we also discussed the main features of a normal hysteresis cycle and mentioned the hysteresis cycles characteristic of certain polycrystalline substances. The magnetic properties of ferromagnetic single crystals were detailed in Chapter IV, and in Chapter X the essentially discontinuous nature of magnetisation processes was emphasised. We will now examine these processes more closely in the light of recent theoretical work, much of which has been described *in extenso* by Becker and Döring.*

We have frequently referred to the Weiss domain theory of magnetisation which postulates that an unmagnetised ferromagnetic crystal consists of a large number of small regions or domains, magnetised practically to complete saturation, with their magnetic vectors orientated in all possible directions. Bulk magnetisation of the material therefore occurs when some of these vectors are caused to align themselves more or less parallel to an applied magnetic field. Referring to Fig. 98, the separate compartments represent a few domains in an unmagnetised specimen of iron, in which the domain vectors set in directions, indicated by the arrows, parallel to the edges of a crystal cube. It is clear that at least three distinct kinds of change might be expected to result from the application of an applied field $\mathscr{H}$ in the direction shown. First, the vector of domain 4 might turn through 180 degrees, so that, while it still remains parallel to a crystal edge, it is in much better alignment with $\mathscr{H}$. In like manner, the vectors of domains 2 and 3 might turn through 90 degrees. Secondly, the vector of domain 1 might grow at the expense of domain 2 as shown by the dotted line. Finally, after all

* R. Becker and W. Döring, *Ferromagnetismus*, Springer, 1939.

330

the domain vectors have set themselves as nearly parallel to $\mathcal{H}$ as possible, although still remaining parallel to cube edges, they might thereafter rotate bodily through angles less than 90 degrees until they are exactly parallel to the field.

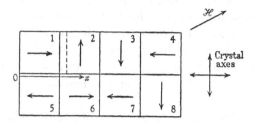

Fig. 98. Illustrating domain structure of a ferromagnetic crystal.

It is frequently assumed that all the spins within a given domain can simultaneously swing through large angles from a position of minimum domain energy to another. The energy needed for such a comparatively large-scale process is so great that the assumption is undoubtedly incorrect, and the instantaneous 180 degree rotation of all the spins in domain 4 and a similar 90 degree rotation of all the spins in one of the domains 2, 3 and 8 is impossible. Consequently, the domain changes must proceed by means of a series of rapid rotations of individual spins. Now, while it is easy to draw domains as if they were of regular shapes situated within a two dimensional frame, it is not easy for an ordinary person to visualise their actual arrangement in three dimensions within a crystal; in any case, in strained materials the vectors will not make angles of exactly 90 and 180 degrees with one another, and, while we have thus far confined attention to single crystals, their properties are not unique, for all ferromagnetic materials must be regarded as collections of such crystals giving rise to hysteresis cycles which must be explicable in terms of the properties of the single crystals of which they are constituted.

Factors affecting Direction of Domain Vectors. It is known that there are three important factors which decide the direction of magnetisation within a particular domain.

331

Experiments with single crystals show conclusively the existence of directions of easy magnetisation, e.g. in iron crystals these are parallel to the [100] directions, and in nickel crystals at room temperatures to the [111] directions. This is because different quantities of energy are associated with different positions of the vector with respect to the crystal axis, and we express this fact by means of the anisotropy coefficients of p. 161. Hence, we may consider that the *crystal energy* per unit volume, E_c (equal to the quantity E of equation (111)), is one of the factors deciding the position of the domain vector.

A second factor must be the effect of strain produced by the magnetostrictive changes of dimensions which accompany magnetisation. It is impossible to prepare a ferromagnetic substance which is perfectly strain-free, because, as it cools through the Curie point, the substance changes from a para-magnetic to a ferromagnetic state, resulting in stresses which cannot be entirely removed, however careful and prolonged any subsequent annealing may be. We know from direct experiment how important such stresses can be; for example, a nickel wire under severe longitudinal stress can be magnetised parallel to its axis only with difficulty, because the domain vectors under these circumstances set perpendicular to the axis of the wire. In fact, it may easily be proved that if we apply a tension of F dynes per sq.cm. to a domain in nickel, for which E is small, we have to supply *strain energy* per unit volume, E_F, equal to $\frac{3}{2}\lambda_s F \sin^2 \phi$ in order to turn the vector through an angle ϕ from the direction of the applied tension, where λ_s is the saturation magnetostriction coefficient.

For, let the vector make an angle ϕ with an applied field $\mathscr{H}$, and let the mechanical couple acting on it be M_F. Suppose that a displacement $d\phi$ is accompanied by a magnetostrictive change of length equal to dl per unit length. If α is the area of the cross-section over which F is applied, we must have, from mechanical considerations,

$$M_F . d\phi = F\alpha dl.$$

But, for nickel, where we may assume that the magneto-striction is isotropic, we may write

$$dl = ld\left(\frac{l-l_0}{l_0}\right) = d\{\tfrac{3}{2}\lambda_s l(\cos_3^2 \phi - \tfrac{1}{3})\},$$

$$= -3\lambda_s l \sin\phi \cos\phi \, d\phi,$$

so that $\qquad M_F = \alpha l F \, 3(-\lambda_s)\sin\phi\cos\phi.$

Now M_F is balanced by the couple $M_{\mathscr{H}}$ which the magnetic field exerts on the domain. If the magnetic moment per unit volume of the domain is I_s, and the volume $V = \alpha l$, we have

$$M_F = M_{\mathscr{H}} = V I_s \mathscr{H}\sin\phi = \alpha l I_s \mathscr{H}\sin\phi.$$

Therefore $\cos\phi = -\dfrac{I_s\mathscr{H}}{3F\lambda_s}$, the linear relation, which changes abruptly when $\cos\phi = 1$ at $I = I_s$, already given on p. 306 for the magnetisation of nickel under severe longitudinal stress. Hence, to find the total mechanical work done in turning the vector from a direction making an angle ϕ with $\mathscr{H}$ to one parallel to it, we write

$$E_F = -\frac{1}{\alpha l}\int_0^{\phi} M_F \, d\phi = -\tfrac{3}{2}(-\lambda_s)\,F\sin^2\phi. \quad \ldots\ldots(197)$$

For nickel, λ_s is negative; for permalloy under tension, it is positive. The above calculation holds only for isotropic magnetostriction, and for the general case, e.g. for iron, where even the sign of λ_s depends upon the direction of the magnetisation with respect to the crystal axes, the calculation of E_F is extremely difficult and does not appear to have led to results of practical value so far.

Finally, in addition to E_c and E_F we have the energy per unit volume $E_{\mathscr{H}}$ which is the energy of position or potential energy of the domain magnetisation *with respect to the applied field*; it is not the energy located in the material itself, as we shall see later. This is clearly equal to $-\mathscr{H}I_s\cos\phi$. Consequently, to sum up, the domain vector must set in such a direction that $E_c + E_F + E_{\mathscr{H}}$ is a minimum. It follows that only when the material is approximately strain-free does E_c

decide the direction of the vector, and that only when $E_{\mathscr{H}}$ is greatly in excess of E_c and E_F can its direction be fixed uniquely. Now, there must always exist a minimum internal stress equal to $F_i = \lambda_s E'$, where E' is Young's modulus for the material, because the latter gains its ferromagnetism, and is correspondingly distorted, by cooling through the Curie point. It is therefore appropriate to divide ferromagnetic materials into two groups according as $E_c \gg E_F$ or not.

Domain Boundary Displacements. We may now give a somewhat more definite picture of what takes place when we magnetise a ferromagnetic substance like iron for which E_c is always $\gg E_F$; where, in fact, tensions considerably above the elastic limit would be needed to make E_F approach E_c in magnitude. Let us consider two adjacent domains, such as 1 and 2 in Fig. 98, whose vectors are at 90 degrees with respect to one another, with what is termed a 90 *degree boundary* between them. Let us plot the difference in the energy minima per unit volume, denoted by $E_1 - E_2$ as we proceed through the two domains in the x direction; i.e. we imagine unit volume around any given point to have its vector pointing first as in 1 and then as in 2. It will be represented by a curve of the form $ABCDEF$, Fig. 99, when no field is

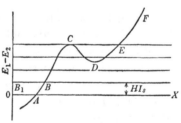

Fig. 99. Illustrating energy changes involved in boundary displacements.

applied, the boundary being located by the point of intersection A on the x axis. On establishing an effective field H parallel to x the energy difference is decreased by HI_s, which may be regarded as equivalent to raising the x axis by this amount, so that it takes up the position $B_1 B$ and intersects the curve at B which now represents the new boundary position. Pro-

334

ceeding in this way with larger values of the applied field successive points of intersection representing a 90 *degree boundary displacement* are obtained. This displacement is magnetically reversible until we reach the point of intersection C, but, once the field is increased beyond the value for C, a region where the direction of the vector is particularly favoured, instability is found and the boundary moves forward rapidly to positions given by points of intersection such as E. The change between C and E is magnetically irreversible and forms a Barkhausen discontinuity. On decreasing the field the minimum D is reached before a similar Barkhausen change takes place in the reverse direction.

When two adjacent domains whose vectors are antiparallel are under consideration, it is necessary to imagine a transition layer between them analogous to those considered in surface tension problems. The experiments of Sixtus and Tonks show that a critical field must be exceeded if a more favoured domain is to grow at the expense of the other. This means that such changes are irreversible. They undoubtedly account for the majority of the Barkhausen discontinuities, particularly in the region below the knee of the hysteresis curve.

The fields necessary to cause the two types of boundary displacement to continue, until all the domain vectors in the iron crystal are set parallel to the cube edge most nearly coincident with the direction of the applied field, are not large. If it is further desired to turn the direction of magnetisation from parallel to the cube edge through an angle less than 90 degrees to exact alignment with the applied field, then, since no further boundary displacements are possible the change may occur only by *rotation of the vectors*; strong fields must now be used, such that $HI_s \gg \lambda_s F_t$. Hence, when the boundary displacements found on the steep part of an $(I, \mathscr{H})$ curve come to an end and rotations begin, a marked kink appears in the curve. Such kinks are shown in Fig. 44. As far as is known these rotations are reversible. It is actually possible to harden* a polycrystalline nickel wire by cold working so that it is isotropically and homogeneously strained to such an extent

* G. Thiessen, *Ann. der Phys.* **38**, 153, 1940.

that $E_F \gg E$, and all magnetisation of the wire occurs by rotations alone. Usually, of course, in polycrystalline materials boundary displacements and rotations are superimposed and their effects are often difficult to disentangle. This is particularly the case with exceptionally hard materials, e.g. steels, where irreversible processes are evident during any change in magnetisation.

Examination of the Hysteresis Cycle. An attempt will now be made to examine how boundary changes and rotations occur in the several portions of a typical hysteresis cycle for a soft material. Referring to Fig. 100, we consider first the initial portion Oa of the virgin curve, whose slope, or initial

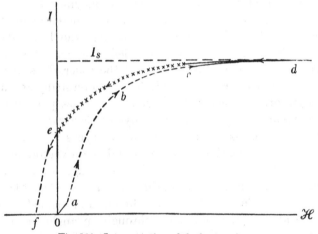

Fig. 100. Interpretation of the hysteresis curve.

susceptibility, is determined by the relative numbers of 90 and 180 degree boundary changes. In general, for annealed specimens the 90 degree changes predominate, and the number of 180 degree boundaries depends on the method by which the specimen was demagnetised, e.g. by reversals with d.c., or by gradually decreasing an a.c. field, prior to the measurement of I and $\mathscr{H}$. We know that the path Oa is magnetically reversible so that the boundary changes there are most probably reversible too, and we therefore conclude that only in exceptional circumstances is the number of 180 degree boundary changes large in this region; indeed, it appears to be large solely in the case of carbonyl iron. Approximate expressions for the initial susceptibility may therefore be deduced.

Thus, for a nickel rod under severe *internal* stress, k_0 can be shown to be approximately equal to $\frac{2}{9} I_s^2 / \lambda_s F_i$. The latter expression, strange as it may seem, differs but little from that for a substance with F_i very small, when 90 degree boundary displacements are considered of overwhelming importance. Assuming, therefore, that the minimum value of F_i may be taken equal to $\lambda_s E'$, the maximum value of the initial permeability could not greatly exceed $1 + 4\pi \frac{2}{9} I_s^2 / \lambda_s^2 E'$, i.e. about 12,000 e.m.u. per c.c.; this is in good accord with some of the experimental values for special materials. Higher values, e.g. 14,000 for hydrogen-treated pure iron[*] and 100,000 for certain specially treated materials[†] are presumably due to 180 degree boundary displacements. Returning to Fig. 100, the steep part ab is mainly due to irreversible 180 degree boundary changes. At b, where the effective field is only slightly greater than the coercivity $\mathscr{H}_c$, is located the bend or "knee" of the curve which is produced by the superposition of huge numbers of the sharp kinks found with single crystals. It represents the transition from the region of 180 degree boundary changes to the region cd in which nearly all changes in magnetisation are due to reversible rotations. If we increase the field beyond the point d we leave the region of technical magnetisation processes and enter the region, so well explored in magnetocaloric experiments, where further magnetisation results from the reorientation of individual electron spins alone.

On decreasing the field, the magnetically reversible path dc is traversed, and the magnetisation thereafter decreases because of reversible 90 degree and, possibly, a few 180 degree boundary displacements, until the point e at the intersection of the curve with the I axis is reached. The field is then

---

[*] P. P. Coiffi, *Phys. Rev.* A, **45**, 742, 1934.

[†] O. L. Boothby and R. M. Bozorth, *Phys. Rev.* A, **71**, 472, 1947.

reversed; a little beyond e the reversible boundary displacements are completed and the curve descends steeply to f because of irreversible 180 degree boundary displacements. There is no advantage in discussing the remaining stages of the cycle, as all the essential features are covered by the above description.

The intercept I_r on the I axis is the retentivity. In the case of magnetically soft materials it is easy to see, from a consideration of the symmetry of the distribution of the vectors about the direction in which the field was formerly a maximum and the absence of vectors pointing in the reverse direction, that I_r should be equal to $I_s/2$. However, there is usually sufficient magnetic interaction between neighbouring domains to make $I_r > I_s/2$. It should be noted that the coercivity $_0f$ cannot be greater than the residual induction, $4\pi I_r$.

In the case of heavily strained materials there should be very little difference between the demagnetised state and the state $I = I_r$, the only real difference being that certain of the domain vectors are oppositely directed in the two states. Hence the differential susceptibility k_r at $I = I_r$ should be equal to the initial susceptibility, k_0, and this is supported by experiment. For lightly strained materials we expect a different relation to hold. Becker calculates that for iron $k_r/k_0 = 1 - 2/\pi = 0\cdot364$, while for nickel the smaller ratio $0\cdot328$ is obtained. In other words, the value of the ratio k_r/k_0 should never fall below $0\cdot3$, but if this should happen, as, e.g. with carbonyl iron, it is easy to explain it by postulating that 180 degree as well as 90 degree boundary displacements are involved.

The coercivity is determined by the field necessary to start a large Barkhausen discontinuity with a 180 degree boundary change. If we postulate that this boundary is of finite thickness, that electron spins therein change their directions in a regular manner as we traverse the boundary from one side to the other, and that a quantity of energy q is associated with each unit area of the boundary surface, then a displacement dx of the latter must satisfy the equation

$$2\mathscr{H}I_s dx = (\partial q/\partial x)\,dx. \qquad \ldots\ldots(198)$$

Now, q may be regarded as made up of the contributions due to the Heisenberg exchange effect q_s, the effect of internal strain q_i and the crystal anisotropy q_c, i.e.

$$q = q_s + q_i + q_c. \qquad \ldots \ldots (199)$$

It is reasonable to suppose that q_s and q_c do not vary appreciably throughout the boundary, so that it is sufficient to write

$$dq/dx = dq_i/dx. \qquad \ldots \ldots (200)$$

But

$$dq_i = \tfrac{3}{2}\lambda_s F_i \sin^2 \theta \, dx,$$

whence $q_i = \tfrac{3}{2}\lambda_s F_i t$, where t is the thickness of the boundary. If the boundary is to be stable it must possess a definite thickness t_0, so that from equation (198) it follows that the maximum value of $\mathcal{H}$ which can be applied before the Barkhausen discontinuity starts is

$$\mathcal{H}_0 = \frac{1}{2I_s}(\partial q_i/\partial x)_{\max} = \frac{3}{4}\frac{\lambda_s}{I_s} t_0 (\partial F_i/\partial x)_{\max}.$$

Therefore, in order to calculate $\mathcal{H}_0$, the measure of the minimum value of the coercivity, we must make special assumptions about the way in which F_i varies from place to place in the ferromagnetic material. Practically nothing is known about the latter variation, but we may certainly take it that $\mathcal{H}_c$ is proportional to $\lambda_s F_i/I_s$, or to I_s/k_0, a statement which is well supported by experiment.

We may confidently assume that the maximum value of $\mathcal{H}_c$ for a given material cannot exceed $\tfrac{3}{2}\lambda_s F_i/I_s$. Assuming that the remanence $B_r = I_s/2$, we may readily estimate the maximum value of $(B_r \times \mathcal{H}_c)$ which is unlikely to be exceeded. Thus, taking λ_s to be 5×10^{-5} and F_i to be the tensile strength, say 200 kg. per sq.mm. or 2×10^{10} dynes per sq.cm., we find $(B_r \mathcal{H}_c)_{\max}$ to be 10×10^6 gauss-oersteds; the value of $(B_r \mathcal{H}_c)$ for alcomax II, the highest for any material in common use in Britain, is about 7×10^6, while the value of $(B\mathcal{H})_{\max}$ is about $5 \cdot 0 \times 10^6$ ergs per c.c.

Adiabatic Temperature Changes accompanying Magnetisation in Moderate Fields. It is generally accepted that the energy liberated in a complete hysteresis cycle is given by the expression $\oint_{+\mathscr{H}}^{+\mathscr{H}} \mathscr{H}dI$, derived on p. 59, a statement which is sometimes known as *Warburg's law*.* The derivation of this expression has been discussed in detail by Becker and Döring,† Bitter,‡ Guggenheim,§ Stoner‖ and others; it is theoretically correct only when the conductors present consist solely of the ferromagnetic specimen under examination and the solenoid which produces the magnetic field applied to it, so that eddy currents are not established in neighbouring conductors. However, all the above-mentioned discussions seek to explain why the energy supplied to a substance during magnetisation is given by $\int\mathscr{H}dI$ and not $\int d(\mathscr{H}I)$ or $\int Id\mathscr{H}$, and are not adequate for an examination of the energy changes associated with small changes in $\mathscr{H}$ in any part of a hysteresis cycle.

Let us assume that $\mathscr{H}dI$ correctly represents the whole energy supplied to a ferromagnetic specimen in any part of a hysteresis cycle and that it is accompanied by an increase in internal energy dE. For the present, we will confine attention to specimens which are not subjected to applied mechanical forces, neglecting those provided by the presence of the atmosphere, so that we may write $dE = d(E_c + E_F)$, where E_F refers only to the energy associated with strain of internal origin. The changes are adiabatic so that we may write

$$\mathscr{H}dI = (\partial E/\partial I)_T\, dI + (\partial E/\partial T)_I\, dT. \quad \ldots\ldots(201)$$

Now, $\mathscr{H}dI$ can be obtained directly from magnetic measurements, while the second term on the right-hand side of equation (201) represents that part of the change in the internal energy which manifests itself as an increase in temperature and can be obtained directly from thermal measurements to be described. In fact, we may write

$$(\partial E/\partial I)_T\, dI = \mathscr{H}dI - dQ_1 \quad \ldots\ldots(202)$$

and we may, therefore, determine $(\partial E/\partial I)_T$ at any stage in the magnetisation of a ferromagnetic, which means that we can

  * E. Warburg, *Ann. der Phys.* **13**, 141, 1881.
  † R. Becker and W. Döring, *Ferromagnetismus*, p. 53, Springer, 1939.
  ‡ F. Bitter, *Introduction to Ferromagnetism*, p. 14, McGraw-Hill, 1937.
  § E. A. Guggenheim, *Roy. Soc. Proc.* A, **155**, 49, 1936.
  ‖ E. C. Stoner, *Phil. Mag.* **23**, 833, 1935.

say whether E increases, decreases or remains constant with change in I at any point on an $(I, \mathcal{H})$ curve, equation (202) being independent of whether the change is reversible or not in the thermodynamic sense. Equation (202) clearly brings out the importance of considering thermal changes with reference to changes in I instead of changes in $\mathcal{H}$.

We must use a more general method of analysis for the examination of ferromagnetic specimens subjected to longitudinal stress. Assuming the phenomena to be thermodynamically reversible we may write

$$dU = T.dS + \mathcal{H}dI + F.dl, \qquad \ldots\ldots(203)$$

where F is the load expressed in dynes per unit area, and dl is the extension per unit length of the specimen. Taking S, $\mathcal{H}$ and F as the three independent variables it follows, e.g. by comparison with analogous gas formulae, that

$$\left(\frac{\partial T}{\partial \mathcal{H}}\right)_{S,F} = -\frac{T}{J\rho C_{F,\mathcal{H}}}\left(\frac{\partial I}{\partial T}\right)_{F,\mathcal{H}}, \qquad \ldots\ldots(204)$$

$$\left(\frac{\partial T}{\partial F}\right)_{\mathcal{H},S} = -\frac{T}{J\rho C_{F,\mathcal{H}}}\left(\frac{\partial l}{\partial T}\right)_{F,\mathcal{H}}, \qquad \ldots\ldots(205)$$

$$\left(\frac{\partial l}{\partial \mathcal{H}}\right)_{S,F} = \left(\frac{\partial I}{\partial F}\right)_{\mathcal{H},S}, \qquad \ldots\ldots(206)$$

where $C_{F,\mathcal{H}}$ is the specific heat of the material under constant F and constant $\mathcal{H}$, and the other symbols have their usual meanings. When $F = 0$, equation (204) reduces to the statement for the magnetocaloric effect, deduced on p. 273, which adequately represents the behaviour in strong fields and will not be considered further here.

In the case of severely loaded nickel specimens we have direct experimental evidence* that the longitudinal magnetostriction changes of length are proportional to I^2. This evidence, together with that of Engler† and of Siegel and Quimby‡ may permit us to regard the length of the specimen

* H. Kirchner, *Ann. der Phys.* **27**, 49, 1936.
† O. Engler, *Ann. der Phys.* **31**, 145, 1937.
‡ S. Siegel and S. L. Quimby, *Phys. Rev.* **49**, 663, 1936.

as prescribed by the value of I^2 alone. If this is correct, then we may write $\mathscr{H} = k_1 I$ and $(l - l_0) = k_2 I^2$, whence $dl = 2k_2 I dI$. Substituting at the outset in equation (203) it follows that

$$\frac{\partial}{\partial S}\left\{-\left(\frac{k_1 + 2Fk_2}{k_1}\right) I\right\}_{\mathscr{H}, F} = \left(\frac{\partial T}{\partial \mathscr{H}}\right)_{S, F}$$

so that the adiabatic change in temperature ΔT for a field change is given by

$$\Delta T = -\frac{T}{J\rho C_{F,\mathscr{H}}}\left(\frac{k_1 + 2Fk_2}{k_1}\right)\left(\frac{\partial I}{\partial T}\right)_{\mathscr{H}} . \Delta\mathscr{H}.$$

Now, Becker's values of k_1 and k_2 for magnetisation processes in nickel under tension are respectively $\dfrac{3F\lambda_s}{I_s^2}$ and $-\dfrac{3\lambda_s}{2I_s^2}$, whence $(k_1 + 2Fk_2)$ is zero and no reversible changes in temperature should accompany changes in magnetisation under these circumstances.

Experimental Studies. There are considerable difficulties in the way of the measurement of the quantity dQ_1 of equation (202), because in the study of soft materials by a step-by-step procedure it is necessary to detect temperature changes of the order of 10^{-6} ° C., and few such measurements have been made. Pioneer studies of hard steel were made by Adelsberger and by Constant. Experiments on carbon steels were made by Ellwood, by Honda, Ôkubo and Hirone and by Okamura, and on nickel by Miss Townsend and later by Hardy and Quimby.* In all these experiments the temperature changes were measured by means of a large number of thermocouples connected in series to a galvanometer. They all suffered from difficulties of change of galvanometer zero, and in most cases there was no means of calibrating the system used for the temperature measurements, apart from the calculation of the total heat on the assumption that Warburg's law is correct.

Bates and Weston* used a new method in which several thermocouple circuits were arranged in parallel. The "hot" junction of each thermocouple was placed in direct contact with the ferromagnetic specimen while the "cold" junction

* Full references and brief descriptions of these experiments are given in the paper by L. F. Bates and J. C. Weston, *Phys. Soc. Proc.* **53**, 5, 1941.

was close thereto, but thermally insulated therefrom, except for conduction along the material of the couple. Each thermocouple circuit was connected to its own insulated primary of low resistance wire wound on a mu-metal spiral core, or ring, which served as the core of a transformer. The secondary coil of many turns of low resistance wire wound on the core was connected to a special form of moving coil galvanometer or fluxmeter. The latter was peculiar in that the magnetic field was provided by an electromagnet, while the slight ferromagnetism normally found in a moving coil was artificially increased to provide a negative restoring couple, i.e. a couple in opposition to that provided by the suspension, so that an instrument of long period and very high sensitivity resulted.

The following theory of the method, though obviously incomplete, is sufficient for present purposes. Let us suppose that only one thermocouple is used. When the temperature of a "hot" junction changes by ΔT, a current proportional to ΔT flows in a primary circuit. It is now assumed that this current remains constant, so that the resulting change in the magnetic induction of the core persists for a reasonable interval of time. Consequently, an unidirectional discharge of electricity occurs in the secondary circuit and results in a fluxmeter deflection θ which is likewise proportional to ΔT. It is evident that a set of P identical thermocouples, each with its own primary winding, produces a deflection $P . \theta$, although each "hot" junction is in direct electrical contact with the specimen.

The sensitivity of the fluxmeter depended upon the electromagnet current and reached a maximum which was constant over a useful current range. Employing twenty thermocouples the sensitivity attained was normally 8×10^{-6} ° C. per mm. deflection with a scale 7 m. distant, and it was easy to read the deflection to 0·1 mm. The specimen to be investigated was a rod some 40 cm. long and 4 or 5 mm. in diameter, to which brass extension pieces were screwed, mounted vertically inside a water-cooled solenoid. The symmetrically arranged thermocouple leads issued from the upper and lower openings of the solenoid. Severe tension could be applied to the specimen, when necessary, by a brass steelyard arrangement.

The temperature measuring system was calibrated by producing standard adiabatic changes of temperature of the order of 0·001° C. by applying known longitudinal stresses to the specimen. Joule* showed that when a tension of F dynes is suddenly applied to a metal rod there occurs an adiabatic fall of temperature $\varDelta T$ given by

$$\varDelta T = -\frac{\alpha T}{J\rho C}\frac{F}{A}, \qquad \dots\dots(207)$$

where α is the coefficient of linear expansion of the rod, T the absolute temperature, J the mechanical equivalent of heat, and ρ, C and A are respectively the density, specific heat and area of cross-section of the rod. We see, then, that the application of F really causes an energy change $J\rho C\varDelta T$ calories per c.c. of the specimen; this is particularly helpful since in hysteresis experiments it is the heat liberated or absorbed per c.c. which is measured. This method of calibration automatically eliminated errors due to loss of heat.

Eddy currents were eliminated by making individual changes in magnetisation small. Effects of stray field upon the thermocouple leads were eliminated by means of a compensating coil C, which was connected to its own primary winding so that it formed a closed circuit with a time factor identical with that of an ordinary thermocouple circuit. The position of C could be adjusted with respect to the specimen. Unwanted induction effects were manifested by sharp or jerky initial movements of the fluxmeter coil, followed by the much more leisurely, true, thermal deflections; their direction depended upon the direction in which the hysteresis cycle was traversed, while that of the true thermal deflections did not. It was therefore easy, even if sometimes very tedious, to see when exact compensation was reached.

Outline of Results. We will now examine some representative results. In Fig. 101 are plotted the data recorded for a hard-drawn rod of nickel, 99·98 per cent pure, with less than 0·001 per cent sulphur, less than 0·02 per cent carbon and

---

* J. P. Joule, *Collected Works* (London: Phys. Soc.), **1**, 426, 1884.

a trace only of other impurities. The data were obtained by starting with the nickel in a field of about 200 oersteds, decreasing the field to zero and then increasing it to the same value in the opposite direction. Starting from the left-hand side of the figure the full-line curve represents the successive values of ΣdQ_1, or Q_1, plotted against the appropriate values of I. The broken line represents the $(\int \mathcal{H} dI, I)$ curve calculated from magnetic measurements made by the ballistic method

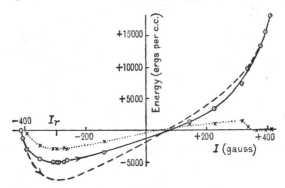

Fig. 101. Hard-drawn nickel, unloaded.
—— $Q_1, I.$ ----- $\int \mathcal{H} dI, I.$ ..... $\int \mathcal{H} dI - Q_1, I.$

and the dotted line represents the $(\int \mathcal{H} dI - Q_1, I)$ curve. In order to save space, the additional data obtained by reducing the field to zero once more and then increasing it to its maximum value in the original direction have not been plotted. They merely give a similar (Q_1, I) curve to that which can be obtained by rotating the printed full-line curve about the axis of ordinates while displacing it parallel to that axis until the first point on the left-hand side rests upon the furthermost point on the right-hand side. The experimental values of the retentivity are indicated by $I_{\text{rem.}}$ or I_r in Fig. 101 and other figures.

In Figs. 102 and 103 are shown the results obtained with specially annealed nickel of the same composition as that used for Fig. 101, when maximum fields of 200 and 400 oersteds, respectively, were used. The contrast between Figs. 101 and 102 brings out the profound effects of annealing, while that

345

between Figs. 102 and 103 indicates the effects of applying fields sufficiently high to magnetise the specimen beyond the regions in which kinks are observed in the $(I, \mathscr{H})$ curves for single crystals of nickel. In all these figures it is seen that

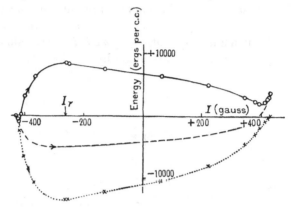

Fig. 102. Annealed nickel, unloaded. Maximum field 192·4 oersteds.
——— $Q_1, I.$ - - - - - $\int \mathscr{H} \, dI, I.$ ..... $\int - Q_1, I.$

when $\mathscr{H} = 0$ and $I = I_r$, $(\partial E/\partial I)$ is always zero. This means that $\partial(E_c + E_F)_T/\partial I = 0$ when $I = I_r$. In other words, in these circumstances the changes in crystal energy are exactly

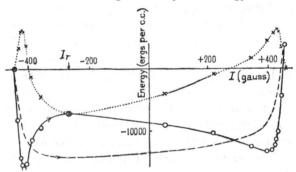

Fig. 103. Annealed nickel, unloaded. Maximum field 411·3 oersteds.

compensated by the changes in strain energy. Indeed, for one specimen of nickel 99·67 per cent pure, containing 0·04, 0·02, 0·04, 0·08, 0·01 and 0·14 per cent of carbon, silicon, copper, iron, manganese and magnesium, respectively, Bates

346

and Weston found that $(\partial E/dI)_T$ was zero for all values of I; but this was exceptional. The lower the coercivity of the nickel specimen the more pronounced was the minimum value of $\int \mathscr{H} dI - Q_1$. Moreover, $(\partial E/\partial I)_T$ was negative whenever the magnetisation processes were magnetically reversible and positive when they were magnetically irreversible, a statement which also applies to magnetisation experiments with virgin material.

For all (hard-drawn) nickel iron alloys studied, $\int \mathscr{H} dI - Q_1$ was always positive for the magnetisation cycles, $(\partial E/\partial I)_T$ being zero when $I = I_r$ and practically zero over a wide range of I extending from this state. There appeared to be no essential difference other than magnitude between the thermal behaviour of mu-metal and permalloy, which show very small magnetostriction, and 42 per cent nickel-iron alloy which has a very large magnetostriction; in these cases, however, the (Q_1, I) curves lie completely below the I axis. In all the experiments Warburg's law was found to hold with an accuracy of less than 1 per cent for a single cycle. The thermal changes for nickel under severe tension were very small, and, consistent with Becker's theory, there was no sign of cooling anywhere in a cycle. Traces of reversible heating and cooling were found in a cycle with a load of 8·57 kg. per mm.^{2}, but there was no cooling with 12·2 kg. per mm.^{2}, in agreement with Kirchner's finding that the magnetostriction varies linearly with I^2 only when the load exceeds 10 kg. per mm.^{2}.

Similar experiments were made by Bates and Healey* with specimens of armco iron containing only 0·11 per cent known impurity. The work was very difficult because of the smallness of the thermal changes, the disturbing effects of mechanical vibrations and the largeness of eddy current effects, so that Warburg's law could not be proved for individual cycles. The Q_1 curves for unannealed iron and unannealed nickel were surprisingly alike, especially in the case of cycles with high values of the maximum field with iron; yet there is a big difference in the magnetostriction behaviour of the two metals.

* L. F. Bates and D. R. Healey, *Phys. Soc. Proc.* **55**, 188, 1943.

Measurements with a maximum field of 400 oersteds were made on iron annealed by maintaining it at a temperature of 950° C. for half an hour *in vacuo*, and then allowing it to cool slowly. Certain asymmetries were found in the thermal measurements, depending on the direction of magnetisation of the specimen, and were interpreted as evidence that annealing does not produce a random distribution of the crystal axes in the specimen, but merely shifts the preferred orientation, and as evidence of the effects of mechanical fatigue on magnetostriction phenomena. An important feature was the lack of definiteness in the values of dQ_1 for stated changes in the field in the region of 350 to 400 oersteds on the ascending portion of the hysteresis curve. There was a systematic fall in the dQ_1 value for a chosen field change as the measurements were continuously repeated until a small constant value was observed. Now, Akulov* pointed out that, theoretically, a discontinuity ought to exist in the $(I, \mathscr{H})$ curve for a single crystal of iron when magnetised parallel to a [111] direction. He showed that the $(I, \mathscr{H})$ curve should bend suddenly towards the I axis when the field reached about 350 to 400 oersteds, which is precisely the region in which Bates and Healey observed the unexpected thermal behaviour. Presumably, the phenomenon has never been observed in ordinary magnetic measurements of I in hysteresis cycles because of the difficulty of obtaining strain-free crystals.

Cobalt, 98·4 per cent pure, in the form of annealed and hard-drawn No. 12 s.w.g. wire, was investigated by Bates and Edmondson.† The heat changes were relatively large and in marked contrast to those observed with iron and nickel. For example, with unannealed iron and nickel there is always cooling followed by heating as a high field on the specimen is reduced, but with cobalt there is always an initial warming; on the whole, cobalt behaves in a contrary manner to iron and nickel. It is important to correlate these differences with differences in crystal structure, e.g. preferred orientation and grain size, magnetostriction and other properties. The collec-

---

* N. Akulov, *Zeit. für Phys.* **69**, 81, 1931.
† L. F. Bates and A. S. Edmondson, *Phys. Soc. Proc.* **59**, 329, 1947.

tion of fundamental information on the energy aspects is a very necessary prelude.

Limitations of the Domain Concept. When we survey the experimental basis of the domain concept,* we see that we require it to explain (*a*) the fact that ferromagnetic substances do not become magnetised except in a magnetic field, (*b*) the Barkhausen effect, (*c*) the Sixtus-Tonks experiments on the growth of large regions in materials under tension in an applied magnetic field, and, possibly, (*d*) the occurrence of Bitter figures on the surfaces of ferromagnetic single crystals. Unfortunately, each of the above phenomena appears to require its own particular type of domain. In fact, the domain concept lacks the precision which can only be given to it when we can find a rigid experimental proof of the existence of domains and also a more fundamental reason for their existence than the above phenomena provide.

Fortunately, we are on the brink of such a discovery, for Néel† has given a theory of domain formation which accounts for types of Bitter figures, mentioned on p. 320. He pictures the 90 degree domains in a rectangular bar cut parallel to the binary axis of an iron crystal as thin fillets with their planes perpendicular to that axis, along which the magnetic field acts. Now, their edges on the bar surface should give very intense surface fields. As these fields are not observed, *closure domains*, of prismatic shape, magnetised in such ways that little flux emerges through the surface, are postulated. There has not been time to test Néel's ideas properly, but we already have some evidence of their essential correctness.‡

* Cf. W. F. Brown, *Rev. Mod. Phys.* **17**, 15, 1945. Industrial aspects of the domain theory are considered very fully by G. C. Richer, *Journ. Iron and Steel Inst.* **2**, 90, 1944.

† L. Néel, *Journ. de Phys.* **5**, 241, 1944.

‡ H. J. Williams, *Phys. Rev.* **71**, 646, 1947.

CHAPTER XII

RECENT WORK ON FERROMAGNETIC SUBSTANCES

Introduction. In Chapter II we studied the production and measurement of magnetic fields and described some of the more important properties of ferromagnetic substances. In the present chapter we will discuss recent additions to our knowledge of these matters. New methods of producing very strong magnetic fields, without the use of iron, have been devised. Great developments have taken place in the manufacture of permanent magnets, and we must now consider the behaviour of these substances more closely than was previously thought necessary. We will also discuss certain hysteresis and magnetoelastic phenomena which have not been treated earlier in this book.

Bitter Magnets. For the production of intense fields, Bitter* has described magnets of new types in which only small quantities of iron are used, and some in which no iron at all is used. The latter must, however, be supplied with currents of the order of 10,000 amperes at less than 170 volts. The field is produced inside an arrangement which may be regarded as a very low resistance spiral of thick copper tape. Actually, it consists of a set of overlapping circular copper discs (or rings with radial slots) separated by thin layers of insulation. The successive overlaps are displaced regularly as one proceeds from the lowest to the uppermost disc, so that the net result is that one has a spiral of copper. The outermost discs are soldered to heavy copper plates which in turn are bolted to the leads from a current generator. The discs are bored and covered with water-resisting paint, so that a stream of cooling water can be forced axially through the system. The water must be distilled, or, at any rate be pure, and may be pumped round the system and a heat exchanger which is cooled in

* F. Bitter, *Rev. Sci. Instr.* **7**, 479 and 482, 1936; **8**, 318, 1937; **10**, 373, 1939.

river water. One of these magnets is capable of giving a field of 100,000 oersteds uniform to 1 per cent over a volume of 25 c.c.

The main feature about the Bitter magnets in which iron is used is that each of a pair of small powerful coils is surrounded by iron. The coil itself provides the major part of the field. For example, a spiral of copper tape of gradually increasing width was wound on a brass tube, as one current terminal, to form a coil with a conical end which was pressed into an iron ring serving as the other terminal. When supplied with 860 amperes at 25 volts and water cooled, a pair produced a total field of 33,000, of which 9000 oersteds was due to the iron. The field was increased to 43,000 oersteds by inserting iron cores into the brass tubes.

Fields up to 200,000 oersteds over a cylindrical space 8 mm. in diameter have been obtained by connecting a very big battery to a coil immersed in liquid hydrogen.* Only a small quantity of liquid evaporates during an experiment, because the coil resistance is very low and the latent heat of liquid hydrogen relatively big. (There seems to be little advantage in using liquid helium below the λ point, since, although the resistance is smaller, so too is the latent heat involved.)

The Properties of Permanent Magnets. The magnetic properties of high coercivity permanent magnet alloys may be investigated in a physical laboratory with ordinary equipment. Manufacturers of such alloys usually employ the magnetic testing gear described in B.S.I. 406. This consists of a soft iron disc, mounted in ball-bearings between a pair of pole-pieces adapted to receive the magnet under test, and driven at a constant speed by a small electric motor. Collecting brushes make contact with the spindle and edge of the disc, and the magnet under test thus provides the field of a simple dynamo whose E.M.F. measures the flux produced by the magnet. The magnetic state of the latter is varied by a magnetising coil of a specified number of turns, which encircles the magnet and which can be supplied with direct current; in the case of

---

* W. J. de Haas and J. B. Westerdijk, *Nature*, **158**, 271, 1946.

horseshoe magnets two magnetising coils, one on each limb, can be used.

Another method for dealing with bars of these alloys has recently been described by Jellinghaus.* The remanent induction in the permanent bar magnet is measured by placing it between the poles of an electromagnet whose base contains a kind of well-made barrelled iron stopper which extends almost across the whole section of the base. In the hollow of the stopper are the windings of a coil which is connected to a fluxmeter. On placing the stopper with the plane of the windings perpendicular to the lines of induction in the iron and then rotating it quickly through 180 degrees, the fluxmeter gives a deflection directly proportional to the flux through the iron, and this is assumed to be the same as the flux through the bar.

The magnetic potentiometer may readily be used in such measurements.† It appears to have been used previously by Lacoste-Tayan‡, but only for the determination of the demagnetisation factors of permanent bar magnets. The potentiometer should be of the small type used by Wolman and by Bates and Lloyd Evans. The most convenient method of manufacture is as follows. A piece of ebonite rod, of circular cross-section for large models and of rectangular cross-section for small models, is used as a former and wound from the centre towards one end with No. 40 s.w.g. d.s.c. copper wire.§ The winding is lightly coated with shellac and then wound back towards the other end to an equal distance from the centre. It is again lightly coated with shellac and wound back to the centre again, so that the two ends of the winding emerge from the middle of the former. The length of the former so wound with two layers of wire is next heated over a flame while it is bent round a metal tube of chosen diameter, an adequate length of bare former material being available, and the uncovered portions are then tied together until the whole is cold. The excess portions are removed and the feet of the

* W. Jellinghaus, *Zeit. für tech. Phys.* **12**, 312, 1942.
† L. F. Bates, *Phil. Mag.* **36**, 297, 1945.
‡ H. Lacoste-Tayan, *Compt. Rend.* **206**, 103, 1938.
§ Cf. also T. A. Margerison and W. Sucksmith, *Journ. Sci. Instr.* **23**, 182, 1946.

former are filed flat until the ebonite is almost down to the wire; the final removal of ebonite is done by careful rubbing on sandpaper until bare wire is just visible. The number of turns in each half-layer of the potentiometer is carefully controlled and care taken to make the windings completely symmetrical. The whole is finally given a heavy coat of shellac. The fine end wires or leads are twisted together and wrapped twice round the middle of the potentiometer to provide material for repairs, which are necessary when the potentiometer is frequently used. The ends of the thin leads are subsequently joined to thicker copper wires and then to copper flex.

It is advisable to fix the two ends or feet of the potentiometer to prevent them from opening, as the ebonite tends to straighten. This is best done in the case of larger potentiometers by inserting the ends through two holes in a flat ebonite plate and cementing them in position. In the case of small potentiometers a thread binding is sufficient. The potentiometer is calibrated during a set of measurements either in a standard solenoid or in a uniform field of an electromagnet with plane pole pieces; the former method is the more reliable.

In the above description of the making of a magnetic potentiometer it has been emphasised that the ends or feet surfaces of the potentiometer are made flat. It is equally important that the surface of any magnetic specimen upon which it is placed should be reasonably plane. When these conditions are satisfied, the deflection, of a sensitive galvanometer joined to the leads, which results on removal of the potentiometer from the surface of the specimen, is a direct measure of the field $\mathscr{H}$ acting upon the specimen in that region in which the potentiometer is placed. Consequently, the complete $(B, \mathscr{H})$ curve for the material of a magnet can be found by the following method.

The magnet, when in the form of a straight bar, is placed between the plane pole pieces of an electromagnet, or of a simple apparatus with adjustable gap designed for the purpose. The middle of the test magnet is covered with a thin layer of (music roll) insulating paper and wound with 40 to 80 turns of thin d.s.c. copper wire, preferably in a single layer, whose ends are connected to a fluxmeter, so that changes in B can be measured directly. The corresponding values of $\mathcal{H}$ are obtained from the galvanometer readings when the potentiometer is placed with its feet astride the B coil clear of insulating material, and then removed quickly. A convenient potentiometer consists of 240 turns wound on a former of 0.95×0.30 cm.^{2} cross-section, bent into a semi-circular arch, with a distance of 1.25 cm. between the mid-points of the feet.

In the case of horseshoe magnets each test magnet is placed to bridge the gap of the electromagnet, its feet resting on the upper portions of the electromagnet pole pieces, which are flattened for this purpose over a reasonable area parallel to the line of electromagnet pole centres. Of course, the maximum fields which can be applied are now considerably weaker than those available for straight specimens. The B coil is wound on the arch of the magnet.

Prior to recording a set of $(B, \mathcal{H})$ data the test magnet is taken through the full magnetisation cycle several times to get the material into a cyclic state. For most purposes, only the demagnetisation portion of the $(B, \mathcal{H})$ curve from $\mathcal{H} = 0$ to $\mathcal{H} = -\mathcal{H}_c$ is required. Hence, at the start of each measurement the electromagnet current is adjusted to its maximum value and the potentiometer reading taken. The magnet current is then switched off and the change of flux recorded. Theoretically, the permanent magnet should now be in the state $B = B_r$ and $\mathcal{H} = 0$. This is, however, rarely found to be the case; for, on taking potentiometer readings, small values of $\mathcal{H}$, sometimes positive and sometimes negative, are obtained. These are due to imperfections of surface and to the way in which the magnet rests upon the electromagnet pole pieces. The condition is, however, reproducible, and in what follows the corresponding induction is denoted by B_r'.

The electromagnet reversing switch is now turned, the circuit resistances fixed at chosen values and the current again switched on. The permanent magnet then suffers partial demagnetisation, the change of flux being recorded by the fluxmeter; the new value of the demagnetising field is found by the potentiometer. It is generally unsatisfactory to continue from this stage and observe further successive changes in B and $\mathscr{H}$. There is always a danger of current fluctuation while $\mathscr{H}$ is being measured, and serious errors in the measurement of B can result. Consequently, after each individual B, $\mathscr{H}$ reading, the magnetisation cycle is completed and the electromagnet current again adjusted to its maximum positive value. It is again switched off, the reversing key turned, the resistance adjusted to a new value and the current switched on. Another change of induction from B'_r to a new value on the demagnetisation curve is thus obtained. It will be seen that the maximum induction, B_m, and the apparent remanent induction, B'_r, have to be found separately. The former is half the sum of all the changes in B recorded as one proceeds from B_m to $-B_m$ without pausing to measure fields. The latter is obtained from the change in induction, B_m to B'_r, which occurs when the current is reduced from its maximum value to zero. All fluxmeter deflections should be calibrated by means of a standard mutual inductance whose secondary is permanently in series with the coil on the magnet. Examples of $(B, \mathscr{H})$ curves so obtained are reproduced in Fig. 104. Curve (a) was obtained with an alloy of 23·3 per cent cobalt and 76·7 per cent platinum, in the form of a polished bar 2·02 cm. long and 1 cm.^{2} in cross-section, and made ferromagnetic by keeping it at 650° C. for three hours. With materials of such high coercivity and comparatively low induction it is necessary to correct for the lines of the applied field which are collected in the space between the B coil and the specimen on which it is wound. Curve (b) was obtained with a bar of sintered "alnico" 3·73 cm. long and 0·90 cm. × 0·85 cm. in cross-section. Curves (c) and (d) were obtained respectively for "alcomax" and "alnico V" ("ticonal") horseshoe magnets of simple design.

Bates* has also shown how the magnetic potentiometer may be used to investigate the circuit relations, p. 76, which permanent magnets obey.

There is an interesting point to note in connection with the definition of coercive force. If we plot B against $\mathscr{H}$ and

* L. F. Bates, *loc. cit.*

355

23-2

$(B - \mathcal{H})$ against $\mathcal{H}$ on the same axes, the intercepts on the axis of ordinates must coincide, but those on the $\mathcal{H}$ axis do not. The coercivity, $_B\mathcal{H}_c$, measured with the $(B, \mathcal{H})$ curve is

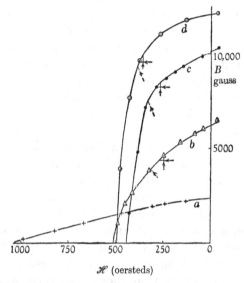

Fig. 104. Demagnetisation curves of permanent magnet materials.

less than the coercivity, $_I\mathcal{H}_c$, measured with the $(B - \mathcal{H}, \mathcal{H})$ curve. The numerical difference is small in the case of soft magnetic materials, but it is really important in the case of permanent magnet materials, when $_B\mathcal{H}_c$ must always be given.

The Economic Utilisation of Permanent Magnets. We have seen on p. 77 how the volume of magnetic material necessary for the maintenance of a stated magnetic field over a given air gap may be calculated. It is desirable to examine this problem more closely in the light of recent work on the subject.*

Let us suppose that we have a horseshoe permanent magnet whose gap is closed by a piece of iron, and that, after energising the magnet, the iron is removed. The magnet will work at point P on the demagnetisation curve determined by the total reluctance, i.e. resistance to the passage of magnetic flux, in

* D. J. Desmond, *Journ. Inst. Elect. Eng.* **92**, 229, 1945.

the circuit. The required flux in the gap will be obtained from
a minimum of magnetic material if each cm. cube is producing
its maximum flux. In addition to the *useful field* in the gap
there is the *leakage field* in parallel with it. Consequently,
it is convenient to work with the reciprocal of the reluctances
associated with these two fields and to treat them from the
viewpoint of each cm. cube of magnet material.

We may write

$$\phi = B_s A_s = \frac{\text{M.M.F.}}{\text{reluctance}} = \frac{\int \mathcal{H}_s dl_s}{\text{total reluctance}} = \frac{\mathcal{H}_s l_s}{r},$$

whence

$$\frac{B_s}{\mathcal{H}_s} = \frac{l_s}{Ar} = p, \text{ say.}$$

The quantity p is called the *total unit permeance*, because it
refers to *one cm. cube* of the material; we have here assumed
that $\int \mathcal{H}_s dl_s = \mathcal{H}_s l_s$ and that each unit volume of the magnet
behaves exactly like its neighbour, and both these assumptions
are only approximately correct. If we further postulate that
all the leakage occurs in the immediate region of the gap, we
may write

Total unit permeance
 = *useful unit permeance + leakage unit permeance*

or
$$p = p_u + p_l.$$

Turning now to the $(B, \mathcal{H})$ curve, Fig. 105, we draw the
line OP through the origin and the *working point* P of the
magnet, so that $\tan POS = B_s/\mathcal{H}_s = p$. The line OP is called
the *total permeance line of the circuit*. Drawing a line OC,
such that $\tan COS = (\text{leakage per c.c.})/\mathcal{H}_s$, we obtain the
leakage permeance line. The total flux supplied per c.c. of the
magnet is PS, of which PX' is useful and $X'S$ is leakage. The
field required to maintain this flux is $\mathcal{H}_s = OS$, so that the
total energy supplied per c.c. of the magnet is $PQOS$ of which
$PQO'X'$ is useful and $X'O'OS$ is leakage. We may note that
$(\tan POS/\tan COS) = q$.

It is evident that, for the magnet to have the maximum
efficiency, P must coincide with the $(B\mathcal{H})_{\max}$ point. This

means that a definite value of p is required for optimum performance, and if for a given magnet we calculate the reluctance of the air gap and assume a value of q, then the total reluctance r can be calculated and the total volume of the magnet found from $q \dfrac{A_s^2 B_a^2}{(B\mathcal{H})_{\max}} r.$

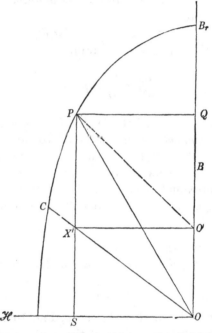

Fig. 105. Permanent magnet working without recoil and with parallel leakage.

The concept of useful flux is rather more helpful when we apply it to magnets which are said to be used under recoil conditions. Let us take a pair of identical horseshoe magnets and place them in contact to form a completely closed magnetic circuit, when they are to be fully energised. In order to use one of these magnets, the pair must be separated, whereupon the magnetisation falls from the point B_r to the point P on the $(B, \mathcal{H})$ curve, Fig. 106. On placing the magnets together again, the magnetisation follows, not the initial path PB_r, but a hysteresis line PR, known as the *recoil line*, which we will assume to be straight. The line OP represents the leakage permeance line *on open circuit*, i.e., with the magnets far apart. If the magnets are now brought together so that a fixed gap exists between one pair of their opposing surfaces

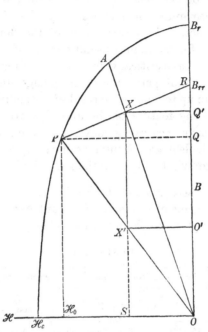

Fig. 106. Permanent magnet working with recoil and parallel leakage.

(while the other gap is closed by iron), the magnetisation rises to some point X. The total flux available is now XS, and the useful flux is given by XX'. The slope of $O'X$ is equal to the *recoil useful unit permeance* and the area $XQ'O'X$ now represents the useful energy. The latter is a maximum when X is halfway between P and R, when it is easy to see that the *maximum recoil useful energy* is half the area of the triangle OPR. It is therefore desirable to make this triangle as large as possible, a condition which will uniquely fix the position of P. The way in which these several quantities may be determined by experiment has been described by Bates.*

* L. F. Bates, *loc. cit.*

359

The slope of the line PR, assumed straight, is equal to $p_u - p_l$, so that the value of p_u for any given value of p_l may be determined. It is easily shown from the geometry of the figure, and assuming that the slope of PR is practically independent of the point P, that the area of the triangle OPR is a maximum when P lies below the $(B\mathscr{H})_{\max}$. point, i.e. at a numerically greater value of the demagnetisation field. Hence, for maximum useful output of a magnet both p_u and p_l must have critical values. Tables of maximum useful energy, useful unit permeance and leakage unit permeance may be drawn up to guide us in the design of efficient magnets. One outstanding feature of the above treatment is that it shows that in some cases one may get a more efficient magnet by altering the leakage. A number of examples of permanent magnet design will be found in the paper by Desmond.*

New Permanent Magnet Materials. The composition and properties of some of the more important permanent magnet materials are given in the following table:†

TABLE XVII

Material	Percentage composition					Properties (average)		
	Al	Ni	Co	Cu	Fe	B_r gauss	H_c oersteds	$(B\mathscr{H})_{\max.} \times 10^6$ gauss oersteds
Alni	13	24	—	3·5	59·5	6,200	480	1·25
Alnico	10	18	12	6	54	8,000	500	1·7
Alnico V	8	14	24	3	51	12,000	525	4·0
Alcomax II	8	11	24	6	51	12,400	570	4·3

It was found, somewhat unexpectedly, that the magnetic properties of permanent magnet materials are greatly influenced by exposing them to a magnetic field while they are under heat treatment. The first experiments were made by Oliver and Shedden,‡ who found that by cooling alnico in a field of 4400 oersteds from 1200° C., $(B\mathscr{H})_{\max.}$ is increased from $1·5 \times 10^6$ to $1·8 \times 10^6$ gauss-oersteds. The changes are even

* D. J. Desmond, *loc. cit.*
† K. Hoselitz, *Journ. Sci. Instr.* **23**, 65, 1946.
‡ D. A. Oliver and J. W. Shedden, *Nature*, **142**, 209, 1938.

more marked with alloys of higher cobalt content, which were previously regarded as uneconomical for the manufacture of permanent magnets, and values of $(B\mathscr{H})_{\text{max.}}$ of up to 5×10^6 gauss-oersteds can now be obtained in one direction; the properties in a direction perpendicular thereto are, of course, less good.

The new materials are usually tempered at about 600° C. after heat treatment; the exact cooling speed and tempering temperature, etc., depend on the composition. Hence the magnets are fairly stable when they leave the makers.

Magneto-resistance. It is desirable to discuss further some of the phenomena described on pp. 293 and 294. This may be done conveniently by reference to some measurements of the magneto-resistance of high coercivity alloys.*

Suppose we take a thin rod of the material whose resistance in the unmagnetised state is R_0, and measure the changes $\Delta R_{\parallel}$ and $\Delta R_{\perp}$ which respectively occur when a magnetic field $\mathscr{H}$ is applied parallel to and perpendicular to the long axis of the rod, and plot the values $\Delta R_{\parallel}/R_0$ and $\Delta R_{\perp}/R_0$ against $\mathscr{H}$. This requires a definition of the unmagnetised state, which we will define as the state of the rod when, by means of a magnetic potentiometer, no difference of magnetic potential can be found over its length. Unfortunately, there is some ambiguity in the matter, since, for example, this result is obtained when the magnetic vectors are either all aligned parallel and antiparallel to the long axis of the rod, or are all aligned perpendicular thereto, in such a way that there is no resultant magnetisation.

Several workers have shown that, for a pure specimen of a ferromagnetic metal, $\Delta R_{\parallel}/R_0$ normally shows a marked initial increase with the applied field; it rises to a maximum and eventually decreases linearly with the field; so that in very high fields it may become zero and, indeed, change sign. In the present work, a specimen of pure cobalt was used for comparison measurements, and the upper curve of Fig. 107 shows the typical variation of $\Delta R_{\parallel}/R_0$. When the direction

* L. F. Bates, *Phys. Soc. Proc.* **58**, 153, 1946.

of $\mathscr{H}$ is perpendicular to the direction of the current in the specimen, the values of $\Delta R_{\perp}/R_0$ are at first subject to complications due to the high value of the demagnetisation field, D.I., but when the demagnetisation field is overcome, the value $\Delta R_{\perp}/R_0$ falls rapidly with increase in $\mathscr{H}$ and finally decreases with the applied field at the same rate as $\Delta R_{\parallel}/R_0$, as shown by the lower curve in Fig. 107. The two points

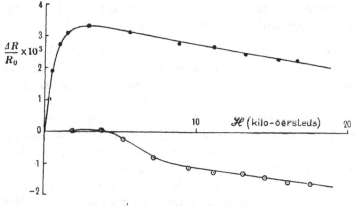

Fig. 107. Magneto-resistance of cobalt.
● Longitudinal field.   ⊙ Transverse field.

nearest the origin on the lower curve lie above the $\mathscr{H}$ axis, but this is merely because the specimen is of finite cross-section, and its long axis cannot be placed absolutely at right angles to the applied field; consequently, a small longitudinal component is always present.

The main features of the curves of Fig. 107 may be explained on the Weiss domain theory of ferromagnetism, if we suppose that magneto-resistance changes occur only when domain vectors turn through angles between 0 and 90 degrees or when boundary displacements take place between neighbouring domains whose vectors are perpendicular to one another, and that they do not occur to any marked extent when vectors as, for example, those parallel to a direction of easy magnetisation in a single crystal, turn through 180 degrees. In order that Ohm's law may be satisfied, the resistance of a polycrystalline

362

specimen of a ferromagnetic metal must be a function of $\overline{\cos^2\theta}$, where θ is the angle between the direction of the current in a portion of the specimen and that of the field when saturation conditions set in; the bar denotes the average taken over all portions of the specimen. Hence the resistivity of the material may in general be written as

$$\rho = A + B\,\overline{\cos^2\theta},$$

where A and B are constants. In an isotropic demagnetised state $\overline{\cos^2\theta}$ would have the value $\tfrac{1}{3}$, giving $\rho = \rho_0 = A + \tfrac{1}{3}B$. Under saturation conditions, with current and field parallel, we have

$$(\rho_\parallel)_{\text{sat}} = A + B,$$

and, with current perpendicular to the field,

$$(\rho_\perp)_{\text{sat}} = A.$$

Since, in practice, we measure relative changes of resistance only, it is more convenient to write for the general case of non-isotropic material, starting from $\rho_0 = A + B\,\overline{\cos^2\theta_0}$,

$$\left(\frac{\Delta R_\parallel}{R_0}\right)_{\text{sat}} = \frac{B(1 - \overline{\cos^2\theta})}{A + B\,\overline{\cos^2\theta_0}},$$

$$\left(\frac{\Delta R_\perp}{R_0}\right)_{\text{sat}} = \frac{-B\,\overline{\cos^2\theta_0}}{A + B\,\overline{\cos^2\theta_0}} = a,$$

and $\quad \left(\frac{\Delta R_\parallel}{R_0}\right)_{\text{sat}} - \left(\frac{\Delta R_\perp}{R_0}\right)_{\text{sat}} = \frac{B}{A + B\,\overline{\cos^2\theta}} = b,$

whence $\quad \overline{\cos^2\theta_0} = 1\Big/\left[1 - \left(\frac{\Delta R_\parallel}{\Delta R_\perp}\right)_{\text{sat}}\right].$

In the case of isotropic material, it follows that $(\Delta R_\parallel)_{\text{sat}}$ should be equal to $-2(\Delta R_\perp)_{\text{sat}}$, as found by Thiessen* for annealed nickel, and, moreover, the last equation means that $(\Delta R_\parallel)_{\text{sat}}$ and $(\Delta R_\perp)_{\text{sat}}$ must have opposite signs or $(\Delta R_\parallel)_{\text{sat}}$ be zero. The value of b, which is the relative change in resistance when the magnetic state of the specimen is changed from transverse to longitudinal saturation, is clearly independent of the initial state of the specimen and, practically, of the value assumed for ρ_0. It should also be the same for

---

* G. Thiessen, *Ann. der Phys.* **38**, 167, 1940.

polycrystalline as for single-crystal specimens. Positive values of b have been found for all materials previously investigated to date.

It is generally agreed that the intensity of magnetisation of each Weiss domain may be increased slightly by the application of a strong field, and this is the reason for the linear rates of change of $\Delta R_{\parallel}/R_0$ and $\Delta R_{\perp}/R_0$ at high values of the field, as depicted in Fig. 107. Theoretically, the two rates should be equal, as is the case for the cobalt specimen within the limits of experimental error. The values $(\Delta R_{\parallel}/R_0)_{\text{sat}}$ and $(\Delta R_{\perp}/R_0)_{\text{sat}}$ refer to saturation conditions extrapolated to zero external field. Consequently, to find the required value of $(R_{\parallel}/R_0)_{\text{sat}}$ from Fig. 107, the linear portion of the upper graph must be produced to cut the axis of ordinates, since we may neglect longitudinal demagnetisation effects. The linear portion of the lower graph must be produced to cut the ordinate at $\mathscr{H} = 2\pi I_s$, where I_s is the value of the intensity of magnetisation, obtained by producing back the $(I, \mathscr{H})$ curve at high field values to cut the I axis, and 2π is the demagnetisation factor for a long narrow cylinder or rod magnetised at right angles to its long axis.

The interesting fact emerged from these experiments that the resistance of every modern high-coercivity alloy investigated is always decreased by the application of a magnetic field, whether parallel or perpendicular to the direction of the current, apart from one or two small exceptions to this rule when very small fields are used, and these are readily understood. A similar behaviour had previously been noted only in the cases of Heusler alloy and copper-rich alloys like constantan. Now, according to the preceding equation, the theory of domain magnetisation requires that $\Delta R_{\parallel}/R_0$ and $\Delta R_{\perp}/R_0$ extrapolated to zero effective field from measurements made in strong fields should have opposite signs, which is in flat contradiction with the experimental data. An obvious suggestion to overcome this difficulty is that the magneto-striction deformations of the domain boundaries or, possibly, of the boundaries between the metal grains, play an important part. Nothing is known about the magnetostriction properties of high coercivity alloys, and it may be that their determination will throw light on the problem. It is also interesting that in all cases the value of b is positive or zero within the limits of experimental error, and we are again forced to the conclusion that the main resistance changes measured may be due to changes in the boundaries between the metal grains. This raises the interesting question of the reason for the marked directional properties of alcomax, and one wonders if they are to be attributed to the boundaries between the grains. No fundamental differences between the nature of the boundary in cast and sintered metals were found in these experiments.

Experiments on Single Crystals. Kondorsky,* and Becker and Döring,† have pointed out that Williams's "picture-frame" experiments on single crystals, p. 159, can be interpreted as showing that there exist preferential directions of spin orientations in the crystal, i.e. the numbers of spins parallel and antiparallel to the three axes are not equal. Denoting the numbers for the directions [100], [110] and [111] by p, q and r respectively, and writing $p = 1$, $q = r = 0$ when $\mathscr{H}$ is parallel to the [100] direction, $p = q$ and $r = 0$ when $\mathscr{H}$ is parallel to the [110] direction and $p = q = r = \frac{1}{3}$ when $\mathscr{H}$ is parallel to the [111] direction, it is easy to show that the initial susceptibilities parallel to these three directions are in the ratio $6 : 3 : 2$, which is in accord with the experiments. Okamura and Hirone‡ made picture frame single crystals of nickel $8 \cdot 0 \times 18 \cdot 6$ mm. with cross-section $(1 \cdot 96 \times 1 \cdot 89)$ mm.², the sides being parallel to the [111] direction. The initial magnetisation and hysteresis curves were found at different temperatures by the ballistic method. The interesting feature of the results is that B_r, H_c and $\int H \, dI$ each rise to a maximum at about $100°$ C., and while H_c is $2 \cdot 4$ oersteds at room temperatures, its value at $-190°$ C. is only $0 \cdot 7$ oersted; but one is inclined to ask whether a trace of some impurity, e.g. copper, might be responsible.

* E. Kondorsky, *Phys. Rev.* **53**, 319, 1938.
† R. Becker and W. Döring, *Ferromagnetismus*, p. 153, Springer, Berlin, 1939.
‡ T. Okamura and T. Hirone, *Phys. Rev.* **55**, 102, 1939.

Becker's Theory and the Kaya Relation. Equation (110), p. 159, has been proved by Becker to follow from his theory that, as we reduce the field on a ferromagnetic single crystal, only rotations occur until, at $I = I_r$, boundary extensions suddenly come into play. Incidentally, Kirkham* showed that at about 200° C. the [100] direction becomes the direction of easy magnetisation in nickel.

Von Engel and Wills† have discussed the behaviour of the component $I_\perp$ of the magnetisation perpendicular to an applied field $\mathscr{H}$. They emphasise that, on the above theories, $I_\perp$ should increase for low values of $\mathscr{H}$, whereas experiment shows that it is zero. They therefore suggest that the magnetisations of the domains must be pictured as distributed continuously in angle, and, by assuming that the proportion of the volume of the domains magnetised in a given direction is the greater the less the energy of magnetisation in that direction, they find reasonable agreement with experiment.

The Torque Magnetometer. The use of a torque magneto-

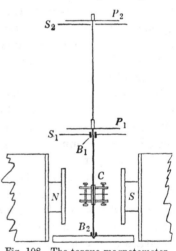

meter was mentioned on p. 162, and one form is shown in Fig. 108. The specimen, in the form of a disc, is clamped on the carrier C, mounted on a phosphor bronze shaft between the jewelled bearings B_1 and B_2. To the upper end of the shaft is fixed a pointer P_1 which moves over the lower semi-circular scale S_1 and fixes the position of the disc with respect to the lines of force. The upper pointer P_2, situated above the upper scale S_2, serves to mea-sure the torque in the suspen-

Fig. 108. The torque magnetometer.

sion. The magnetic field is applied between two flat pole-faces, NS, each attached to an iron core 5 cm. sq. and 30 cm. long. Flat

* D. Kirkham, *Phys. Rev.* **52**, 1163, 1937.
† A. von Engel and M. S. Wills, *Roy. Soc. Proc.* A, **188**, 464, 1947.

pole-faces are not strictly necessary, as the torque approaches a limit asymptotically at high fields, a field of 2000 oersteds giving a torque of about 80 per cent of the maximum value.

The torque per c.c. of crystal is given by the general expression

$\Gamma = -dE/d\alpha$ in the limit where I approaches $\mathscr{H}$ when $\mathscr{H}$ is large,

$$= -K_1(2\sin 2\alpha_0 + 3\sin 4\alpha_0)/8$$

(where α_0 is the angle between the applied field and the [100] direction). The value of K_1 can then be obtained by trial to make the theoretical curve fit the experimental one.

In the case of an unannealed disc cut from hard rolled sheets we find anisotropy similar to that for a single crystal, because rolling causes the crystals to become orientated in a special manner. For example, an unannealed nickel disc may be regarded as made up of about equal numbers of crystals in the two possible positions with their (110) planes in the plane of the sheet and their [111] directions coincident with that of rolling. The theoretical expression for the torque per c.c. in this case is

$$\Gamma = K_1'(7\sin 4\beta + 2\sin\beta)/24,$$

where β is now the angle between the direction of rolling and that of the applied field, and K_1' is not, of course, the anisotropy constant K_1. Again, an annealed nickel disc gives rise to a torque $\Gamma = -K_1'(\sin 4\beta)/2$, similar to that for a single crystal disc cut parallel to the (100) plane, the direction of rolling being [010]. Hence, annealing does not produce a random crystal distribution but merely changes the preferred orientation of the crystals. Bozorth and Williams* find that certain apparent anomalies, obtained with discs of low demagnetisation factor cut parallel to the (110) and (100) planes, the torque passing through a maximum before the final approach to saturation, are readily explained on the domain theory.

Rotational Hysteresis. If we place a single disc of magnetically isotropic material in a uniform magnetic field $\mathscr{H}$

* R. M. Bozorth and H. J. Williams, *Phys. Rev.* **59**, 827, 1941.

whose lines of force are parallel to the plane of the disc and rotate the latter about an axis through the mid-point and perpendicular to the plane of the disc, then the intensity of magnetisation I will lag behind $\mathscr{H}$ by an angle α. Consequently, a uniform torque $\mathscr{H}I\sin\alpha$ per c.c. of material will be needed to rotate the disc, so that the energy wasted in *rotational hysteresis* will be $2\pi\mathscr{H}I\sin\alpha$ per cycle, independent of the rate of rotation of the disc. Theoretically, it is therefore possible by measuring the torque for very small rates of rotation to avoid eddy current disturbances and so determine the relation between $\mathscr{H}$ and I.

We saw, however, that rolled sheets have pronounced directional properties, so that a disc cut from rolled sheet will experience a torque which changes its direction as the disc rotates. The latter torque is often about one hundred times as great as the hysteresis torque in the case of magnetically soft materials, and so measurements on a single disc do not enable us to obtain reliable information about hysteresis losses. We therefore use a composite disc made up of three identical discs so superimposed that corresponding diameters on the discs are relatively displaced through angles of 0, 120 and 240 degrees respectively.*

The general expression for the torque per c.c. on a single thin disc of polycrystalline material is, *neglecting hysteresis,*

$$\Gamma = A\sin 2(\alpha+\delta_2) + B\sin 4(\alpha+\delta_4) + C\sin 6(\alpha+\delta_6),$$

where A, B, C, δ_2, δ_4 and δ_6 are constants. In the case of the composite disc this expression reduces to $C\sin 6(\alpha+\delta_6)$, a value which is usually so small that it is unnecessary to reduce it further by making the composite disc of more components. Introducing the torque $\Gamma_{\mathscr{H}}$ per c.c. due to rotational hysteresis, we have for one direction of rotation of the composite disc

$$\Gamma_1 = C\sin 6(\alpha+\delta_6) + \Gamma_{\mathscr{H}}$$

and for the opposite direction

$$\Gamma_2 = C\sin 6(\alpha+\delta_6) - \Gamma_{\mathscr{H}}.$$

so that $\Gamma_{\mathscr{H}} = \frac{1}{2}(\Gamma_1 - \Gamma_2)$.

* F. Brailsford, *Journ. Inst. Elect. Eng.* **83**, 566, 1938.

In practice, the terms in $2(\alpha + \delta_2)$ and $4(\alpha + \delta_4)$ are not entirely absent when a composite disc is used. Γ_1 and Γ_2 are measured with a torque magnetometer as follows. The applied field is fixed and the composite disc rotated in a clockwise direction through one revolution to occupy a standard position in the field. The values of B and the effective field $\mathscr{H}$ are obtained by ballistic methods. Curves of rotational hysteresis loss and of $\mathscr{H}$ plotted against $4\pi I$ are shown in Fig. 109;

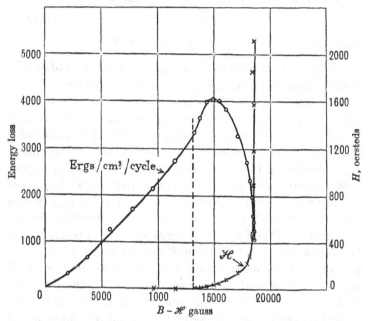

Fig. 109. Rotational hysteresis loss and magnetisation curve
for silicon steel transformer.

these were obtained with a composite disc $1\frac{1}{4}$ in. in diameter. There appears to be a sharp rise in loss when I reaches about 0·7 of its technical saturation value, i.e. in the region of the "knee" of the I, $\mathscr{H}$ curve. It is important that these experiments should be extended in order that we may obtain an accurate knowledge of the exact relation between rotational hysteresis loss and magnetisation, and, hence, of the fundamental mechanism of hysteresis in general.

Further Studies on the Approach to Saturation. We will now turn to some experiments which have bearing on current theories of ferromagnetism, and we will first consider an elaboration of equation (167), p. 242. For low fields we find

$$\sigma_{\mathcal{H},T} = \sigma_{\infty,T}(1 - b/\mathcal{H}^2). \qquad \ldots\ldots(167a)$$

Polley* has shown by experiments on polycrystalline nickel that equations (167) and (167a) may be combined, and we may write for the differential mass susceptibility

$$\chi = A/\mathcal{H}^3 + B/\mathcal{H}^2 + \chi_c \qquad \ldots\ldots(167b)$$

which becomes $(\chi - \chi_c) = (A + B\mathcal{H})\,\mathcal{H}^{-3}$,

where χ_c is the practically constant value of the differential mass susceptibility in very high fields at a given temperature.

The term $A\mathcal{H}^{-3}$ is of particular interest, for, if rotational magnetic processes alone were present in the domains, it would entirely represent the variation of χ. From the value of A this conception allows us to calculate the crystal energy K_1 defined on p. 161; A is a function of T, increasing rapidly with approach to absolute zero; K_1 also increases with decrease in temperature and reaches a saturation value at absolute zero. It is difficult to find satisfactory explanations for the occurrence of B and χ_c. B increases with fall in temperature, but below 20° C. variations in B can no longer be detected and the term is unimportant compared with that in A; it cannot be explained by thermal agitation of the domains. χ_c decreases with fall in temperature but below $-20°$ C. its behaviour cannot be followed. It is some ten times larger than would be expected if it were due to incomplete saturation within the Weiss domains, and this parasitic magnetisation, cf. p. 251, seems to depend very much on the presence of impurities.

We may add that Sanford and Bennett† found $\sigma_{0,298} = 218\cdot0$ e.m.u. per gm. for a specimen of very pure iron.

Ferromagnetic Gadolinium. Ferromagnetic gadolinium is extremely interesting in that the value of p_{eff} for $T > \theta_f$

* H. Polley, *Ann. der Phys.* **36**, 625, 1939.
† R. L. Sanford and E. G. Bennett, *Journ. Bur. of Stds.* **26**, 1, 1941.

is found equal to 7·95, compared with the calculated value from $\sqrt{4S(S+1)} = \sqrt{4 \times \frac{7}{2} \times \frac{9}{2}} = 7.94$, while its saturation moment for $T < \theta_f$ is found by direct experiment to be $7.15\mu_B$, compared with $2S\mu_B = 7.0\mu_B$. The agreement is most striking, and this metal provides the only known case where the experimental values of p_{eff} above and below the Curie point agree with theory.

Néel's Work on Alloys. The theoretical importance of the work described on p. 252 becomes more clear when we discuss the algebraic increase in the number of Bohr magnetons when one atom of the solvent metal is replaced by an atom of the dissolved metal. For the undermentioned metals dissolved in nickel* we have

TABLE XVIII

Metal	Zn	Al	Ti	Si	Sn	V	Sb	Mo	W
$\Delta\mu$	−2·1	−3·0	−3·8	−3·4	−4·2	−5·2	−4·8	−5·4	−5·8
Metal	Cr	Pt	Au	Cu	Mn	Ru	Fe	Co	
$\Delta\mu$	−4·4	−0·6	−0·8	−1·0	+2·4	−2·8	+2·8	+1·2	

From this table it follows that for solid solutions of non-ferromagnetic metals in nickel $-\Delta\mu$ *is equal to the number p of the external or valency electrons of the dissolved metal,* within the limits of experimental error. In the case of ferromagnetic metals $-\Delta\mu$ *is equal to the difference in the total number of external electrons of the dissolved metal and nickel atoms.* Thus, palladium which has the same external electron configuration as nickel, gives $\Delta\mu = 0$. This rule is, however, approximate only. In the case of metals dissolved in iron a large number behave simply as diluents, viz. Si, Al, Cr, V, Au, Sn, Mn and Zn, with $\Delta\mu = -2.2$ (approximately), which is in marked contrast with their behaviour in nickel.

An Extension of Sadron's Work on Alloys. More recent work† has shown that certain elements of the platinum group

* L. Néel, *Le Magnétisme*, II, 97, Strasbourg, 1940.
† M. Fallot, *Ann. de Phys.* **10**, 291, 1938.

24-2

are very effective in increasing the magnetic moment of the iron atom as shown in the following table.

TABLE XIX

Metal	Ru	Rh	Pd	Os	Sr	Pt
Increase in Bohr magnetons	0·0	+1·0	~ −0·2	−1·6	+0·9	+2·0
Per cent atoms added (max)	6·5	20	5	6·5	7·0	10·0

The maximum moments observed with 7·5 per cent iridium, 20 per cent rhodium and 12·4 per cent platinum are 2·28, 2·42 and $2·43\mu_B$ respectively compared with $2·46\mu_B$ for 25 to 30 per cent cobalt.

Explanation of the Magnetic Properties of Alloys. Farineau* examined the width of the L X-rays emitted by the transition elements and proved that the 4s electrons behaved as more or less free, so that their properties would be those of an energy band very similar to the parabolic band of free electrons. Hence, the 4s electrons can be pictured as forming an atmosphere common to the whole volume of the alloy; at the same time the saturation value of its ferromagnetism can be obtained from the total number of holes in the 3d bands of all the atoms in the alloy. Actually, when an alloy is formed of a magnetic element (nickel) and a non-magnetic element (copper) the number of positive holes is referred to the magnetic atoms only.

Néel has calculated on these lines the average number D of positive holes in the 3d band per magnetic atom, and the average number S of 4s electrons which is found from the constitution of the alloy and the number of positive holes in the 3d bands. On plotting the values of S against D, the points corresponding to the same families of alloys, e.g. nickel-cobalt, fell on separate curves. The points characteristic of iron, nickel and cobalt fell on the lines $S - D = 2$, $S - D = 0$, $S - D = 1$. To the left of the line $S - D = 0$ were the nickel

* J. Farineau, *Ann. de Phys.* **10**, 20, 1938.

alloys with non-ferromagnetics; here S remains practically constant, as if the valency electrons of the added metal filled up the positive holes in the nickel band. Again, in the Ni-Co and Ni-Fe alloys, S changes slowly as a function of D and depends to some extent upon the nature of the alloy. On the whole, then, for the alloys just mentioned, the density of the 4s electron atmosphere may be considered practically constant. But iron-rich alloys give quite different results and there is a rapid fall in S with increase in D reaching a minimum for pure iron itself.

Néel therefore pictures a transition metal as a mixture of ions—i.e. atoms surrounded by their 3d electrons—immersed in an atmosphere of 4s electrons whose electron pressure is measured by S. Thus, in the case of nickel, one ion possesses zero magnetic moment and is a neutral $3d^{10}$ ion, while the other is a $3d^9$ or a more highly ionised carrier. There is an equilibrium between the two which can be influenced by the addition of s electrons. For example, Pd is made less and less paramagnetic by the occlusion of hydrogen until 0·65 atom of hydrogen per atom of Pd is added. This is because S for Pd is about 0·6, and the added hydrogen electrons fill up the positive holes. Rado and Kaufmann* point out that many elements, e.g. Al, V, Cr, Sn and Au have loosely bound electrons which do not go into the 3d band of iron, so that the initial change of saturation magnetic moment of the iron atom with increase in concentration of the alloyed metal is zero.

If we dissolve a nickel atom in a metal in which $S > 0·6$, the nickel will tend to form $3d^{10}$ ions with zero magnetic moment. This is what happens when nickel is dissolved in the monovalent metals Cu, Ag and Au, when paramagnetism varying with the temperature does not arise. Bates and Baker† found that, in freshly prepared dilute solution in mercury, nickel exhibits a large paramagnetism more or less independent of the temperature, i.e. it behaves as antiferromagnetic. At 225° C., however, a sudden change sets in and the nickel becomes ferromagnetic;‡ simultaneously, resistance

* G. I. Rado and A. R. Kaufmann, *Phys. Rev.* **60**, 336, 1941.
† L. F. Bates and C. J. W. Baker, *Phys. Soc. Proc.* **52**, 436, 1940.
‡ L. F. Bates and J. H. Prentice, *Phys. Soc. Proc.* **51**, 419, 1939.

changes occur and these are consistent with the view that $3d^9$ or $3d^8$ ions are formed in place of $3d^{10}$ ions, the electrical resistance increasing by the passage of conductivity electrons from the s band to holes in the d band. In frozen amalgams at $-78 \cdot 5°$ C. the metals iron, nickel and cobalt behave as if they possess specific magnetisations of 130, 45·5 and 190, instead of the values 220, 57·6 and 163·9 c.g.s. units per gm., respectively, found for the pure metals in bulk.

Depression of the Curie Point. On dissolving certain metals in nickel there results a loss in the saturation atomic moment which is directly proportional to the number of the external or valency electrons of the dissolved metal. Manders* showed that a similar linear relation exists between the lowering of the Curie point of nickel per atom of dissolved metal and the number of its valency electrons. Néel considers that this is because the dissolved atoms, in replacing those nickel atoms which have no magnetic moment, set with their electron spins antiparallel to those of nickel, so that we are justified in assuming that the carriers responsible for ferromagnetism are also responsible for paramagnetism. The law which these alloys obey is $\chi = a + C/(T - \theta)$, and the constant a may be regarded as a measure of the deformation of the couples of antiparallel spins, resulting in a paramagnetism independent of the temperature.

A very striking result is obtained in the case of zinc dissolved in iron, even up to a concentration of over 16 atoms per cent; the Curie point decreases regularly at the rate of $6°$ C. for each atom per cent of added zinc. In the case of elements of the platinum group we have the results in the following table for the lowering of the Curie point per atom per cent dissolved.

TABLE XX

Metal	Ru	Rh	Pd	Os	Ir	Pt
Lowering of Curie point	16	2	3	11	4	0
Maximum conc. of dissolved metal	10	5	5	5	4·5	5

* R. S. Manders, *Ann. de Phys.* 5, 167, 1936.

New Saturation Measurements with Alloys. Sucksmith* has used his ring balance, p. 105, for rapid measurements on the saturation intensity of a large number of ferromagnetic alloys, for the determination of their equilibrium phase boundaries. The specimen is roughly cut to the form of an ellipsoid $4 \cdot 0 \times 1 \cdot 5 \times 1 \cdot 5$ mm. and held in a platinum-iridium holder, which does not contaminate. The holder is attached to a vertical molybdenum rod connected to the balance, the latter being suitable for forces of the order of 5 to 10 gm. weight. The specimen occupies a position of maximum $d\mathscr{H}/dx$ between the poles of an electromagnet whose conical poles-tips are machined to have three plane faces for maximum efficiency. The balance records a small deflection proportional to $\sigma m \cdot d\mathscr{H}/dx$ when a strong field $\mathscr{H}$ is excited, m being the mass of the specimen and σ its saturation intensity per gm.

Fluctuations of the Internal Field. We have to provide an explanation of the fact, mentioned on p. 257, that the saturation magnetisation of nickel at very low temperatures is usually much lower than that corresponding to the magneton number deduced from experiments above the Curie point. It would seem from the experimental data that the number of magnetic carriers increases as a metal passes from the ferromagnetic to the paramagnetic state, but this is a conclusion which must not be accepted without rigorous examination. We must also explain the occurrence of the two Curie points θ_f and θ, and see what place they have in the formulation of a satisfactory magnetic equation of state.

Néel† has discussed these matters on the hypothesis that the interactions between magnetic carriers in a ferromagnetic substance depend on their proximity in such a way that we may distinguish between long-range and short-range inter-actions which give rise to the internal field and to purely paramagnetic behaviour, respectively. In the usual $(1/\chi, T)$ curve a continuous curved line represents the somewhat complicated relation between $1/\chi$ and T which is attributed

* A. T. Pickles and W. Sucksmith, *Roy. Soc. Proc.* A, **175**, 331, 1940; K. Hoselitz and W. Sucksmith, *Roy. Soc. Proc.* A, **181**, 303, 1942.
† L. Néel, *Le Magnétisme*, II, 67, Stras., 1940; *Journ. de Phys.* **5**, 104, 1934.

to short-range interactions. Since the effect of the Weiss field is to introduce a constant contribution equal to $\nu_0\rho_0$ to $1/\chi$, the net result of its presence is that we have to draw a new axis of temperature at $1/\chi = \nu_0\rho_0$. We then find that θ_f and θ are respectively given by the intersections of the continuous curved line, and of the straight line asymptotic to it at high temperatures, with this new temperature axis.

For the purposes of ease of calculation, Néel assumes that the magnetic carriers in a ferromagnetic substance occur in groups of N with equal interactions between any two of them. Let us suppose that p of these carriers set parallel to the direction of an applied field while the remainder q set antiparallel thereto. Assuming that the magnetic moment of each carrier is μ and that a field $\mathscr{H}$ is acting, the potential energy of the group is given by

$$W = pq\omega - (p-q)\mu\mathscr{H},$$

where ω is the mutual potential energy of any two antiparallel carriers. Hence, the probability $P_{(p,q)}$ that any particular arrangement (p,q) may exist is given by

$$P_{(p,q)} = \frac{\lfloor N}{\lfloor p \; \lfloor q} e^{-W/kT} \Big/ \Sigma \frac{\lfloor N}{\lfloor p \; \lfloor q} e^{-W/kT},$$

where Σ includes all the positive values of p and q which satisfy the relation $p + q = N$. Since the magnetic moment of the (p,q) configuration is $(p-q)\mu$, and that of the group is $\Sigma(p-q)\mu P_{(p,q)}$, it follows that the magnetic moment per gm. is equal to

$$\sigma = \sigma_0 \Sigma \frac{p-q}{N} P_{(p,q)},$$

where σ_0 is the saturation value when all the carriers are parallel to the field. The last equation may be written as

$$\sigma = \sigma_0(a\mathscr{H} - \lambda a^3 \mathscr{H}^3),$$

and, by comparison with the experimental results for nickel, the above constants may be found. For example, it is easy to show that the initial susceptibility χ_0 is equal to $\alpha\sigma_0$, and that the value of θ is given by $\omega(N-1)/2k$. The Curie constant is equal to $\sigma_0\mu/k$. By plotting the values of $\partial(1/\chi)/\partial T$ against T for chosen values of N, a set of curves is obtained, and the value of N which best fits the experimental results can then be picked out. The value thus found for N in the case of nickel is 750. With $\sigma_0 = 57\cdot5$ e.m.u. per gm. there is excellent agreement between the calculated and experimental $(1/\chi, T)$ curves. The paramagnetic Curie point θ is thus much more fundamental than θ_f, for the difference $\theta-\theta_f$ is fixed entirely, on this hypothesis, by the number N of interacting carriers in the group; when it is large, θ and θ_f coincide. The number of carriers in the group thus represents a kind of coefficient of fluctuation; if it is small the fluctuations produce marked effects, if it is large a pure Weiss field obtains.

Returning to the first equation we have, below θ_f and in its neighbourhood,

$$W = pq\omega - (p-q)\mu(\mathscr{H}\nu\rho\sigma),$$

so that, using the expression for σ, we may calculate W, ω and σ/σ_0 as a function of $(\mathscr{H}+\nu\rho\sigma) = m$, to which we may give chosen values. Taking $N = 750$, $\theta = 642\cdot8°$ K., $\nu\rho = 2400$, $k\theta/\sigma_0\mu = 12,000$, $\mathscr{H}\mu = m - 0\cdot02\theta\sigma/\sigma_0$, we may plot a magnetic isotherm as a function of σ/σ_0 and $\mu\mathscr{H}$. Its intersection with the line $\mu\mathscr{H} = 0$ will give the value $\sigma_{0,T}/\sigma_{\infty,0}$ for the spontaneous magnetisation at the temperature of the isotherm. The curve of $\sigma_{0,T}/\sigma_{\infty,0}$ so deduced is in very good agreement with the experimental results of Weiss and Forrer.

The internal field decreases with rise in temperature because of an increase in the mean distance between the carriers due to expansion and to atomic vibrations. Néel considers that it varies according to the expression

$$(\nu\rho)_T = \nu_0\rho_0(1 - 2\cdot23 \times 10^{-4}T)$$

in the case of nickel. Taking account of these features Néel calculated and plotted the values of $\sigma_{0,T}/\sigma_{\infty,0}$ as a function of T/θ for the three cases $J = \frac{1}{2}$, $J = 1$ and $J = \frac{3}{2}$. The $J = 1$ and $J = \frac{3}{2}$ curves are much more in accord with experiment than the $J = \frac{1}{2}$ curve. If the carriers are distributed in an

RECENT WORK ON FERROMAGNETIC SUBSTANCES

energy band, it can be shown that $J = \frac{3}{2}$ fits the facts better than $J = 1$. The curve for $J = \frac{1}{2}$ in Fig. 78, p. 265, was, of course, computed without taking account of the fluctuation or thermal variation of the internal field. The value $J = \frac{3}{2}$ is consistent with the specific heat data and with the value of the Curie constant. For, the latter should be given by $C = (\gamma\mu_B^2/3R)\,4J(J+1)$, where γ, the number of magnetic carriers per atom of nickel, takes the values 0·606, 0·303 and 0·202 for $J = \frac{1}{2}$, $J = 1$ and $J = \frac{3}{2}$ respectively, so that the corresponding values of C are 0·197, 0·262 and 0·328, while the experimental value is $0·323 \pm 0·002$.

Order and Disorder Phenomena. We can much more appropriately treat the problem of the transition from ferromagnetism to paramagnetism as one of transition from an ordered to a disordered state according to the theory of Bragg and Williams* for a simple A, B-type alloy. In the general case the numbers of A and B atoms are respectively represented by rn and $(1-r)\,n$ where $r \leqslant \frac{1}{2}$, and, in a state of complete order they occupy lattice points which are termed α and β sites. As some atoms may take no part in an order-disorder transition, n need not be equal to the total number of atoms in the lattice; for example, iron atoms always occupy the cube corners in the Fe_3Al lattice. The degree of order S is dependent on the fraction p of A atoms occupying α sites at a given instant, and Bragg and Williams define S by the equation $S = (p-r)/(1-r)$. Hence $S = 1$ corresponds to $p = 1$, i.e. to perfect order, while $S = 0$ means that $p = r$ and the atoms are randomly distributed. In the case of ferromagnetism we have only spins which can be directed parallel or antiparallel, $r = \frac{1}{2}$, and from p. 240, we have

$$S = 2p - 1 = \frac{\bar{x} - \bar{y}}{\bar{x} + \bar{y}} = \frac{\sigma_{0,T}}{\sigma_{\infty,0}} = \tanh\left(\frac{V}{4kT}\right),$$

where V is the increase in energy when an A atom on an α site changes places with a B atom on a β site, thus producing a pair of disordered atoms. Now, Bragg and Williams assume

* W. L. Bragg and E. J. Williams, *Roy. Soc. Proc.* A, **114**, 340, 1930; cf. also F. C. Nix and W. Shockley, *Rev. Mod. Phys.* **10**, 1, 1938.

that V is a function of S, steadily increasing from the disordered, i.e. demagnetised, state with $S = 0$ to a maximum V_0 when $S = 1$, so we may write $V = V_0 S$, whence we deduce that

$$\frac{\sigma_{0,\,T}}{\sigma_{\infty,\,0}} = \tanh\left(\frac{V_0 S}{4kT}\right).$$

Consequently, we find that the critical temperature θ is equal to $V_0/4k$; above θ, S is always equal to zero, i.e. the substance is not spontaneously magnetised, while at temperatures below θ, S has a value between 0 and 1, being 1 at $0°$ K. We note that at $T = \theta$ the tangent to the curve of $\sigma_{0,\,T}/\sigma_{\infty,\,0}$ against T/θ should be vertical. It is easy to see that there should be a difference between the ferromagnetic and the paramagnetic Curie point, because of the effects of atoms other than immediate neighbours; in other words we have effects of *short distance* and *long distance*. Long distance order disappears at θ_f, but short distance or local order may gradually decrease as the temperature rises above θ_f.

The Specific Heats of Ferromagnetic Metals and their Alloys. The specific heat of very pure iron (99·99 per cent) has been measured by Awbery and Griffiths[*] over the range 0 to $950°$ C. The curve obtained is similar in shape to Grew's curve for nickel, p. 272. Indeed, thermal effects of the magnetic transformation at $755°$ C. with this iron are apparent even when the $\alpha - \gamma$ transformation sets in at $903°$ C., where there is a true phase-change with an associated latent heat of 3·9 cal. per gm. There is also a small secondary peak in the specific heat curve at $908°$ C., corresponding to about 0·39 cal. per gm.

The specific heat of iron over the range 1·5 to $20°$ K. was measured by Duyckaerts[†] who gives the expression

$$C_p = 464·5(T/\theta)^3 + 1·20 \times 10^{-3}T \text{ cal. per gm. atom per } °\text{C.,}$$

θ being equal to $465°$ K. The linear term represents the specific heat of the electrons and the T^3 term the (Debye) specific heat of the lattice. The contribution of the electrons to the specific heat at these low temperatures is greater than the theory,

[*] J. H. Awbery and E. Griffiths, *Roy. Soc. Proc.* A, **174**, 1, 1940.
[†] G. Duyckaerts, *Physica*, **6**, 401, 1939.

based on the density of the energy states being assumed proportional to the square root of the energy, would lead us to expect.* Duyckaerts found the same value for the linear term in the case of cobalt over the range 2° to 18° K.

Keesom and Kurrelmeyer measured the specific heats of some alloys of nickel with copper (and iron) in order to find how γ varied with concentration. The saturation magnetisations of these alloys vary linearly with the concentration, because a certain percentage of the s electrons of the copper atoms go into the "holes" in the d levels of nickel, until, at 60 per cent copper, these "holes" are all filled. The number of levels per unit energy range at the top of the filled levels is much larger for the 3d band of nickel than for the 4s band of copper, as shown by the fact that γ, the coefficient of T in the last equation, is $1 \cdot 75 \times 10^{-3}$ cal. per mol. per deg.² for nickel and $1 \cdot 78 \times 10^{-4}$ for copper. The experimental results showed that γ, instead of decreasing uniformly to low values at 60 per cent copper, actually showed a maximum near that composition, and that even for 80 per cent copper the value of γ is $2 \cdot 5$ times the value for pure copper.

The Significance of Ni_3Fe. Measurements of the specific heats of Ni-Fe-Co alloys over large ranges of temperature by Kaya and Nakayama† by Sykes' method have shown that a maximum in the specific heat curve becomes the more pronounced the closer the alloy approaches in composition to Ni_3Fe. It is thought that this arrangement of atoms is of great importance in all the ternary alloys. In the case of the perminvars the maximum is sharp when the specimen is annealed, but with quenched specimens it is flattened and spread over a wide temperature range.

Low Temperature Ferromagnetic Alloys. Many Cu-Ni alloys exhibit ferromagnetism at low temperatures only. An example is provided by the commercial material "K" Monel, a modification of standard monel metal‡ containing aluminium, with

---

* Cf. F. Simon, *Zeit. für Elektrochem.* **34**, 528, 1928.
† S. Kaya and M. Nakayama, *Zeit. für tech. Phys.* **112**, 420, 1939.
‡ *Ni Bulletin,* **12**, 9, 1939.

the constitution nickel 66, copper 29, aluminium 2·75, silicon 0·25, manganese 0·4, iron 0·9 and carbon 0·15 per cent, respectively.

The ΔE Effect. When a mechanical stress is applied to a ferromagnetic material in a magnetising field, in general a change of magnetisation occurs; the magnitude of the change depends upon the total intensity of magnetisation and is zero at saturation. This indicates that a stress may change the orientation of the vectors, since the phenomenon of magnetostriction may be considered as a change of dimensions consequent upon changes of the orientations in the domains. Thus, when a stress is applied to a ferromagnetic substance, the resulting strain may be considered to have two components, one of purely elastic origin and the other due to the above magneto-mechanical process. The latter component is a function of the total intensity of magnetisation, and one would expect that the values of the elastic moduli of a ferromagnetic would also depend upon the state of magnetisation. In particular, the variation of Young's modulus, E, with intensity of magnetisation is known as the *ΔE effect.*

Young's modulus may be determined by either static loading or dynamic oscillatory methods, leading to values of the isothermal and adiabatic Young's modulus, respectively. The first method was employed by Honda and Terada.* In later work the dynamic method of measurement has usually been adopted. If the specimen is in the form of a rod of length l and is set into longitudinal resonant oscillation at its fundamental frequency, f, the adiabatic Young's modulus is given by $E = 4\delta f^2 l^2$ dyne cm.$^{-2}$, where δ is the density of the material.

Several methods of exciting longitudinal oscillations and detecting resonance have been employed; e.g. composite oscillators of quartz and ferromagnetic materials excited by electrostriction in the quartz have been used by Zacharias,† Siegel and Quimby, Cooke and Williams, Bozorth and

* K. Honda and T. Terada, *Phil. Mag.* **13**, 36, 1907.

† K. Zacharias, *Phys. Rev.* **44**, 116, 1939; S. Siegel and S. L. Quimby, *Phys. Rev.* **49**, 663, 1936; W. T. Cooke, *Phys. Rev.* **50**, 1159, 1936; H. J. Williams, R. M. Bozorth and H. Christensen, *Phys. Rev.* **59**, 1005, 1941.

Christensen. Self-exciting oscillators were employed by Auwers* and Engler. In experiments made by Yamamoto† oscillatory strains were set up in rod specimens by the magnetostriction changes in length due to alternating current flowing in a small coil encircling the rod.

Brown,‡ in a theoretical study showed that, owing to the "magnetic skin effect", measurements of Young's modulus obtained by a dynamical method give approximately the value at constant flux density, while static methods give the value for constant magnetising field. The differences in the observed values of E for annealed nickel using static and dynamical methods of measurement are thus theoretically explained. In dynamic measurements macro-eddy currents are produced at the surface of the specimen, and the magnetic fields thus set up serve to neutralise the tendency of the stress to change the flux density. The results of Brown's analysis for annealed polycrystalline nickel specimens may be expressed by the following formulae

$$\frac{E_s - E_0}{E_0} = \frac{a}{1-a} \qquad \qquad \text{......(207)}$$

and
$$\frac{E - E_0}{E_0} = 54a/35(I/I_s)^2, \qquad \text{......(208)}$$

where $a = 3\lambda_S^2 \chi_0 E_0/5I_s^2$; λ_S is the saturation magnetostriction coefficient for the direction of easy magnetisation in the crystal; χ_0 is the corresponding initial susceptibility of a single crystal of nickel; I is the appropriate intensity of magnetisation, while I_s is the saturation intensity. E_0, E, E_s are the values of the adiabatic Young's modulus at the constant flux densities 0, I and I_s respectively.

The first equation gives the theoretical value of the total change in Young's modulus from $I = 0$ to I_s. The second represents the behaviour of the specimen from zero magnetisation up to the "knee" of the magnetisation curve, the

* O. v. Auwers, *Zeit. für Phys.* **94**, 707, 1935; O. Engler, *Ann. der Phys.* **31**, 145, 1937.
† M. Yamamoto, *Sci. Rep. Tok. Univ.* **27**, 115, 1938; *Phys. Rev.* **59**, 768, 1941.
‡ W. F. Brown, *Phys. Rev.* **52**, 325, 1937.

assumption being made that rotation of the domain vectors does not occur over this range of magnetisation. The experimental results of Siegel and Quimby are in satisfactory quantitative agreement with these equations; the linear relation between $(E - E_0)/E_0$ and $(I/I_s)^2$ holds for temperatures up to 300° C., and for values of I/I_s equal at least to 0·4.

Quantitative results for the ΔE effect in a few ferromagnetic materials are given in the following table.

TABLE XXI

Material	$\dfrac{E - E_0}{E_0}$	I gauss	Authority
Iron:			
(unannealed)	+0·0020	1500	Cooke
(annealed)	+0·0031	1500	Cooke
	+0·0034	1500	Yamamoto
Nickel (annealed):			
At 23° C.	+0·067	485 (sat.)	Siegel and Quimby
At 185° C.	+0·187	415 (sat.)	Siegel and Quimby
At 353° C.	Zero	0	Siegel and Quimby
Electrolytic cobalt	+0·00168	($\mathscr{H} = 900$)	Yamamoto
Annealed Co-Fe:			
50 per cent	+0·22	($\mathscr{H} = 900$)	Yamamoto
68 Permalloy:			
(unannealed)	+0·00088	About 1000 (sat)	Williams, Bozorth and Christensen
(Cooled rapidly)	+0·104	About 1000 (sat)	Williams, Bozorth and Christensen

The dynamic oscillatory methods for the measurement of elastic moduli may also be used to determine the decrement of the oscillations. In this way the magnitude of the *ferromagnetic internal friction* may be found. For nickel, Siegel and Quimby showed that the internal friction decreased with increasing intensity of magnetisation, and at saturation intensity the value was of the same order of magnitude as that for non-ferromagnetic materials. A detailed explanation of the variation of ferromagnetic internal friction with magnetisation is not yet available, but a qualitative analysis has been attempted for iron by Brown.*

* W. F. Brown, *Phys. Rev.* 50, 1165, 1936.

Incidentally, the effects of heat treatment can be very important, as we know from other experiments, for example, Bozorth and Christensen* showed that the longitudinal magnetostriction of 68 permalloy varies with heat treatment from $2 \cdot 5 \times 10^{-6}$ to 22×10^{-6}, under standard conditions.

Masiyama† found that an alloy of 58 per cent iron and 42 per cent nickel has a volume magnetostriction some ten times greater than that of pure iron and some 60 times greater than that of pure nickel; Döring‡ suggests that this explains the large changes in the elasticity which occur at the Curie point of this alloy.

The Earth's Magnetism. By applying Clapeyron's equation, Slater§ finds that the Curie point of nickel increases by approximately 5×10^{-5} deg. C. per atmosphere, and he therefore concludes that a ferromagnetic explanation of the earth's magnetism is improbable.

* R. M. Bozorth and H. Christensen, *Phys. Rev.* **59**, 1005, 1941.
† Y. Masiyama, *Sci. Rep. Tok. Univ.* **20**, 574, 1931.
‡ W. Döring, *Ann. der Phys.* **62**, 465, 1938.
§ J. C. Slater, *Phys. Rev.* **58**, 54, 1940.

CHAPTER XIII

RECENT WORK ON WEAKLY MAGNETIC SUBSTANCES

Introduction. It would have been possible to enlarge the subject matter of Chapters III and IV by the provision of many notes and addenda at the end of this book, but it was clear that most of these notes could be collected in one chapter under the above heading. Consequently, while there follow sections of general interest in the experimental and theoretical study of weakly magnetic substances, there are also given sections which form a collection of new and important magnetic data.

Feeble Paramagnetism, Antiferromagnetism and Metamagnetism. When θ, which occurs in the statement of the Curie-Weiss law on p. 50, has an extremely large negative value, the susceptibility may be paramagnetic and practically independent of the temperature. This is presumably the case with metallic chromium, metallic manganese and metallic aluminium under normal conditions, where we may picture the electron spins as being oppositely directed, or, where we may say that the exchange integral is negative. The spin of each individual atom is presumably as high as that in a highly magnetic metal, but the exchange effects are demagnetising and so make the resultant spin of the entire metal crystal small. For example, the ground state of the aluminium atom is $^2P_{\frac{1}{2}}$, yet the metal is only weakly paramagnetic. When chromium and manganese are dissolved in copper, silver or gold,* their atoms are exposed to much lower internal fields and they then obey the Curie-Weiss law. Similar paramagnetism is shown by Fe_2Cl_6 and by organic molecules containing three atoms of iron or chromium, such as $Fe_3(CH_2CO_2)_6(OH)_2Cl \cdot 7H_2O$.

Again, feeble paramagnetism may be caused by the spreading of energy levels due to the migration of electrons

* Cf. L. Néel, *Ann. de Phys.* **5**, 258 and 272, 1936.

in a solid conductor. In the absence of exchange coupling this is sufficient to remove the spin paramagnetism, particularly if only s and p electrons, with their wave functions overlapping adjacent atoms, are present. The feebly magnetic transition metals may owe their paramagnetism to the fact that the wave functions of their d electrons overlap other atoms less than those of s and p type electrons.* Distortion due to interatomic forces may produce feeble paramagnetism in an ionic salt whose isolated ions are diamagnetic, and this may be the explanation of the feeble paramagnetism shown by Sc_2O_3, TiO_2, CeO_2 and some of the cobaltamines.

It is, clearly, very difficult to say whether any particular case of feeble paramagnetism is due to a negative exchange integral or to electron migration or to ionic deformation. But there is a class of materials which appears to stand apart and to which we may apply the term *antiferromagnetic*. They are distinguished by having a susceptibility which passes through a maximum as the temperature is raised, whereas ordinary feebly paramagnetic substances have susceptibilities which decrease gradually with rise in temperature. MnO, MnSe, Cr_2O_3 and CrSb are examples of antiferromagnetic substances. This strange behaviour can be explained if we have a crystal whose constituent atoms may be distributed into two sub-lattices A and B, and, further, if the mean direction of spin in A is directed oppositely to that in B under the influence of the resultant internal field. At low temperatures this field becomes more and more effective and the resultant spin less; at high temperatures the field must disappear. Consequently, there is a temperature for which the number of spins pointing in the direction of an applied field is a maximum. It follows, from the above explanation, that the effect of an applied field on an antiferromagnetic should be much less pronounced than that on a ferromagnetic substance, and this is supported by experiment. Perhaps the substances which have been described above as feebly paramagnetic, instead of antiferromagnetic, are so because their crystal structures cannot be

* Cf. Mott and Jones, *Theory and Properties of Metals and Alloys*, Oxford, pp. 189–200; Seitz, *Modern Theory of Solids*, New York, Section 29.

resolved into sub-lattices, or because pronounced electron migration occurs in them.

It has been known for a long time that anhydrous salts of the iron group, such as $NiCl_2$, $CoCl_2$, etc., exhibit a susceptibility which is a function of the applied field at liquid hydrogen temperatures. At these temperatures and in fields greater than 800 oersteds, I increases more rapidly than $\mathscr{H}$, hysteresis and traces of remanence are observed, but the magnetisations are only of the order of those found with paramagnetic substances, although the temperatures are well below the Curie points, and there are no signs of saturation phenomena; the coercivities are large and of the order of 2000 oersteds. Such substances are sometimes termed *metamagnetic*; the same term has previously been wrongly applied to the anomalous behaviour of weakly magnetic bodies containing traces of ferromagnetic impurity, and to crystals which are paramagnetic in one direction and diamagnetic in another. Schubnikow found that they exhibit a pronounced maximum specific heat in the neighbourhood of the Curie point. Starr* also explains their behaviour on the assumption that the atoms in these compounds form groups with either a parallel or an antiparallel alignment of magnetic moments. Of course, such groups can only be formed below the temperature at which the magnetic properties change sharply, where there is a specific heat anomaly.

The Internal Field of Antiferromagnetic Substances. Néel† has obtained the constant $\nu\rho$ of the internal molecular field for a series of antiferromagnetic substances using the relation $-2\nu\chi_a = 1$, a formula which has a reasonable theoretical basis. On plotting $\nu\rho/N$, the ratio of the Weiss constant to the number of atoms with which each atom of the substance interacts, against $D - 2r$, Néel obtains a smooth curve which shows a progressive change with increase in $D - 2r$ from constant paramagnetism to variable paramagnetism with a negative internal field and thence to ferromagnetism.

Metallic chromium shows such a constant paramagnetism

* C. Starr, *Phys. Rev.* **58**, 984, 1940.
† L. Néel, *Le Magnétisme*, III, 65, Strasbourg, 1940.

25-2

or antiferromagnetism, over a wide temperature range. The distance between the atoms is in this case only 2·49 Å., so that even a large rise in temperature cannot break down the interaction between neighbouring atoms. However, in Cr_2O_3 and CrSb the interatomic distances are greater, and here there is a very marked decrease in antiferromagnetism with rise in temperature; the graphs of $(1/\chi, T)$ show very pronounced minima, after which in the case of Cr_2O_3 the graph is linear corresponding to the normal magnetic behaviour of the Cr^{+++} ion. In the case of vanadium dioxide, Horcheck* has shown that a sudden increase in the temperature-independent paramagnetism occurs at about 70° C. although no corresponding change in the crystal lattice occurs. Squire† finds similar behaviour in the case of the Mn^{++} compounds, MnO, MnS, MnSe and MnTe, each of which shows an antiferromagnetic-paramagnetic transition point at low temperatures; in addition he finds evidence of hysteresis and of change in susceptibility with field strength.

Cylindrical Pole-pieces of Special Design. In the measurement of susceptibilities by the Curie method, p. 104, using spherical pole-faces, $\mathscr{H} d\mathscr{H}/dx$ can be made constant in one direction only, so that non-homogeneous specimens cannot be used unless they are extremely small. Bates, Baker and Meakin‡ used one convex-cylindrical and one plane pole-face, so that a relatively long specimen could be placed within a vertical tube whose axis was perpendicular to the lines of force in such a way that the mechanical force acting on each unit volume of the specimen was independent of its position from the end of the tube. Further improvements resulted from a theoretical discussion of the shape of the pole-pieces by Davy,§ who showed that the force on each element of the specimen would be entirely independent of its position in the tube, provided that the pole-faces had sections which satisfied the equation $r^{\frac{3}{2}} \cos \frac{3}{2}\theta = $ constant, where r and θ are

* Cf. W. Klemm, *Zeit. Elektrochemie*, **45**, 587, 1939.
† C. F. Squire, *Phys. Rev.* **56**, 922, 1939.
‡ L. F. Bates, C. J. W. Baker and R. Meakin, *Phys. Soc. Proc.* **52**, 425, 1940.
§ N. Davy, *Phil. Mag.* **33**, 575, 1942.

the polar coordinates of a point in the z plane. Such pole-faces were also used by Bates and Somekh* in their investigation of the properties of non-homogeneous mercury amalgams, and a simple proof of the above equation is given in their paper. A very cheap and inexpensive method of evacuating and sealing a straight tube provided with a simple stopper at one end was used in these experiments.

A New Electrodynamic Balance. A small electrodynamic balance was used to investigate the effects of cold-working on weakly magnetic metals, by Hutchinson and Reekie.† Its most important feature is a moving coil consisting of two sets of 500 turns, wound on a small cylindrical former and joined in series opposition. The moving coil is so situated that the field of a larger, fixed, coil acts more effectively upon the nearer set of turns than upon the more remote, while any stray, but reasonably uniform fields, act equally and oppositely upon the two sets. Consequently, the balance may be used with d.c., and in either the horizontal or vertical position.

The Magnetic Properties of Copper. Pure copper is undoubtedly diamagnetic at ordinary temperatures, but it has been shown‡ that it tends to become paramagnetic at low temperatures. Most specimens of copper contain iron, but they may be rendered completely non-magnetic by heating them in hydrogen for 2 hours at 900° C., when the magnetic impurity dissolves in the copper lattice and can be fixed by quenching.§ Annealing at 850° C. supplemented by cold work restores the ferromagnetism. Smith‖ found that copper containing 2·3 per cent of iron showed little trace of ferromagnetism, until it was cold worked so that a 24 per cent reduction in cross-section was caused, when the iron was supposed to be in the body-centred cube form.

The General Theory of Ferromagnetic Impurity. A more general theory of the behaviour of ferromagnetic impurities

* L. F. Bates and E. M. Somekh, *Phys. Soc. Proc.* **56**, 182, 1944.
† T. S. Hutchinson and J. Reekie, *Journ. Sci. Instr.* **23**, 209, 1946.
‡ F. Bitter, A. R. Kaufmann, C. Starr and S. T. Pan, *Phys. Rev.* **60**, 236, 1941.
§ F. W. Constant, R. E. Faires and H. E. Lenander, *Phys. Rev.* **63**, 441, 1943.
‖ C. S. Smith, *Phys. Rev.* **57**, 337, 1940.

than that on p. 115 has been given by Jaanus and examined in detail by Bates and Baker.* Jaanus assumes that each particle of ferromagnetic has a demagnetisation coefficient D and a coercivity h, and Bates and Baker show that when measurements are made by the Gouy method the Jaanus theory leads to the formula

$$k_{\mathscr{H}} = k_{\infty} + \frac{2c'\sigma_s}{\mathscr{H}} - \frac{c'\sigma_s}{\mathscr{H}}\frac{DI_s}{\mathscr{H}},$$

where I_s is the saturation intensity of magnetisation of the ferromagnetic impurity, provided that $(3h + DI_s) < \mathscr{H}$. It is therefore only when $DI_s/\mathscr{H} \ll 2$ that a linear relation between $k_{\mathscr{H}}$ and $1/\mathscr{H}$ should be found.

The Graphical Representation of the Curie Law. If we plot χT against T we get a horizontal line when the Curie law is obeyed, but we get a sloping line when a law of the type $(\chi + \text{const.})\, T = \text{const.}$ is followed; the latter line will be curved† when the law is of the Curie-Weiss form

$$(\chi + \text{const.})\,(T + \varDelta) = \text{const.}$$

In the case of many metallic alloys a law of the form $\chi = a + C/T - \theta$ is followed, and Néel‡ has devised a method of successive approximations for finding the three constants a, C and θ from a series of experimental observations with suitable range of T.

The Curie-Weiss Law. Föex§ has proved that expression (88), p. 126, must be used in cases where a Curie-Weiss law holds, by comparing the two sets of values of the constant C found for a particular ion, from experiments on its compounds which follow the Curie law with $\theta = 0$ and on its compounds which follow the Curie-Weiss law with a marked value of θ. While the experimental errors are about the same for all the compounds, the products $\chi_M T$ are very different, the differences being the more marked the greater the value

* L. F. Bates and C. J. W. Baker, *Phys. Soc. Proc.* **52**, 443, 1940.
† B. Cabrera and A. Duperier, *Phys. Soc. Proc.* **51**, 843, 1939.
‡ L. Néel, *Ann. de Phys.* **5**, 232, 1936.
§ G. Föex, *Le Magnétisme*, III, 203, 216, Strasbourg, 1940.

of θ, but the values of C obtained from $\chi_M(T-\theta)$ are constant. For example, for the Cr^{+++} ion in $Cr_2(SO_4)_3 \cdot 16H_2O$ where $\theta = 0$, and for the same ion in $Cr_2O_3 \cdot MgO$ where $\theta = -363$, we find C equal to 1838 and 1844, respectively, from the expression $C = \chi_M(T-\theta)$.

New Data for Rare Earth Ions. In order to study the magnetic properties of the rare earth metals themselves, Klemm and Bommer* reduced the pure chlorides in liquid sodium. The reduced metal was, of course, in each case mixed with $NaCl$, but allowance can easily be made for the diamagnetic properties of the latter. The values of p_{eff} for the metals, obtained from Curie-Weiss law expressions with θ positive, all fall very close to the Van Vleck curve of Fig. 35, p. 129, except in the case of Eu and Yb, for which p_{eff} is 8 and 0 respectively. This means that in the metallic state we have to deal with Eu^{++} and Yb^{++} ions, and the data illustrate *Kossel's magnetic displacement rule,* viz. *ions of adjacent elements in the periodic table which have the same number of electrons have similar magnetic properties.* Thus, Eu^{++} and Gd^{+++} both contain 61 electrons and both have $p_{\text{eff}} = 8$, while Yb^{++} and Cp^{+++} both have 68 electrons and $p_{\text{eff}} = 0$. Metallic cerium has been found by Starr and Kaufmann† to exhibit traces of ferromagnetism of an unusual kind, but their work needs confirmation with specimens free from iron.

The Significance of Odd and Even Numbers of Electrons. An example of the significant difference which exists between systems with odd and even numbers of electrons is given on p. 129. Kramers‡ proved that in the case of odd numbers of electrons the energy levels fall together in pairs and remain at least two-fold degenerate so that a reasonable magnetic moment may occur. With even numbers of electrons the crystalline field, if it is sufficiently asymmetric, will completely remove the degeneracy. Nd^{+++} with an odd number of electrons should have an infinitely great susceptibility close

---

* W. Klemm and H. Bommer, *Zeit. für anorg. allg. Chem.* **231**, 138, 1937.
† C. Starr and A. R. Kaufmann, *Phys. Rev.* **58**, 657, 1940.
‡ H. A. Kramers, *Acad. Sci. Amsterdam Proc.* **33**, 959, 1930.

to absolute zero, while Pr^{+++} with an even number should possess a finite susceptibility.

Ions of the Iron Group. To the first line of Table V, p. 130, may be added the ion $Mn^{+++++++}$ which exists in silver permanganate, $AgMnO_4$, for which $\chi = -0.28 \times 10^{-6}$, so that the above ion has zero magnetic moment, in accord with theory.[†]

The Platinum and Palladium Groups. The ions of the Pt and Pd groups are interesting because of the smallness of the experimental values of χ compared with the values calculated on the Hund rules. Cabrera and Duperier[‡] give the following values.

TABLE XXII

Ion	Configuration	$2\sqrt{S(S+1)}$	p_{eff} (expt)	$\dfrac{p_{\text{eff}} \text{ (expt)}}{p_{\text{eff}} \text{ (calc.)}}$
Ru^{+++}	$^6S_{\frac{5}{2}}$	5·92	2·09	0·35
Rh^{+++}	$^5D_4$	4·91	0·062	0·013
Pd^{++}	$^3F_4$	2·83	0·07₆ } *	0·027
			0·13₅ }	0·048
Cs^{++}	$^5D_4$	4·91	0·27₈ } *	0·057
			0·50₉ }	0·104
Ir^{+++}	$^5D_4$	4·91	0·112	0·040
Pt^{++}	$^3F_4$	2·83	—	—

* Depending on the value taken for K in the equation $(\chi + K)(T + \Delta) = C$.

Metal Ions in Dilute Solution in a Solvent Metal. The magnetic properties of a metal in dilute solution in another metal may be considerably different from those of the solute metal in bulk. Examples of metals dissolved in ferromagnetics are described on p. 253. The results obtained for metals in very dilute solution in mercury at room temperature, obtained by Bates and Somekh,[§] show that in nearly all cases the residual paramagnetism of the metals after correction for the diamagnetism of their cores, is due to electrons set free by solution of the metals. Freed and Sugarman[||] made similar measurements on K, Ce, Ca and Ba in dilute solution in liquid

† S. S. Bhatnagar, *Ind. Nat. Acad. Sci. Proc.* **10**, 154, 1939.
‡ B. Cabrera and A. Duperier, *Phys. Soc. Proc.* **51**, 845, 1939.
§ L. F. Bates and E. M. Somekh, *Phys. Soc. Proc.* **56**, 182, 1944.
|| S. Freed and N. Sugarman, *Journ. Chem. Phys.* **11**, 354, 1943.

ammonia, and drew the same conclusion, viz. that the free electrons may be treated as an electron gas.

Standardising Solutions. Nettleton and Sugden* give as a standardising liquid a solution of 23·13 per cent $NiCl_2$ in water, the density of the solution being 1·255 gm. per c.c. at 20° C., when $\chi = 7·363 \times 10^{-6}$ e.m.u. per gm.

Anantakrishnan† suggests the use of $\chi_{HN_4^+} = -13·8 \times 10^{-6}$ and $\chi_{NO_3^-} = -18·5 \times 10^{-6}$ e.m.u. per gm. ion, for checking the ionic susceptibilities of other ions. He points out that, while a free proton makes no contribution to the susceptibility of a substance, it is invariably present in the solvated OH_3^+ state in the solution of an acid. The whole question of the values of ionic susceptibilities of the diamagnetic ions is still a very open one. The values given in the Table on p. 132 differ considerably, for example, from those given in Table II of Selwood's *Magnetochemistry* (New York, 1943, p. 36) and there is no reliable method of choosing between the two sets of values.

The Properties of Dialogite. Krishnan and Banerjee‡ used some specially good crystals of dialogite (rhodochrosite) and found very feeble magnetic anisotropy, the law $\chi = C/(T + 13)$ being followed with $p_{\text{eff}} = 5·56$. Van Vleck and Penney§ deduced from the temperature variation of the Verdet constant that the law should be $\chi = C/(T + 11)$; the agreement is very satisfactory, and shows clearly that the Mn^{++} ion is in an S state with a Stark separation produced by the crystalline field of only 0·07 cm.$^{-1}$.

Data for Rare Earth Crystals. Krishnan and Chakravorty‖ showed that the absorption spectra of the rare earth salts themselves indicate anisotropy departures from cubic symmetry, for many of the absorption lines are strongly polarised, some being almost entirely produced by vibrations parallel to one or other of the principal axes of the optical ellipsoid

* N. Nettleton and S. Sugden, *Roy. Soc. Proc.* A, **173**, 313, 1939.
† S. V. Anantakrishnan, *Ind. Nat. Acad. Sci. Proc.* **21**, 120, 1945.
‡ K. S. Krishnan and S. Banerjee, *Zeit. für Kryst.* A, **99**, 499, 1938.
§ J. H. Van Vleck and W. G. Penney, *Phil. Mag.* **17**, 961, 1934.
‖ K. S. Krishnan and D. G. Chakravorty, *Journ. Chem. Phys.* **6**, 224, 1938.

of the crystal, and variations in the direction of polarisation occur even among lines of the same group, i.e. among the Stark components originating from the same absorption line of the free ion. The absorption spectra also contain weak, sharp, lines which cannot be allocated to stated electron transitions alone; but Ewald* found that, in the case of neodymium salts, in which these lines are relatively well pronounced, they arise from combination transitions of the electron of the Nd^{+++} ion and the characteristic frequencies of a neighbouring anion with respect to the water of crystallisation. Jackson† has investigated the principal magnetic susceptibilities of the monoclinic crystals of neodymium sulphate octahydrate $Nd_2(SO_4)_3.8H_2O$ over the temperature range 14 to 200° K. All three susceptibilities obeyed the law $\chi(T+\Delta) = $ const. above about 120° K., the three $(1/\chi, T)$ curves being parallel straight lines, but at low temperatures χ_2 tends to become independent of T while χ_1 and χ_3 increase more rapidly with decrease in T than at higher temperatures. At 17° C. the magnetic anisotropy $\Delta\chi/\chi$ is 11 per cent, increasing to 70 per cent at 14° K. The magnetic ellipsoid whose principal axes are formed of the principal susceptibilities of the crystal, rotates about the symmetry axis through a large angle as the temperature falls from 290 to 14° K. Penney and Kynch‡ show that it is impossible to reconcile these magnetic properties with current views on the absorption spectra and specific heat data,§ for the magnetic data require a very much larger splitting (20 cm.$^{-1}$) than is found in the absorption spectra, and the specific heat data over the region 3 to 40° K. require a still larger splitting.

The Internal Field of Copper Sulphate Crystals. Some interesting results have been obtained with the triclinic crystal $CuSO_4.5H_2O$ by Krishnan and Mookherji.|| In each unit cell of this crystal there are two Cu^{++} ions, each of which is the

---

* H. Ewald, *Ann. der Phys.* **34**, 209, 1939.

† L. C. Jackson, *Roy. Soc. Proc.* A, **170**, 266, 1939.

‡ W. G. Penney and G. J. Kynch, *Roy. Soc. Proc.* A, **170**, 112, 1939.

§ J. E. Ahlberg, E. R. Blanchard and W. O. Lundberg, *Journ. Chem. Phys.* **5**, 552, 1937.

|| K. S. Krishnan and A. Mookherji, *Phys. Rev.* **50**, 660, 1936; **54**, 533, 1938.

centre of an octahedron formed by six negatively charged oxygen atoms (O^-). This octahedron is not regular, but may be pictured as formed by pulling out a regular octahedron symmetrically along one of its diagonals, which forms an axis of tetragonal symmetry for the electric field acting on the Cu^{++} ion at its centre, and also for the magnetic field. The very extensive measurements of de Haas and Gorter* showed that the powdered crystal obeyed the Curie-Weiss law

$$\chi = \frac{\chi_1 + \chi_2 + \chi_3}{3} = 0.456/(T + 0.70),$$

giving $\mu_{\text{eff}} = 1.92\mu_B$, instead of $1.73\mu_B$ for spin contributions only. Now the small value of θ would indicate that the internal fields are feeble, and we should expect the magnetic anistropy to be small and all three principal susceptibilities to follow Curie-Weiss laws with θ very small. It is found by experiment, however, that the magnetic anisotropy is not small. The crystal is nearly uniaxial as far as its magnetic properties are concerned, and, if we denote the molar susceptibilities of the Cu^{++} ion and its accompanying O atoms parallel and perpendicular to this axis by $\chi_\parallel$ and $\chi_\perp$ respectively, we find

$$\chi_\parallel = 0.572/(T + 4.5) \quad \text{and} \quad \chi_\perp = 0.399/(T - 2.0).$$

Leaving out of account the small Curie temperature corrections as equal and unimportant, we see that the values of μ_{eff}, viz. 2·13 and $1.80\mu_B$, are widely different, although both are practically independent of temperature. We therefore conclude that while a large value of θ in the Curie-Weiss law means that a large internal field is present, a small value of θ does not necessarily mean a weak field.

Reekie† measured the mean mass susceptibilities of

$$CuSO_4.5H_2O, \quad CuSO_4.K_2SO_4.6H_2O$$

and $$CuSO_4(NH_4)_2SO_4.6H_2O$$

down to 1·6° K. The first follows a Curie-Weiss law with $\theta = -0.7°$ over the whole temperature range; the two double sulphates follow a similar law with $\theta = -0.3°$ down to 14° K., but obey the simple Curie law at liquid helium temperatures.

* W. J. de Haas and J. G. Gorter, *Leid. Comm.* 210d, 1930.
† J. Reekie, *Roy. Soc. Proc.* A, **175**, 367, 1939.

It is clear that the Cu^{++} ion must be under different crystalline fields in the two cases, unless magnetic interaction effects become important in copper sulphate at relatively high temperatures.

It is interesting that the susceptibility of a powder obeys the Curie law much more exactly than the susceptibility along a particular axis. Actually, quantum mechanics requires that the mean susceptibility shall be given by

$$\chi = CT^{-1} + aT^{-3} + bT^{-4},$$

where we observe that a term in T^{-2} is absent.

Large Orbits in Graphite. An interesting example of large orbits is provided by graphite, which is a hexagonal crystal with perfect basal cleavage. The carbon atoms are arranged in layers parallel to the basal plane, and the atoms in each layer form a hexagonal network. The distance between adjacent layers is 3·4 Å., while that between adjacent atoms in the same layer is only 1·42 Å. The susceptibility in the basal plane $\chi_\perp$ is only -0.5×10^{-6} e.m.u. per gm. while that parallel to the hexagonal axis $\chi_\parallel$ is -22×10^{-6} at room temperature, and the latter susceptibility changes appreciably with T. It is suggested[*] that the three electrons which assist in attaching each carbon atom to its neighbours in the same layer may be considered bound, while the fourth may move about more or less freely in the basal plane, giving rise to large orbits, large diamagnetism and high electrical conductivity (a kind of superconductivity in the basal plane which is at least 10,000 times greater than the conductivity normal thereto). This view is strongly supported by experiments in which attempts are made, deliberately, to restrict the motion of these "free" electrons. Thus, if the crystal forms the anode in a bath of strong sulphuric acid, a current of about 1 milliampere for a few minutes reduces the magnetic anisotropy practically to zero, because HSO_4 groups creep in between and separate the adjacent carbon layers still further, and convert the crystal to a single crystal of bisulphate of graphite. The anisotropy may be partially restored by washing in water, by immersion

* K. S. Krishnan, *Le Magnétisme*, III, 312, Strasbourg, 1940.

in ferrous sulphate solution, or by electrolysis. Again, when a graphite crystal is exposed to potassium vapour, a metallic, copper-coloured alloy is formed; the potassium atoms increase the distance between the adjacent layers and the magnetic anisotropy is again reduced.

The Effect of Impurities on Crystal Anisotropy. In the case of very pure bismuth* (99·998 per cent) $\chi_\| = -1·05 \times 10^{-6}$ and $\chi_\perp = -1·45 \times 10^{-6}$ at 27° C. so that at this temperature $\chi_\| - \chi_\perp = 0·402 \times 10^{-6}$. The latter quantity decreases linearly with rise in temperature to $0·09 \times 10^{-6}$ at 260° C., whereupon it remains constant up to the melting point, 271·1° C. Less pure bismuth also showed a linear change up to 260°, but thereafter $\chi_\| - \chi_\perp$ decreased very rapidly and reached very low values just before the crystal melted. Rao and Narayanaswamy† found that the magnetic anisotropy of thallium crystals at 30° C. was increased by the addition of small quantities of Cd, but decreased by small quantities of Pb, Sn and Bi. There was evidence that the anisotropy of thallium was not entirely determined by the number of the valency electrons of the added metal, for tin produced twice as great an effect as lead for the same atomic concentration, although both possess the same number of valency electrons, and bismuth with 5 valency electrons produces about the same effect as lead which has only 4 valency electrons.

Lane and Browne‡ used single crystals of Sb alloyed with Sn, Ge, Pb and Te. They found that $k_\|$ decreased rapidly with the percentage of the added metal, even changing from dia-magnetic to paramagnetic in the case of Sn and Ge, although $k_\perp$, due mainly to holes in the first zone, remained practically unchanged. Calculation showed that the number of electrons which overlap into the second Brillouin zone in antimony is 10^{-2} per atom. Moreover, for Sb-Sn alloys $k_\perp$ was practically temperature independent from 77° K. to room temperature, but $k_\|$ decreased in magnitude with increase in temperature both for alloys for which $k_\|$ was paramagnetic and those for which it was diamagnetic.

* W. J. John, *Zeit. für Kryst.* A, **101**, 337, 1939.
† S. R. Rao and A. S. Narayanaswamy, *Phil. Mag.* **26**, 698, 1938.
‡ C. T. Lane and S. H. Browne, *Phys. Rev.* **60**, 895, 899, 1941.

CHAPTER XIV

RADIOFREQUENCY SPECTROSCOPY

The g-Factor of the Nucleus. Some very spectacular results
have recently been obtained by the new experimental tech-
niques for the examination of molecular beams. As a prelude
to their description we will briefly review and extend some of
the concepts dealt with in Chapter VI.

In general we may consider the magnetic moment of
a nucleus to be given by

$$\mu_I = \frac{e}{2M}\frac{h}{2\pi}Ig_I = Ig_I\mu_n,$$

where M is the mass of a proton or neutron and g_I is the
g-factor of the nucleus which takes account of the complex
structure of the nucleus in a way similar to the splitting factor
of an electron system. Now, between the nuclear spin I and
the angular momentum J of the electron system there must
exist a weak coupling. Assuming that the field produced by
the electron system in the region of the nucleus is $\mathscr{H}_0$, then
we must take into account an energy of magnetic interaction
equal to $-\mu_I\mathscr{H}_0\cos(I,J)$; we neglect, for the present, any
electrostatic effects due to asymmetry of the nuclear charge.
Hence, the energy of a spectral term consists of two parts, one
being the value of the term assuming the nucleus to possess
zero magnetic moment, the other being $-\mu_I\mathscr{H}_0\cos(I,J)$ or

$$-(\mu_I\mathscr{H}_0/2IJ)\left[F(F+1)-I(I+1)-J(J+1)\right]$$
$$= A[F(F+1)-I(I+1)-J(J+1)],$$

where F is the resultant of I and J, and may take the values
$I+J$, $I+J-1$, ..., $I-J$. The existence of the second part
therefore results in a *hyperfine structure of the multiplet term.*
From it we deduce the *hyperfine structure interval rule,* viz. the
intervals of a hyperfine structure multiplet term are in the
ratio $(I+J):(I+J-1),\,...\,:(I-J+1)$. The *overall separation*

398

of the hyperfine structure, i.e. the energy difference of the term with $F = (I + J)$ and $F = (I - J)$, is therefore equal to

$$AI(2J + 1) \text{ for } J \geqslant I \quad \text{and} \quad AJ(2I + 1) \text{ for } I \geqslant J.$$

The Formal Theory of Hyperfine Structure. Goudsmit and Bacher* first showed how the results previously obtained in the case of the Zeeman effect may be carried over or applied to the problem of the hyperfine structure. This may be done by replacing, in the Zeeman formulae, the electron spin S by the nuclear spin I; the electron orbital momentum L by the resultant electron momentum J of the whole electron system; the resultant of S and L, namely J, by F; and the component M_J of J parallel to the direction of the field by M_F, the component of F parallel to the field. In addition, the magnetic moment $g_S S$ associated with the resultant electron spin is replaced by $g_I . I$, the magnetic moment of the nucleus; and the resultant magnetic moment $g_L . L$ of the orbital motion is replaced by $g_J . J$, the magnetic moment of the electron system. It was emphasised in Chapter VI that on occasion terms in g_I can be neglected, because of its smallness.

New Measurements. We will first deal with new measurements carried out by improved methods similar to those already described in Chapter VI.

The resolving power of the method described on p. 190 was much increased by Millman, Rabi and Zachariast† who produced $\mathscr{H}$ by what was in effect a metre tube of Armco iron with a longitudinal gap. Equations (128) and (129) are no longer sufficiently accurate and we have, when $J = \frac{1}{2}$,

$$\mu_{+\text{eff}} = -g_I . M_F \mu_n \mp 2 \frac{y + (2M_F/2I + 1)}{[1 + (4M_F y/2I + 1) + y^2]^{\frac{1}{2}}} \cdot (g_J - g_I) \mu_B,$$
$$\dots\dots(128a)$$

where
$$y = \frac{(g_J - g_I) \mu_B \mathscr{H}}{\Delta E}.$$

Equation (128a) reduces to (128) when we write $g_I = 0$. Whereas, for a given negative value of M_F, $\mu_{+\text{eff}}$ and $\mu_{-\text{eff}}$

* S. Goudsmit and R. F. Bacher, *Zeit. für Phys.* **66**, 13, 1930.
† S. Millman, I. I. Rabi and J. R. Zacharias, *Phys. Rev.* **53**, 385, 1938.

were both equal to zero for the same values of $\mathscr{H}$, this is no longer the case, and we find that they are zero at two slightly different values of $\mathscr{H}$, say, for example, $\mathscr{H}_1$ and $\mathscr{H}_1 + \varDelta\mathscr{H}_1$, for the first position, and so on. In other words, instead of a single peak maximum being found by the receiver, a double peak maximum is now found. It can be proved that

$$\frac{\varDelta\mathscr{H}_1}{\mathscr{H}} = \frac{2I+1}{I} \cdot \frac{\mu_I}{\mu_J} \left[1 - \left(\frac{2M_F}{2I+1} \right)^2 \right]^{\frac{1}{2}},$$

where μ_I and μ_J are the nuclear and the extranuclear moments respectively. The last expression is extremely interesting in that it contains μ_J independent of $\varDelta E$, etc. Measurements on these lines were successfully made with In^{113} for which were found $I = \frac{9}{2}$ and $\mu_I = 6\cdot40\mu_n$.

A further point of interest arose in these experiments. The normal state of the In^{115} atom is $^2P_{\frac{1}{2}}$, but, at the temperature of the furnace used, some 19 per cent of the atoms were in the $^2P_{\frac{3}{2}}$ state, in which they had a mean life of 1 sec. Hamilton* investigated their zero moment peaks in detail and obtained interesting information of the electric quadrupole moment of In^{115} which he found was $0\cdot84 \times 10^{-24}$ cm.^{2}

Radiofrequency Spectroscopy. While the foregoing experimental method is applicable to atoms possessing an electronic magnetic moment, it must be replaced by another† when we wish to deal with individual molecules which are diamagnetic. In this case the molecular beam is passed through a strong non-uniform field A, wherein the nuclear spins are completely decoupled from one another and from the molecular rotation. It is then passed through a uniform field B, with a strength $\mathscr{H}$ of some few thousand oersteds, upon which is superimposed an alternating field $\mathscr{H}_1$ of radiofrequency ν perpendicular to the steady field B. Finally, after passage through another equally non-uniform field C, which serves to refocus the particles, the latter fall upon a Pirani collector. The alternating

* D. R. Hamilton, *Phys. Rev.* **56**, 30, 1939.

† I. I. Rabi, S. S. Millman, P. Kusch and J. R. Zacharias, *Phys. Rev.* **53**, 318 and 495, 1938; **55**, 526 and **56**, 729, 1939.

field $\mathscr{H}_1$ is produced by a high frequency current in a "hair-pin" loop of wire, placed in the gap of the electromagnet providing the field B, the plane of the "hair-pin" being parallel to the beam and to $\mathscr{H}$. Hence the resultant magnetic field oscillates slightly about the fixed field.

If now we picture the effect upon an ideal system with a single angular momentum vector J and a magnetic moment μ parallel to J, then we see that the small oscillating field $\mathscr{H}_1$ must cause J to precess about $\mathscr{H}$ with the Larmor precession frequency, $f = \mu\mathscr{H}/hJ$, the orientation of J being given by one of the $(2J+1)$ possible values of M_J, the projection of J along $\mathscr{H}$. In fact, the field $\mathscr{H}_1$ provides an oscillating torque on J about an axis perpendicular to $\mathscr{H}$.

Hamilton* has illustrated its effect by reference to the following mechanical model. When a top precessing about a vertical axis under the action of gravity is subjected to an oscillating torque about a horizontal axis, e.g. by a sinusoidal displacement of the point of support of the top, the latter will be forced during part of the torque cycle to become more vertical and at other times to become less vertical. It is, however, only when the precession and oscillation frequencies are equal that pronounced changes will occur and the top will either "stand up" or "lie down" depending on the phase relation between the oscillating torque and the precession. This corresponds in the atomic system case to an increase or decrease in M_J, which means that transitions between states of different M_J occur, when the frequency ν approximates to that of the Larmor precession, i.e. $\nu = \mu\mathscr{H}/Jh$. As the energy of our simplified atomic system is $-\mu\mathscr{H}M_J/J$, we see that the energy change in a possible transition between two M_J states is numerically equal to $\mu\mathscr{H}/J$ which is h times the frequency of the electromagnetic radiation emitted or absorbed according to the Bohr theory.

Returning to our beam experiment, when the above frequency condition obtains, the reorientation of the nuclear spin and magnetic moment with respect to the steady field takes place, the beam is no longer focused and the detector

* D. R. Hamilton, *Amer. Journ. of Phys.* 9, 319, 1941.

records a decrease in the number of particles incident upon it. The state of affairs is depicted in Fig. 110, where the two full-line curves running from the oven O to the detector D show the paths of two particles, having different angular

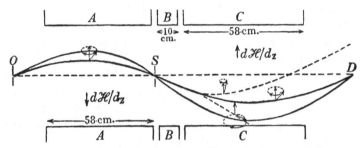

Fig. 110. Paths of molecules in the radiofrequency spectrograph.

moments and translational velocities, whose effective magnetic moments are not changed during passage through the apparatus. This condition is represented by the small gyroscopes drawn on one of these paths with a fixed or constant projection of the magnetic moment parallel to the lines of force. The two dotted curves in the region of the C magnet represent the paths of two particles for which the projections of the nuclear magnetic moments along the field have been changed by the oscillating field within the B magnet region. Small gyroscopes are also drawn on these dotted curves, one showing that the projection of the magnetic moment along the field has been increased and the other that it has been decreased. If $Fh/2\pi$ represents the total angular momentum of a molecular system, it should be realised that we do not measure F directly by these experiments, because

$$Fh/2\pi\mu = \frac{2}{g}\frac{M}{e},$$

so that
$$g = 2\frac{M}{e}.n\mu_n\bigg/\frac{Fh}{2\pi} = 4\pi\frac{M}{e}\frac{n\mu_n}{Fh}$$

$$= 4\pi\frac{M}{e}\frac{\nu}{\mathscr{H}} = 1{\cdot}3122\times10^{-3}\,\nu/\mathscr{H}$$

is the quantity actually measured; here M is the mass of the proton and $n\mu_n$ is the nuclear magnetic moment.

The curve of detector current plotted against $\mathscr{H}$ should have a marked depression at resonance, i.e. whenever the product $h\nu$ is equal to the difference in energy of two molecular states. It is analogous to finding an unusual form of Zeeman pattern by measuring the absorption of a particular narrow spectral line as a function of the magnetic field in which the absorber is placed. It was found by experiment that the magnetic effects of the ends of the hair-pin conductor are such that an additional resonance minimum was formed on one side or the other of the principal minimum according to the direction of $\mathscr{H}$ and the sign of μ. The radiofrequency method was first successfully used for measuring the moments of Li^6, Li^7, N^{14} and F^{19}, and later used for measurements with HD, CCl_2F_2, H_2, D_2, CsCl, CsF, NaF, NaCN, Na_2, KCN, K_2, etc. It is important to use at least two molecules containing the same atom. For example, when using both LiCl and LiF a minimum of the same type was found at the same value of $\nu/\mathscr{H}$, showing that the resonance arose from the Li^7 nucleus. The main experimental error resided in the measurement of $\mathscr{H}$, but in later experiments radiofrequency measurements on atoms provided an accurate direct calibration of $\mathscr{H}$ in terms of frequency, provided that adequate knowledge of the Paschen-Back spectrum of one of the atoms was available.

The H_2 experiments were made with a beam from a source cooled in liquid nitrogen, which caused the parahydrogen molecules to behave as inert because they were all in the rotational state $R = 0$ with total nuclear angular momentum $I = 0$. At the same time the orthohydrogen molecules were in the state $R = 1$, $I = 1$, giving $(2R+1)(2I+1) = 9$ energy levels from which six transitions under the rule $\Delta M_I = \pm 1$ were important in these experiments. These six transitions give six minima as shown in Fig. 111. If additional rotational states were excited, as by using a higher temperature source, then so many of these fine structure minima would be produced that they would merge into one another.

The experiments made with D_2 and HD molecules have led to the spectacular discovery of the existence of a *quadrupole moment in the deuteron*, which means that the forces between

26-2

the neutron and proton in the deuteron nucleus are not strictly central. The low temperature D_2 beam for the transitions $\Delta M_I = \pm 1$ gave six minima grouped about a much more pronounced minimum due to the Larmor precession of the deuteron. The HD beam gave two distinct sets of minima.

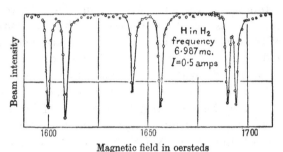

Magnetic field in oersteds

Fig. 111. Resonance curve arising from reorientation of the resultant nuclear angular momentum in H_2.

One, of nine subsidiary resonances with a very deep single resonance at the Larmor frequency of the proton, arose from reorientation of the proton in the HD molecule. The other, of twelve subsidiary resonances with a deep single resonance at the Larmor frequency of the deuteron, arose from reorientation of the deuteron in HD. The deep minima fixed the magnetic moments of the proton and deuteron as $2 \cdot 785 \pm 0 \cdot 02$ and $0 \cdot 855 \pm 0 \cdot 006 \mu_n$ respectively, while the detailed analysis of the resonance curves gave the value of the electric quadrupole moment as $Q = 2 \cdot 73 \pm 0 \cdot 05 \times 10^{-27}$ cm.^{2}

The method has been extended* to the examination of the hyperfine structure intervals of the normal states of the atoms Li^6, Li^7, K^{39} and K^{41}, etc., by observing transitions produced by the oscillating field in external fields ranging from 0·5 to 4000 oersteds.

The Magnetic Moment of the Neutron. The most accurate determination of the magnetic moment of the neutron to date has been made by Alvarez and Bloch† using the system employed by Rabi and his collaborators. The experimental

* P. Kusch, S. Millman and I. I. Rabi, *Phys. Rev.* **57**, 765, 1940.
† L. W. Alvarez and F. Bloch, *Phys. Rev.* **57**, 111, 1940.

arrangements are shown in Fig. 112. A powerful beam of neutrons, from a cyclotron in which high speed deuterons bombarded a Be target, passed through a polariser, then through an oscillating field $\mathscr{H}_1$ of frequency ν and a steady uniform field $\mathscr{H}_0$, which caused them to precess, and then through the analyser to a BF_3 counting chamber. Large blocks of paraffin were used to slow down stray neutrons which had not passed down the collimating cadmium tube.

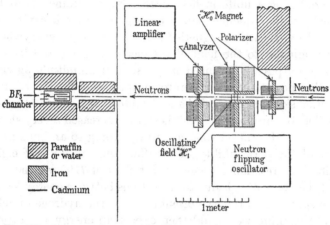

Fig. 112. Apparatus for measuring the magnetic moment of the neutron.

If the neutron has no magnetic moment then there should be no difference in the number of neutrons counted per sec. when this field is on or off, but, if it has a magnetic moment μ, then a decrease in the counting rate should be recorded when the relation between the steady field and the frequency of the oscillating field is $\mathscr{H}_0 = h\nu/2\mu$. This is because the precession of the neutron is now markedly affected by the oscillating field, and the state of polarisation of the beam is thereby altered. Such decreases were observed, and although it was difficult to measure $\mathscr{H}_0$ with accuracy, a clever comparison measurement with respect to the cyclotron field driving a proton beam gave reliable values, the final result for μ being $-1 \cdot 93 \pm 0 \cdot 02 \mu_n$. This means that within the limits of experimental error the magnetic moment of the deuteron is equal to the sum of the magnetic moments of the proton and neutron.

Resonance Absorption due to Nuclear Magnetic Moments. Let us imagine that a high frequency resonant cavity, made in the form of a short section of coaxial line loaded heavily by the capacity of an end plate, and adjusted to resonate at about 30 megacycles per sec., is provided with input and output coupling loops. Let the inductive part of the cavity be filled with solid paraffin at room temperature, and let the resonator be placed in the gap of an electromagnet capable of producing a uniform field over a large volume. Now let (weak) radiofrequency power be supplied to the cavity, so that the alternating magnetic field in the cavity is everywhere perpendicular to the field of the electromagnet, and let the cavity output be balanced in phase and amplitude against another portion of the generator output, and, finally, let any residual signal be amplified and detected. It is found* that as the field of the electromagnet is slowly increased a very sharp resonance absorption sets in, corresponding to an imaginary part of the permeability of paraffin equal to 3×10^{-6} c.g.s. units. This resonance occurs with a field of 7100 oersteds.

If the absorption is due to transitions between energy levels corresponding to the two orientations of the hydrogen nuclei in the paraffin, we should then expect an energy separation corresponding to $\nu = 30$ megacycles per sec., from the relation $h\nu = 2\mu\mathscr{H} = 2\mu \times 7100$, and the imaginary part of the permeability of the order $h\nu/kT$. From the above experimental data, knowing ν and $\mathscr{H}$, we deduce that $\mu = 2 \cdot 75$ nuclear magnetons. It is found in these experiments that the relaxation time is very much shorter than current theories of spin-lattice coupling would indicate, and the method seems likely to be very useful in investigating this coupling. (See also p. 422.)

Arnold and Roberts† have used the method to measure the ratios of the magnetic moments of the deuteron and neutron to that of the proton. With water in a resonant cavity, the electromagnet field was adjusted and fixed to give proton resonance at about 30 megacycles per sec. The water was removed, a powerful beam of neutrons was sent through the

---

* E. M. Purcell, H. C. Torrey and R. V. Pound, *Phys. Rev.* **69**, 37, 1946.
† W. J. Arnold and A. Roberts, *Phys. Rev.* **70**, 776, 1946.

cavity, and the radiofrequency was adjusted to give neutron resonance; the latter was shown by a drop of some 5 to 8 per cent in the intensity of the neutron beam. The ratio of the magnetic moments of the proton and neutron was given by the ratio of the two radiofrequencies, which was found to be 0·68479.

In the case of the deuteron, resonance was obtained with heavy water in position, followed by similar proton resonance with ordinary instead of heavy water; the ratio of the two radiofrequencies was then the ratio of the magnetic moment of the deuteron to that of the proton, and was found to be 0·30702. Assuming that the magnetic moment of the proton is $2·7896\mu_n$, the magnetic moments of the neutron and deuteron are therefore $-1·9103 \pm 0·0012\mu_n$ and $0·85647 \pm 0·0003\mu_n$, respectively.

Nuclear Induction. An entirely new method of attack has been discovered by Bloch* and his collaborators, based on the phenomenon of *nuclear induction*. Suppose we place a sample of matter, containing the nuclei we wish to investigate, in a steady field $\mathscr{H}$ along the z axis, so that it acquires a magnetic moment $M = \chi\mathscr{H}$. If we now superimpose an oscillating field $\mathscr{H}_x = \mathscr{H}_1 \cos \omega t$ parallel to the x axis, we may, by suitable choice of ω, disturb the settings of the nuclear magnetic axes parallel to $\mathscr{H}$, and so produce a magnetisation along the y axis. This will be a maximum when $\rho\mathscr{H} = \omega$, where ρ is the gyromagnetic ratio of the nuclei under discussion. Bloch showed that, although the amplitude of the field associated with the latter magnetisation was only about 5×10^{-6} oersteds, the magnitude of ω (8×10^6 cycles per sec.), and hence the rate of change of flux, was so large that measurable induction effects were produced in a properly designed coil.

The new method is not limited, as molecular beam methods are, by choice of detector, and Bloch suggests that it may be extremely useful for the comparison of nuclear magnetic moments, isotope analysis, the measurement of relaxation times of nuclei, and even for the calibration of intense magnetic fields.

* L. Bloch, *Phys. Rev.* **70**, 460, 474, 1946.

CHAPTER XV

ADIABATIC DEMAGNETISATION
AND ITS APPLICATIONS

Theoretical Aspects. On p. 284 we discussed some theoretical aspects of the production of very low temperatures by adiabatic demagnetisation. Now, nickel sulphate heptahydrate, $NiSO_4.7H_2O$, is interesting because of its high value of θ_m. This is clear from the results of Stout and Giauque* shown in Fig. 113 obtained with the amorphous carbon thermometer, described below. In the same figure the way in which the Curie

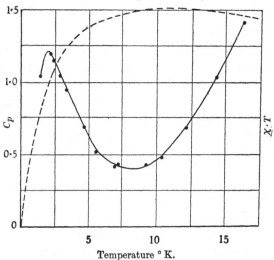

Fig. 113. —— Heat capacity in cal. per deg. per mole of $NiSO_4.7H_2O$.
– – – – Values of $\chi.T$ for $NiSO_4.7H_2O$.

law breaks down at low temperatures may be seen from the sudden bend in the curve of $(\chi.T)$ against T, which should, of course, be a horizontal line when the Curie law is obeyed.

Equation (188), p. 285, may be more effectively written as $(T_f/T_i)\,\mathscr{H} = \alpha = $ constant, where the precise significance of the constant need not be discussed. Bleaney showed that, in

* J. W. Stout and W. F. Giauque, *Journ. Amer. Chem. Soc.* **63**, 714, 1941.

the particular case of potassium chrome alum, α was constant over a relatively wide range.

Low temperature heat capacity measurements provide information from which we may deduce the magnetic energy spectrum in the solid. For example, measurements on anhydrous manganous fluoride show clearly how the random distribution of the magnetic ions among the six states of the Mn^{++} ion possible at high temperatures changes as the temperature is lowered, a very striking maximum being found in the heat capacity curve at 66·5° K. The changes in magnetic entropy are not, however, confined to the restricted temperature range in which this anomalous behaviour is observed, but must extend to higher temperatures where the magnetic contributions to the heat capacity cannot be disentangled from those due to crystalline vibrations.

Thermal Insulation. Cooke and Hull* showed that the thermal insulation of a sample is greatly improved by suspending it between two taut fibres of artificial silk, one attached to each end. In this way the heat flow may be reduced to 5 ergs per min., and this may be further reduced to 1 erg per min. by using "guard rings", formed of samples maintained at the same low temperature, interposed on similar supports, between the main sample and containing vessel or frame. This suggests that the residual heat flow of 1 erg per min. is due to transfer by helium gas "desorbing" from the wall of the cryostat, at 1° K.

Experiments with Solid Hydrogen. Helium must not be allowed to condense as helium II in the inner space of *B*, p. 287, or it will escape through the ground joint and spoil the outer vacuum; this is known as the λ-leak phenomenon. Incidentally, in more modern forms of the apparatus of Fig. 86, the guard tube *S* is omitted.

Kürti, Lainé and Simon† have shown that it is possible to produce low temperatures with iron alum by starting from temperatures reached with solid hydrogen, which is relatively

---

* A. H. Cooke and R. A. Hull, *Roy. Soc. Proc.* A, **181**, 83, 1943.
† N. Kürti, R. Lainé and F. Simon, *Compt. Rend.* **208**, 173, 1939.

easy to produce by evaporation of liquid hydrogen under low pressure. The main advantage of demagnetising from the higher hydrogen temperatures is that the thermal insulation is increased owing to the much lower vapour pressure of hydrogen. The entropy temperature diagram of iron alum shows that lattice effects are relatively unimportant for the magnetisation provided one uses temperatures below 9° K. A strong magnetic field is necessary. Starting from a temperature of 8·5° K. and using a field of 28,800 oersteds, a temperature of 0·36° K. was reached; a field of 42,000 oersteds should enable 0·15° to be attained. In this connection Bitter magnets are likely to be important.

The Definition of $T^\circledast$. Kürti and Simon* pointed out that correct allowance for the demagnetisation factor is most important in calculating T^*, and they used $T^\circledast$ to denote temperatures based on measurements made with, or referred to, specimens of spherical shape.

The field $\mathscr{H}_i$ acting on a particle within the sample, when the latter is exposed to an external field $\mathscr{H}$ may be expressed in the form

$$\mathscr{H}_i = \mathscr{H} + \left(\frac{4\pi}{3} - D\right) M/V,$$

where M is the magnetic moment acquired by the specimen of volume V, and D is the ordinary demagnetisation factor for the specimen. Using the Oxford compensation method of measuring M by a ballistic galvanometer deflection δ, resulting from breaking a current i in the primary of a mutual inductance, we may write $\delta = aM$, where a is a constant. If Curie's law is obeyed, $M = \dfrac{VC}{T}\mathscr{H}_i$, and, consequently, it follows that $M = CV\mathscr{H}/T - \Delta$, where $\Delta = C\left(\dfrac{4\pi}{3} - D\right)$. Now, $\mathscr{H}$ is equal to ai, so that

$$\delta/i = M/\mathscr{H} = CV/T - \Delta.$$

Hence, on plotting δ/i against $1/T - \Delta$, a straight line should be obtained. By extrapolation outside the calibration range one may define temperatures on a scale which depends upon

* N. Kürti and F. Simon, *Phil. Mag.* **26**, 849, 1938.

the value taken for Δ. The value chosen above is that for which the Lorentz and the ordinary demagnetisation fields cancel one another; this is the case for a spherical specimen, and temperatures so defined are, as indicated above, denoted by $T^{\circledast}$. If the specimen is a loosely packed powder, for which f is the ratio of the mass of unit volume of the specimen to the mass of unit volume of the solid material, then

$$\Delta = Cf\left(\frac{4\pi}{3} - D\right).$$

Bleaney and Hull* determined Δ experimentally and compared it with the value calculated from the last expression, with $f = 0{\cdot}55$ and $D = 0{\cdot}95$; the two values were respectively $0{\cdot}038 \pm 0{\cdot}002$ and $0{\cdot}037°$. Although Δ is small, its value can be very important in very low temperature calculations, and in any such calculations it is imperative to state the value of Δ which has been used.

The Carbon Thermometer. Giauque, Stout and Clark† showed that at very low temperatures an excellent resistance thermometer can be made from amorphous carbon. Two platinum strips deposited on the specimen tube formed electrical connections. A single layer of lens paper was then applied to the tube with ethyl alcohol and a camel hair brush; while still wet, lampblack mixed with a large quantity of alcohol was painted into the paper, which was later coated with a weak solution of collodion in alcohol and ethyl ether. Amorphous carbon exhibits a very high specific resistance and a small change of resistance with magnetic field so that accurate temperature measurements can be made. Hence the resistance of the electrical leads, made very fine to prevent heat losses, may be neglected, and the heat produced by eddy currents is extremely small. The resistance of one thermometer was found to be 28,062 ohms at 293° K. and 58,640 ohms at $0{\cdot}129°$ K., and, after ageing, such thermometers gave reproducible results far better than one would expect.

---

* B. Bleaney and R. A. Hull, *Roy. Soc. Proc.* A, **178**, 86, 1941.
† W. F. Giauque, J. W. Stout and C. W. Clark, *Journ. Amer. Chem. Soc.* **60**, 1053, 1938.

The Measurement of the Absolute Temperature T. Kürti and Simon* have suggested another important method of finding T by means of heat supplied by γ-rays. A specimen is first cooled by demagnetisation to a point A in zero field. When heated by γ-rays, its temperature rises, and a (χ, t) curve, AF, Fig. 114, shows how T varies with the time, t. If the experiment is now repeated and a field $\mathscr{H}$ is switched on when the substance reaches a standard temperature T_0 at E, the curve GJ is subsequently traced out after removal of the field.

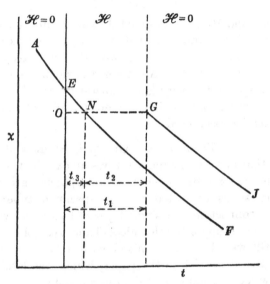

Fig. 114. Determination of T.

Drawing the horizontal line GO we see that the states of the specimen at N and G are the same. We can thus suppose that these states were reached (a) by supplying a quantity of heat αt_3 from E to N, thus increasing the entropy by $\alpha t_3/T_0$, and (b) by heating the specimen in the field $\mathscr{H}$ for a time t_1 from E to G, thus increasing the entropy by $\alpha t_1/T_\mathscr{H}$. But, these entropy changes are equal; therefore $T_\mathscr{H}/T_0 = t_1/t_3 = t_1/t_1 - t_2$. Now, the absolute value of $T_\mathscr{H}$ can be found by taking $\mathscr{H}$ to be the same as the initial field which produced the demagnetisa-

* N. Kürti and F. Simon, *Phil. Mag.* **26**, 840, 1938.

tion cooling. The great advantage of this method is that the susceptibility measurements can be made in zero field.

Bleaney used potassium chrome alum and evaluated $(\partial S/\partial T^*)_{\mathscr{H}}$ from the T_f/T_i relation mentioned in the opening section of this chapter. He also measured $(\partial Q/\partial T^*)_{\mathscr{H}}$ using γ-ray heating, and found the relation between T and T^* from the expression $T = (\partial Q/\partial T^*)_{\mathscr{H}}/(\partial S/\partial T^*)_{\mathscr{H}}$ given on p. 291. The specific heat $(\partial Q/\partial T)_{\mathscr{H}=0}$ was found to be equal to $0\cdot031/T^2$ cal. per gm. ion per $^\circ$K., in good agreement with Hebb and Purcell's results. The difference between $T^\circledast$ and T is a measure of the departure from Curie's law due to crystalline splitting and magnetic interaction. Bleaney also showed that the latter interaction is not adequately expressed by the Lorentz $4\pi/3$ factor, but is better represented by the Onsager or Van Vleck local field.*

When considering the processes involved in an adiabatic demagnetisation one feels on comparatively safe ground as long as one is sure that the magnetic moment of a specimen remains constant as the field is decreased adiabatically from an initial high value to a low one, i.e. the Curie law holds accurately, as de Haas and Wiersma† found in the case of caesium titanium alum. This is not so for iron alum, where the magnetic moment of the specimen in a field $\mathscr{H}$ is given by an expression of the type $M = M_0(1-b/\mathscr{H}^2)$. This case has been discussed very carefully by Sauer‡ who shows, fortunately, that for fields between 10,000 and 5000 oersteds little error is made in temperature determination, but appreciable errors are found with fields below 5000 oersteds.

Sauer and Temperley§ suggested a simple model of freely suspended magnets to account for the behaviour of alums at very low temperatures, only dipole-dipole interactions between the paramagnetic ions being considered, to which the theory of order and disorder is applied. On this basis they showed that ferromagnetism may arise in an ellipsoidal specimen with axes in the ratio of 6 to 1.

* Cf. H. B. G. Casimir, *Magnetism and Very Low Temperatures*, p. 56, Cambridge, 1940.
† W. J. de Haas and E. C. Wiersma, *Physica*, **3**, 491, 1936.
‡ J. A. Sauer, *Phys. Rev.* **64**, 94, 1943.
§ J. A. Sauer and H. N. V. Temperley, *Roy. Soc. Proc.* A, **176**, 203, 1940.

Returning to the expression $T = \partial Q/\partial T^{\circledast}/\partial S/\partial T^{\circledast}$, we observe that what we really want is a $(Q, T^{\circledast})$ curve and an $(S, T^{\circledast})$ curve to which we may draw tangents at chosen values of T. Hull* discusses this mode of attack in detail. Assuming that the specimen obeys Curie's law, we can calculate the values of S under known fields and temperatures; we can measure the values of $T^{\circledast}$ which result from adiabatic demagnetisations from these initial conditions, and so build up an $(S, T^{\circledast})$ curve.

The $(Q, T^{\circledast})$ curve is obtained by supplying γ-ray heat to the specimen, which is almost the only method of generating heat *uniformly throughout the specimen*. This rate of supply must be calibrated by making similar experiments at higher temperatures where the specific heat is established by other methods and absolute temperatures are likewise known. In effect, the $(Q, T^{\circledast})$ curve is really a curve of U, internal energy, against $T^{\circledast}$, obtained by plotting time of γ-ray heating against $T^{\circledast}$. A clever artifice, which amounts to removing adsorbed helium, or baking it out at $1 \cdot 5^{\circ}$ K., is employed to reduce temperature drift, and enables accurate specific heat measurements to be made at temperatures up to $0 \cdot 8^{\circ}$ K.

These measurements fix the temperature scale and heat capacity for any chosen paramagnetic substance, so that the examination of the physical properties of this and other substances is rendered possible. The great difficulty is to maintain temperature equilibrium between such other substances and the poorly conducting paramagnetic salt. To overcome it, the capsule technique mentioned on p. 290 is used. The specially designed capsule consists of a hollow metal cylinder containing the powdered salt and enough helium to cover the salt as liquid helium, the good conducting properties of helium II being utilised. For example, by joining two such capsules at different temperatures by a wire of any metal, the thermal conductivity of the latter can be studied; much more is likely to be heard of this technique.

* R. A. Hull, *Phys. Soc. Proc.*, in press. (My thanks are due to Dr Hull for much helpful information.)

The Lowest attainable Temperature. The lowest temperature likely to be reached depends very much on the actual conditions, such as the interval of time for which it is to be maintained and on the degree of thermal insulation, etc. There is no reason why nuclear spins should not be used in extending the regions already reached by adiabatic demagnetisation, provided that it can be proved that equilibrium between the nuclear spins and the lattice is set up quickly enough and that the relevant θ_m is sufficiently low. The transition between the method based on electron spins and that on nuclear spins should occur about $10^{-4°}$ K. We must remember, however, that the study of the specific heats of substances at very low temperatures has revealed the existence of energy levels unknown in spectroscopy and which have not yet been explained, e.g. there is a marked anomaly* in the specific heat of $CuSO_4 . 5H_2O$ in the region of $0.5°$ K.

* Cf. G. Duyckaerts, *Thèse*, Liége, 1942.

CHAPTER XVI

PARAMAGNETIC ABSORPTION
AND DISPERSION

Theory of the Phenomena. If the high frequency susceptibility χ of a paramagnetic substance may be treated as made
up of two terms, according to the relation $\chi = \chi' + j\chi''$, it
follows that *paramagnetic dispersion* may be observed when
χ' is not identical with the static susceptibility χ'_0 in constant
magnetic fields. It also follows that *paramagnetic absorption*
must arise from the presence of χ'' when alternating fields
are employed, and many workers have studied the consequent dissipation of heat in paramagnetic specimens
exposed simultaneously to a constant steady field $\mathscr{H}_c$ and
an alternating field of amplitude $\mathscr{H}_0$ and pulsatance ω
(frequency ν).

The theory of the phenomena has been discussed by Debye*
and by Starr† and others mentioned below. If it is assumed
that electron spins are responsible for paramagnetism, it may
be postulated that there exists at any instant a temperature
difference θ between the temperature T of the crystal lattice
and that of the spins, an idea first advanced by Casimir and
du Pré.‡ Hence, in an interval of time dt, the heat imparted
to the lattice will be $dQ = \alpha . \theta . dt$, where α is a constant.
Let σ be the specific magnetisation and U the total energy of
the system, then,

$$dQ = dU - \mathscr{H}\, d\sigma. \qquad \qquad \ldots\ldots(209)$$

When a small alternating field h is superimposed upon
a steady field $\mathscr{H}_c$, so that $\mathscr{H} = \mathscr{H}_c + h$, we may obtain θ as
a function of t by writing, from (209),

$$[\partial U/\partial T - \mathscr{H}\,\partial\sigma/\partial T]\,\dot{\theta} + \alpha\theta = -[\partial U/\partial\mathscr{H} - \mathscr{H}\,\partial\sigma/\partial\mathscr{H}]\,\dot{h},$$
$$\ldots\ldots(210)$$

* P. Debye, *Phys. Zeit.* **39**, 616, 1939.
† C. Starr, *Phys. Rev.* **60**, 241, 1941.
‡ H. B. G. Casimir and F. K. du Pré, *Physica*, **5**, 507, 1938.

whence, replacing h by $\mathscr{H}_0 e^{j\omega t}$, we have

$$\theta = -j\omega[\partial U/\partial\mathscr{H} - \mathscr{H}\partial\sigma/\partial\mathscr{H}]h/(\alpha + j\omega[\partial U/\partial T - \mathscr{H}\partial\sigma/\partial T]).$$
$$\ldots\ldots(211)$$

Since, for very small values of h and θ we may write

$$\sigma = \sigma_0 + h\partial\sigma/\partial\mathscr{H} + \theta\partial\sigma/\partial T,$$

$$(\sigma - \sigma_0)/h = [\partial\sigma/\partial\mathscr{H} + j\omega/\alpha\{(\partial U/\partial T - \mathscr{H}\partial\sigma/\partial T)\,\partial\sigma/\partial\mathscr{H}$$
$$- (\partial U/\partial\mathscr{H} - \mathscr{H}\partial\sigma/\partial\mathscr{H})\,\partial\sigma/\partial T\}]$$
$$\div [1 + j\omega/\alpha\{\partial U/\partial T - \mathscr{H}\partial\sigma/\partial T\}].$$
$$\ldots\ldots(212)$$

The coefficient of $j\omega/\alpha$ in the denominator of (212) is $C_{\mathscr{H}}$, the specific heat of the material in a constant field, vide p. 273. The coefficient of $j\omega/\alpha$ in the numerator is, somewhat less obviously, C_σ, the specific heat of the material at constant magnetisation.

We may now replace $\partial\sigma/\partial\mathscr{H}$ by χ_0', the mass susceptibility in a steady field, and $(\sigma - \sigma_0)/h$ by χ, the incremental susceptibility in a high frequency field of small amplitude, so that

$$\chi/\chi_0' = (1 + j\omega C_\sigma/\alpha)/(1 + j\omega C_{\mathscr{H}}/\alpha)$$
$$= \left\{1 + \frac{\omega^2}{\alpha}C_\sigma C_{\mathscr{H}} - j\omega(C_{\mathscr{H}} - C_\sigma)\right\}\Big/(1 + \omega^2 C_{\mathscr{H}}^2/\alpha). \quad\ldots\ldots(213)$$

When Curie's law is obeyed, i.e. $\chi = C/T$ (or is a function of C/T alone), $C_\sigma = 0$, the real part of (213) becomes

$$\chi'/\chi_0' = 1/(1 + \omega^2 C_{\mathscr{H}}^2/\alpha^2)$$
$$= 1/(1 + \nu^2\rho^2), \qquad\ldots\ldots(214)$$

where $\rho = 2\pi C_{\mathscr{H}}/\alpha$. Equation (214), which shows the presence of paramagnetic dispersion, was obtained by Casimir and du Pré. The constant ρ is termed the *relaxation constant*, and its value determines the possibility of paramagnetic dispersion at the radio-frequencies available.

Measurements of paramagnetic dispersion have so far been made on powders, where χ_0' must be isotropic, and confined almost entirely to the case of longitudinal steady fields. χ' may readily be found by noting the frequency change of an oscillatory circuit when the powdered material is placed within a coil of suitable self-inductance, due allowance being made for any changes of capacity caused thereby. Details are given by Gorter and Kronig.* Results are termed *positive when χ' is not equal to χ'_0*, and such results are found with salts containing the hydrated ions Cr^{+++}, Fe^{+++} and Mn^{++}. The other common ions of the iron group and non-hydrated ions, e.g. Fe^{+++} in $FeNH_4(SO_4)_2$, give *negative* results. The ions which give positive results are known to occupy ground states in the crystalline field which are separated from higher energy states by considerable intervals, and to obey Curie's law excellently. The fact that non-hydrated ions of the same kind do not show paramagnetic dispersion may be due to the smaller distances between the neighbouring ions, whereby ρ may be much diminished. Few results have yet been obtained for ions of the rare earths. Gorter† found positive results with the hydrated sulphate, oxalate and acetate of gadolinium at low temperatures, but negative results with gadolinium oxide and hydrated dysprosium sulphate (and with hydrated gadolinium oxalate at room temperature).

It is important to know how the dispersion depends on the values of $\mathscr{H}_c$ and ω. Measurements are most fruitful when $\mathscr{H}_c$ and ω are high and the temperature is low. Iron ammonium alum, $FeNH_4(SO_4)_2 . 12H_2O$ has been most thoroughly investigated; a typical set of results is shown in Fig. 115, χ'_0 being the constant value of χ' obtained in small alternating fields of low frequency at high temperatures. They satisfy the relation

$$\chi' = \chi'_0 F/(1 + \rho^2 \nu^2) + \chi'_0(1 - F),$$

where F and ρ are constants which depend on the temperature and on $\mathscr{H}_c$. Both increase with decrease in temperature and with increase in $\mathscr{H}_c$. The values of F are practically independent of T at very low temperatures, where ρ appears to vary as $T^{-1.4}$. In the region 64 to 90° K., ρ appears to vary as T^{-5}, and roughly as T^{-3} in the intervening temperature range.

* C. J. Gorter and R. Kronig, *Le Magnétisme*, III, 65, Strasbourg, 1940.
† C. J. Gorter, *Physica*, **10**, 621, 1942.

Returning to equation (213), we may now express the specific magnetisation of a material in a field $\mathscr{H}_c + \mathscr{H}_0 \cos \omega t$ by the relation

$$\sigma = \chi_0' \mathscr{H}_c + \chi' \mathscr{H}_0 \cos \omega t + \chi'' \mathscr{H}_0 \sin \omega t, \quad \ldots\ldots(215)$$

whence it follows that the energy dissipated per gm. of material per cycle is given by

$$W_1 = \int_0^{2\pi} \sigma \, d\mathscr{H} = \pi \chi'' \mathscr{H}_0^2,$$

or, the energy dissipated per gm. per sec. is

$$W = \pi \chi'' \mathscr{H}_0^2 \nu = [\pi \mathscr{H}_0^2 \rho \nu^2 / 1 + \rho^2 \nu^2] / (C_{\mathscr{H}} - C_\sigma)/C_{\mathscr{H}} \quad \ldots\ldots(216)$$

since

$$\chi'' = 4\pi \chi_0 \omega (C_{\mathscr{H}} - C_\sigma)/\alpha(1 + \omega^2 C_{\mathscr{H}}/\alpha^2).$$

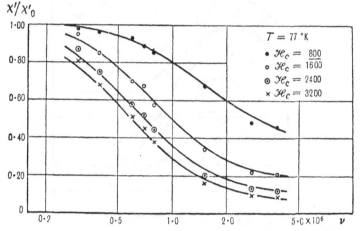

Fig. 115. Paramagnetic dispersion of $FeNH_4(SO_4)_2.12H_2O$ at 77° K.

The absorption or dissipation of energy in this manner is termed *paramagnetic absorption*. Since it depends on the squares of the amplitude and frequency of the alternating field, a high power oscillator and sensitive thermal measuring instruments have to be used to record it. In spite of the practical difficulties thereby introduced, experiments have shown conclusively that χ'' is independent of $\mathscr{H}_0$ and is usually proportional to ν and inversely proportional to T, whence we

419

deduce that ρ is independent of $\mathcal{H}_0$, ν and T when $\rho \ll 1$. The variation of χ'' with frequency is shown in Fig. 116; it satisfies the equation

$$\chi'' = \chi' F \rho \nu / 1 + \rho^2 \nu^2.$$

It follows from Fig. 116 that ρ can be deduced from the frequency ν_m at which the absorption is a maximum, where $\rho \nu_m = 1$ and $\chi'' = \frac{1}{2}\chi' F$. In more recent work* the variation of χ'' is given by

$$\chi'' = \chi_0 F \rho_1 \nu / (1 + \rho_1^2 \nu^2) + \chi_0 (1 - F) \rho_2 \nu,$$

where $F = C \mathcal{H}_c^2 / b + C \mathcal{H}_c^2$, ρ_1 and ρ_2 are the relaxation constants of the lattice and spins, respectively, and b and C are constants.

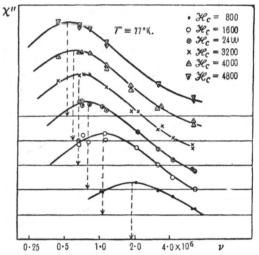

Fig. 116. Paramagnetic absorption of $FeNH_4(SO_4)_2 . 12H_2O$ as a function of frequency at 77° K.

In effect, it is postulated that direct interaction may take place between the magnetic field and the spin system and that this is independent of the crystal lattice; we have already discussed the interaction between the spin system and the lattice. It is considered that the two absorption processes can usually be separated readily. Thus, the absorption per

* Cf. L. J. Dijkstra, C. J. Gorter and J. Volger, *Physica*, **10**, 337, 1943; and F. W. de Vrijer, J. Volger and C. J. Gorter, *Physica*, **11**, 412, 1946.

cycle in the case of lattice relaxation is a maximum at frequencies of the order of 10^6 per sec. at liquid air temperatures, and at 10^2 per sec. at liquid helium. Spin absorption, however, is independent of temperature and is a maximum at frequencies of about 10^9 per sec. In the case of caesium titanium alum, there is very little dispersion, and absorption cannot be measured over the above ranges of temperature, presumably because $\rho < 6 \times 10^{-9}$ sec. for this alum.

The absorption appears to be inhibited by the action of a strong transverse magnetic field. Teunissen and Gorter* found that paramagnetic dispersion, which is pronounced in manganese ammonium sulphate at room temperatures, disappeared on the application of a strong transverse field. De Vrijer, Volger and Gorter† showed that the small absorption of $Gd(SO_4)_3 . 8H_2O$ at $77°$ K. fell off rapidly in a transverse field.

The mechanism of the interaction responsible for these phenomena has been discussed by Gorter‡ and by Temperley.§ The latter shows that a new kind of mechanism of interaction between the spins and the lattice, requiring the simultaneous transition of two or more spins with the emission or absorption of one quantum of energy to or from the lattice vibrations, provides a time of relaxation of the correct order of magnitude and explains why caesium titanium alum gives such a small effect; for, on the supposition of single transitions, the time calculated for the latter substance is much too large. However, Van Vleck† has shown that at liquid helium temperatures the lattice vibrations cannot possibly play the part of a thermostat thus assigned to them, and he has suggested a substitute conduction mechanism¶ which satisfies the thermodynamic formulae of Casimir and du Pré. It is assumed that when a paramagnetic ion reverses its spin it emits a "virtual" vibrational quantum which is absorbed in the

* P. Teunissen and C. J. Gorter, *Physica*, **7**, 33, 1940.
† De Vrijer, Volger and Gorter, *loc. cit.*
‡ C. J. Gorter, *Le Magnétisme, loc. cit.*
§ H. V. N. Temperley, *Camb. Phil. Soc. Proc.* **35**, 256, 1939.
‖ J. H. van Vleck, *Phys. Rev.* **59**, 724, 1941.
¶ J. H. van Vleck, *Phys. Rev.* A, **59**, 938, 1941.

surrounding liquid helium. This means that conductivity is pictured as taking place by virtual rather than by real lattice waves. Starr finds that the relaxation time for iron ammonium alum fits the empirical formula derived from Van Vleck's treatment; however, Gorter and his co-workers obtain values for this time which do not agree with those of Starr.

Hysteresis Phenomena. At the very low temperatures produced by adiabatic demagnetisation a different kind of absorption occurs with iron ammonium alum in alternating fields of frequencies of the order 50 to 75 cycles per sec. It appears to be unconnected with the relaxation phenomena described above and to be more akin to ferromagnetic hysteresis processes. It has been investigated by Giauque and McDougall,* Cooke and Hull† and Shire and Barkla.‡ The latter showed that the absorption was proportional to the frequency of the alternating field. They found no trace of remanent magnetisation. The heat dissipated increased markedly on the application of a steady field perpendicular to the alternating field, and increased somewhat less markedly when the steady field was applied parallel. For a specified field $\mathcal{H}_c$ the heating was proportional to $\mathcal{H}_c^2 / T^3$, and a maximum value of the specific heat was found at $T = 0 \cdot 09°$ K., which is rather higher than the temperature found by Kürti, Lainé and Simon; the difference is perhaps due to differences in the chemical purity of the specimens of salt.

Temperley§ has shown that the classical theory of action between magnetic dipoles, i.e. the Ewing theory of hysteresis, appears to apply to these phenomena. It is only necessary to assume that a definite fraction of the spins are locked, and, consequently, are unable to follow small changes in the alternating field.

Nuclear Absorption Phenomena. Rollin‖ carried out nuclear absorption measurements by a method similar to that of

* W. F. Giauque and D. P. McDougall *Journ. Amer. Chem. Soc.* **58**, 1032, 1936.
† A. H. Cooke and R. A. Hull, *Roy. Soc. Proc.* A, **162**, 404, 1937.
‡ E. S. Shire and H. M. Barkla, *Camb. Phil. Soc. Proc.* **35**, 327, 1939.
§ H. N. V. Temperley, *Camb. Phil. Soc. Proc.* **36**, 79, 1940.
‖ B. V. Rollin, *Nature*, **158**, 670, 1946.

Purcell, Torrey and Pound, described in Chapter XIV. The material was placed inside the coil of a tuned circuit within a transverse magnetic field which was modulated at 750 cycles per sec. Radiofrequency power from a crystal oscillator was fed through the tuned circuit to an amplifier followed by a detector and tuned audiofrequency amplifier. When the mean value of the applied field is adjusted to a value near resonance, the radiofrequency power reaching the detector may be modulated at 750 cycles per sec. if there is any variation in the loss in the tuned circuit due to nuclear absorption.

Measurable absorptions have so far been found in substances containing proton or fluorine nuclei. The resonance line appeared stronger and sharper for substances in the liquid than in the solid state, although the relaxation time for the transfer of energy from the spin system to the lattice appeared shorter. The relaxation time was about a second for most proton-containing liquids. Measurable absorption was not found with ice or with heavy water. There appeared to be no marked change in relaxation time with temperature, apart from the change noted on solidification.

Absorption in Ferromagnetic Substances. An anomaly in the high frequency resistance of ferromagnetic metals has been found by Griffiths.* At very high frequencies the product of the resistivity and the differential permeability rises to a well defined, broad maximum at a certain value of a strong steady field on which is superimposed the alternating field; the position of the maximum depends on the frequency. The values of $\mathscr{H}_{max.}$ λ, $\mathscr{H}_{max.}$ being the effective field at which the maximum occurs for a wave-length λ, suggest that the phenomenon is due to resonant absorption by magnetic particles with moments of the order of $2\mu_B$.

Absorption of Microwaves by Oxygen Molecules. See Note 10.

* J. H. E. Griffiths, *Nature,* **158,** 670, 1946.

NOTES

The following notes do not readily fit into the framework of Chapters XI to XVI, and are therefore given separately.

NOTE 1, p. 39. *The Quadratic Zeeman Effect*

Kapitza, Strelkov and Laurman (*Roy. Soc. Proc.* A, **167**, 1, 1938) studied the Paschen-Back effect for a Be doublet using a field of 300,000 oersteds and found that the splitting followed the theoretical predictions accurately.

In discussing the Lorentz procession on p. 11 a second order term was neglected. If we take account of this term in the quantum mechanics theory, we find that there results a frequency *displacement*

$$\Delta \nu_Q = \frac{e^2 \mathscr{H}^2}{8mc^3h} a_1 n_{\text{eff}}^4 (1 + m_l^2),$$

where $a_1 = h^2/4\pi^2 e^2 m$ is the radius of the first Bohr circular orbit, n_{eff} is the effective quantum number of the upper energy level and m_l is the change in the orbital magnetic quantum number, equal to ± 1 for the σ components and to 0 for the π components. Since the displacement is proportional to $\mathscr{H}^2$, it is known as the *Quadratic Zeeman Effect*. It becomes important when $\mathscr{H}$ and n_{eff} are large; e.g. when $\mathscr{H} = 27,000$ oersteds and $n_{\text{eff}} = 25$, $\Delta \nu_Q$ is of the same order as the separation of the triplet. It has been found by F. A. Jenkins and E. Segrè (*Phys. Rev.* **55**, 52, 1939) with the absorption lines of sodium and potassium, which can be observed as far as $n_{\text{eff}} = 40$ or 50, using a cyclotron magnet. Agreement with theory was excellent for $\mathscr{H} = 27,000$ oersteds and $n_{\text{eff}} = 21$, but beyond the latter value the displacement increased more rapidly than n_{eff}^4. (Cf. also H. E. White, "Some Recent Developments in the Zeeman Effect in Atomic Spectra", *Rep. on Progress in Physics*, **6**, 137, 1939.)

NOTE 2, p. 47. *The Jahn-Teller Rule*

In the case of polyatomic molecules, i.e. those which contain more than two atoms, no multiplet structure exists, because of the *Jahn-Teller rule* (H. A. Jahn and E. Teller, *Roy. Soc. Proc.* A, **161**, 220, 1937 and H. A. Jahn, *Roy. Soc. Proc.* A, **164**, 117, 1937) which states that a degenerate state in a polyatomic molecule is never stable since the molecule is automatically deformed to give an asymmetry sufficient to remove the degeneration. Hence the susceptibility should follow the Curie law with $\chi_M = \dfrac{N^2 \mu_B^2}{3RT} [4S(S+1)]$, due to spin only. The only gas on which tests can be made is nitrous oxide, NO_2, for which $S = \frac{1}{2}$, and the measurements made by G. C. Havens (*Phys. Rev.* **41**, 337, 1932) are in good agreement with theory.

NOTES

NOTE 3, p. 55. *Uniform Fields in Solenoids*

It is not generally realised that the field inside a short, thick, solenoid can be made much more nearly uniform over an axial region near its centre by supplying a single small correcting coil, placed equatorially, with reversed current. Thus, L. W. McKeehan (*Rev. Sci. Instr.* **10**, 371, 1939) shows that a winding of less than 50 turns is sufficient to correct for a main winding of about 8000 turns.

S. W. Rubens (*Rev. Sci. Instr.* **16**, 243, 1945) has described a system of five equally spaced square coils, forming a cubical coil system, containing respectively 19, 4, 10, 4 and 19 turns, which provides a uniform magnetic field over a considerable portion of the cube volume, which is, moreover, accessible from outside.

NOTE 4, p. 62. *Demagnetisation Factors*

J. A. Osborn (*Phys. Rev.* **67**, 351, 1945) has prepared charts and tables which enable one to find the demagnetising factor for any principal axis of an ellipsoid of any shape. He also gives formulae for the general ellipsoid and for a number of special cases; E. C. Stoner (*Phil. Mag.* **36**, 816, 1946) has prepared tables of the demagnetisation factors of ellipsoids of revolution; and R. M. Bozorth and D. M. Chapin (*Journ. Appl. Phys.* **13**, 320, 1942) have dealt thoroughly with the demagnetisation factors of rods. To the references on p. 62 should be added the experimental study by E. Dussler (*Ann. der Phys.* **86**, 66, 1928).

NOTE 5, p. 67. *Very Pure Iron*

Very pure iron has been made by J. G. Thompson and H. E. Cleener (*Bur. of Stds. Journ. of Res.* **23**, 163, 1939) by reducing purified iron oxide to sponge iron in hydrogen, later melting the sponge in hydrogen and remelting it *in vacuo*.

NOTE 6, p. 86. *The Fluxmeter as a Null Instrument*

The fluxmeter may be used as a null instrument. H. R. Nettleton and S. Sugden (*Roy. Soc. Proc.* A, **173**, 314, 1939) arrange that the removal of the search coil from the field gap breaks the primary current in a mutual inductance whose secondary is in series with the fluxmeter and search coil. The flux in the secondary opposes that in the search coil and, by adjustment of the primary current, the two can be made to annul each other. The primary current, which can be measured accurately by a potentiometer method, is then a direct measure of the field strength. A fluxmeter may also be combined with a thermocouple and used to measure the average power of irregular a.c. waves, or currents arising from speech, music, etc. (vide H. K. Dunn, *Rev. Sci. Instr.* **10**, 362, 1939). Special care should always be exercised in levelling a fluxmeter, as, by proper adjustment of levelling screws, much troublesome zero drift can be avoided.

When a condenser of capacity K whose plates are charged to a potential difference of E is suddenly discharged through a fluxmeter whose terminals are joined by a resistance r, the deflection ϕ which

425

results is given by $NAH\phi = rKE$, provided that r is not so large that the instrument is insufficiently damped. For, if the potential difference between the ends of r corresponding to the passage of a current i is V_r at any instant, the latter must be balanced by the E.M.F. generated by the moving fluxmeter coil, so that

$$\int_0^\phi NAH\, d\phi/dt\, dt = \int V_r dt = \int ri\, dt = r \int_0^q dq$$

or $NAH\phi$ is equal to $r.KE$ c.g.s. units when r, K and E are expressed in c.g.s. units, and to $100r.KE$ when they are expressed in ohms, microfarads and volts respectively. This provides an easy and accurate method of finding NAH, using a 120 volt H.T. battery, a 10 ohm resistance and a capacity of 5 microfarads.

NOTE 7, p. 118. *Furnace Materials*

A valuable analysis of the properties and sources of supply of heating elements, refractories, cements and thermal insulating materials for use in the construction of furnaces, is given by L. Walden, *Journ. Sci. Instr.* **16**, 1, 1939.

NOTE 8, p. 221. *Gyromagnetic Data for Alloys*

F. Galavics (*Helv. Phys. Acta*, **12**, 581, 1939) measured g for a number of alloys of manganese and antimony and of iron and selenium; in the latter the sulphur atom of pyrrhotite is replaced by selenium. The method used was that of Coeterier and Scherrer, considerable attention being paid to the elimination of disturbances along the lines described on pp. 223 *et seq.* The g values for the Mn-Sb alloys lay between 1·9 and 2·0, while those for Fe-Se alloys lay between 0·35 and 0·88. While the values are regarded as provisional only, they do in a way provide support for the unusual value of g for pyrrhotite. The wide spread of the experimental values was caused by the strong permanent magnetisation of the specimens. It is to be hoped that these results will be checked by an independent method, e.g. magnetisation by rotation.

NOTE 9, p. 229. *g-Values for Paramagnetic Salts*

Presumably, a result similar to that for Eu^{+++} would be found with $TiCl_4$ and $Ce(SO_4)_2$, as the Ti^{++++} and Ce^{++++} ions have a rare gas configuration and yet possess a paramagnetism independent of temperature.

NOTE 10, p. 423. *Absorption of Microwaves by Oxygen*

J. H. Van Vleck (*Phys. Rev.* **71**, 413, 1947) predicted that the paramagnetic O_2 molecule might absorb microwaves by means of transitions for magnetic dipole radiation between closely-spaced components in the ground state; and marked absorption of wavelengths about 0·5 cm. has been found.

SUBJECT INDEX

427

SUBJECT INDEX

Dispersion, paramagnetic, 416 *et seq.*
Displacement rule, Kossel's, 391
Domains, effects at boundaries of, 232, 330, 334, 362, 364
  limitations of concept, 349
  Néel's theory of, 349
  size effects of, 249 *et seq.*, 279
  surface effects of, 320
  theory, 160, 238, 305, 330 *et seq.*
  threadlike, 63
  Weiss, definition of, 238
Dysprosium oxide, gyromagnetic ratio of, 228
Dysprosium sulphate, 418

Earth's magnetism, 384
Eddy currents, effects of, 113, 276
Einstein-de Haas effect, 205 *et seq.*
Electromagnet, condition for maximum field of, 71
  condition for uniform field of, 72
  design of, 70
  for special purposes, 74
  reluctance of, 69
Electron, charge on, 54
  magnetic properties of free, 50, 392
  specific charge of, 54
Electron energy bands, 150, 266, 372, 386
Electron spin, 23
Electron states of atom, 24
Electron states of molecule, 44
Elements, diamagnetic, 124
  ferromagnetic, 266
  paramagnetic, 124
  susceptibilities of, 30, 31
Energy, crystal, 332, 346
  strain, 332, 346
Entropy, zero point, 284
Erbium oxide, use of, 119
Europeum, doubly ionised, 391
  trebly ionised, 41, 43, 128 *et seq.*, 228, 229
Exchange forces, 239, 385
Exchange integral, 239

Ferromagnetics, ballistic investigation of, 56
  for special purposes, 66
  high frequency absorption in, 423
  internal friction in, 383
  magnetometer investigations of, 59
  properties of, 56 *et seq.*

Ferromagnetics, resistance of, 292 *et seq.*, 361
  saturation of, 57, 237 *et seq.*, 241, 305
  specific heat of, 268 *et seq.*, 379
  surface fields on, 320
  thermoelectric properties of, 296
  Young's modulus of, 381
Ferromagnetism, collective electron treatment of, 257
  condition for existence of, 239
  correction for, 114
  definition of, 1, 235
  Heisenberg's theory of, 239, 242 *et seq.*
  Weiss theory of, 231, 233 *et seq.*
Ferrous ammonium sulphate, 132, 289
Field, crystalline, 49, 129, 135, 143, 394
  electrostatic, magnetic effects of, 49
  in antiferromagnetics, 387
  in crevasse, 6
  in solenoid, 5
  in spherical cavity, 233, 291, 410
  intense, 80, 113, 407
  leakage, 357
  measurement of, 81 *et seq.*
  molecular, 233 *et seq.*, 268 *et seq.*, 394
  self-consistent, 52
  stabilisation of, 87 *et seq.*
  symmetry, 49
  uniform, definition of, 3
  Weiss internal, 233 *et seq.*, 268 *et seq.*, 375
Flux density, definition of, 7
Fluxmeter, 82 *et seq.*, 425
Force, line of, definition of, 2
  tube of, definition of, 2
Furnaces, 117
  materials for, 426

Gadolinium, ferromagnetic, 252, 370
  trebly ionised, 127, 284
Gadolinium phosphomolybdate, 291
Gadolinium sulphate, 17, 42, 286, 289, 418, 421
  specific heat of, 286
γ-rays, use of, 290, 414
Gas, paramagnetic, theory of, 14 *et seq.*
Gases, diatomic, experiments with, 177 *et seq.*
  susceptibilities of, 118 *et seq.*
  test bodies for, 122
Gauss, 2

429

431

NAME INDEX

435

Masiyama, Y., 301, 308
Masumoto, H., 79, 160, 162
Mathews, B. H. C., 299
Matteucci, C., 310
Maxwell, J. Clerk, 14, 155, 202
Meakin, R., 388
Meissner, W., 169, 172 *et seq.*
Michaleck, J. C., 270
Milikan, R., 53
Millman, S., 187, 399, 400, 404
Mishima, T., 79
Montgomery, C., 147
Mookherji, A., 143, 394
Mott, N. F., 239, 256, 264, 386

Nagaoka, H., 308
Nakayama, M., 380
Narayanaswamy, A. S., 397
Néel, L., 123, 349, 371, 374, 385, 387, 396
Nettleton, H., 393, 425
Nix, F. C., 378

Ochsenfeld, R., 258
Okamura, T., 342, 365
Ôkubo, J., 342
Oliver, D. A., 78, 360
Onnes, Kamerlingh, 42
Onsager, L., 291, 413
Osborn, J. A., 425
Owen, M., 111, 115

Pacault, A., 132
Pan, S. T., 389
Pascal, P., 12, 97
Paschen, F., 37 *et seq.*
Pauli, W., 21
Pauling, L., 52, 144
Pearce, R. R., 257
Peierls, R., 150
Penney, W. G., 49, 129, 142 *et seq.*, 393
Perry, J., 202
Pettersson, H., 139
Pfaffenberger, J., 322
Phipps, T. E., 178, 197
Piccard, A., 101
Pickles, A. T., 375
Pidgeon, H. A., 312
Pierce, G. W., 315
Pol, B. van der, 323
Pollard, A. F., 299

Polley, H., 370
Potter, H. H., 90, 109, 118, 154, 159, 262, 269, 276, 293, 295
Pound, R. V., 406, 423
Powers, H. N., 201
Preisach, F., 306, 322, 324
Prentice, J. H., 373
Preston, T., 35
Purcell, E. M., 291, 406, 413, 423

Quimby, S. L., 341, 381
Quincke, G., 95, 100

Rabi, I. I., 139, 167, 182, 188, 195, 197, 198, 399, 400, 404
Rado, G. I., 373
Ramsey, N. F., 199
Rankine, A. O., 96, 110
Rao, S. R., 397
Ray-Chaudhuri, D. P., 216
Rayleigh, Lord, 58, 62
Reddi-Pantulu, D. V., 118, 258
Reekie, J., 389, 395
Richardson, O. W., 202
Richer, G. C., 349
Richter, G., 63
Roberts, A., 406
Rodebush, W. H., 270
Rogowski, W., 63
Rollin, B. V., 422
Roscoe, R., 144
Rosenblum, S., 75
Rubens, S. W., 425
Ruedy, R., 148
Ruhemann, M., 245, 286
Runge, C., 35
Russell, H. N., 25
Rutherford, Lord, 88

Sadron, C., 252, 253 *et seq.*, 371
Sanford, R. L., 65, 370
Sauer, J. A., 413
Saunders, F. A., 25
Scharff, Gertrud. 293
Scheffers, H., 169, 172
Scherrer, P., 220
Schlapp, R., 49, 129, 142 *et seq.*
Schmidt, Th., 199
Schneiderhan, K., 292
Schnurmann, R., 178
Schoen, E., 242
Schubnikow, L. W., 199, 387

438

Williams, H. J., 158, 163, 349, 367, 381
Williams, S. R., 299, 312
Wills, M. S., 366
Wolf, A., 303
Wolman, W., 63
Woltjer, H. R., 42
Wood, A. B., 318
Wood, R. W., 177
Wrede, E., 177

Würschmidt, J., 62
Wynn-Williams, C. E., 88, 89

Yamamoto, M., 382

Zacharias, J. R., 195, 197, 199, 399, 400
Zeeman, P., 35 *et seq.*
Zuithoff, A. J., 272

Printed in the United States
By Bookmasters